Rhythmic Stimulation Procedures in Neuromodulation

Rhythmic Stimulation Procedures in Neuromodulation

Edited by

James R. Evans
Sterlingworth Center, Greenville,
SC, United States

Robert P. Turner
University of South Carolina School of
Medicine, Columbia;
Network Neurology LLC, Charleston,
SC, United States

Academic Press is an imprint of Elsevier
125 London Wall, London EC2Y 5AS, United Kingdom
525 B Street, Suite 1800, San Diego, CA 92101-4495, United States
50 Hampshire Street, 5th Floor, Cambridge, MA 02139, United States
The Boulevard, Langford Lane, Kidlington, Oxford OX5 1GB, United Kingdom

Notices
Knowledge and best practice in this field are constantly changing. As new research and experience broaden our understanding, changes in research methods, professional practices, or medical treatment may become necessary.

Practitioners and researchers must always rely on their own experience and knowledge in evaluating and using any information, methods, compounds, or experiments described herein. In using such information or methods they should be mindful of their own safety and the safety of others, including parties for whom they have a professional responsibility.

To the fullest extent of the law, neither the Publisher nor the authors, contributors, or editors, assume any liability for any injury and/or damage to persons or property as a matter of products liability, negligence or otherwise, or from any use or operation of any methods, products, instructions, or ideas contained in the material herein.

Library of Congress Cataloging-in-Publication Data
A catalog record for this book is available from the Library of Congress

British Library Cataloguing-in-Publication Data
A catalogue record for this book is available from the British Library

ISBN: 978-0-12-803726-3

For information on all Academic Press publications visit our website at
https://www.elsevier.com/books-and-journals

Working together
to grow libraries in
developing countries

www.elsevier.com • www.bookaid.org

Publisher: Mara Conner
Acquisition Editor: Natalie Farra
Editorial Project Manager: Kathy Padilla
Production Project Manager: Julia Haynes
Designer: Mark Rogers

Typeset by Thomson Digital

Dedication

This book is dedicated to:

The ancients of Greece and other early civilizations who recognized and practiced the healing powers of rhythmic stimulation and harmony in music, poetry and dance.

The late 19th and 20th century writers, practitioners and creative thinkers who, despite frequent ridicule and resistance, brought renewed attention to the frequency and vibratory aspects of human function and dysfunction, and their potential for healing and self-regulation.

The neuroscience researchers of the 21st century using objective measures of brain function to investigate relationships between rhythmic stimulation techniques and neuromodulation.

Today's clinical practitioners who combine neuroscience with personal acumen to regulate neural timing, rhythmicity and internal harmony for the benefit of clients and, perhaps ultimately, the benefit of society as a whole.

James R. Evans
Robert P. Turner

Contents

CHAPTER 5   Nexalin and related forms of subcortical
electrical stimulation 131

Nancy E. White, Leonard M. Richards

CHAPTER 6   The use of music for neuromodulation...............................159

Eric Miller, Lynn Miller, Robert P. Turner, James R. Evans

List of Contributors

Marvin H. Berman   Quietmind Foundation, Elkins Park, PA, United States
Paul Chazot   Durham University, Durham, United Kingdom
Tom Collura   Brain Enrichment Center, Bedford, OH, United States
James R. Evans   Sterlingworth Center, Greenville, SC, United States
Michael R. Hamblin   Harvard Medical School, Wellman Center for Photomedicine, Massachusetts General Hospital, Boston, MA, United States
Berthold Langguth   University of Regensburg, Regensburg, Baveria, Germany
Eric Miller   Montclair State University, Montclair, NJ, United States
Lynn Miller   Expressive Therapy Concepts, Phoenixville, PA, United States
Siegfried Othmer   The EEG Institute, Woodland Hills, CA, United States
Susan F. Othmer   The EEG Institute, Woodland Hills, CA, United States
Leonard M. Richards   Unique MindCare, Houston, Texas, United States
Dirk De Ridder   University of Otago, New Zealand
Dave Siever   Mind Alive Inc., Edmonton, AB, Canada
Theresia Stöckl   Neurodevelopmental Clinic, Gauting, Germany
Wing Tin To   Center for Brain Health, The University of Texas, Dallas, TX, United States
Robert P. Turner   University of South Carolina School of Medicine, Columbia; Network Neurology LLC, Charleston, SC, United States
Sven Vanneste   The University of Texas, Dallas, TX, United States
Nancy E. White   Unique MindCare, Houston, Texas, United States
Udo Will   Ohio State University, Columbus, OH, United States

Preface

This book's origins date back over several decades. This preface summarizes historical highlights of the first editor's personal experiences on the long path to the decision to publish it. Hopefully, at least some readers will find this adds a useful personal touch to the book, thus being worthy of inclusion.

My maternal grandmother was a Christian Scientist. This may have been the source of a life-long interest in "mind over matter" and mind/body healing, and the eventual decision to become a clinical psychologist. But, it does not explain development of an obsessive interest in the power of tone and other rhythmic stimuli. The latter had a clear beginning, when, at the age of 22, a man of similar age told me of a book he had been reading in which it was claimed that ancient Egyptians used tones to open entrances to pyramids. For reasons, which remain unknown to me, I began searching for related topics in the library books and journals of the small college I was attending. Although I now know that some related literature existed at that time, this was long before the Internet, and the searches were in vain. The interest persisted, however, and frequently was reinforced by chance events. I began noticing the distinct rhythms of the voices of charismatic ministers, and wondering if that played a causal role in the apparent "miracle" cures they seemed to achieve. While serving in the Army in Turkey and on a tour of ancient ruins, I visited the site of an Aesculapia (a type of ancient Greek "health resort"). The guide pointed to what amounted to a music therapy room, and to another where clients were exposed to rhythmic, healing sounds of trickling water. I came to believe that tone might do more than open pyramids! Later, while an intern at a state hospital for persons with all levels of intellectual disability, I had occasion to observe from a balcony an entire auditorium of residents moving in synchrony with the live music of an on-stage rhythm band. The power of rhythmic stimulation in the form of music was reaching the central nervous systems of even the most severely intellectually disabled. During that same time period a "neurological organization" program, often referred to as the Doman–Delacato method, was quite popular and involved rhythmic stimulation of brain-injured persons via creeping/crawling exercises and "patterning". There were many reports of effectiveness, and I wondered, "can it be that rhythmic stimulation other than sound also can be healing?"

Events such as these motivated further searching. I discovered some literature on radionics (discussed further in Chapter 1). However, when I discovered that radionics practitioners claimed to be able to heal at long distances this was so far from my science-oriented university training that I dismissed it as quackery. Then I came across an older article in Science by a psychologist, T.C. Kahn (1952) titled Audio–Visual–Tactile Rhythm Stimulation in which he discussed equipment he had devised to enable persons to adjust independently the frequencies of each of the three types of sensory stimulation until they discovered their "preferred" combination. He cited evidence that preferred frequency combinations differed among patients with differing psychiatric-type disorders, and speculated on the treatment potential of stimulating patients with their own preferred frequencies. Finally, I had found a reputable behavioral scientist with similar interests! Dr. Kahn was still professionally active, and, upon being contacted, expressed pleasure that his ideas, which he had never pursued in detail were of interest to me. He even sent his original equipment for my use. However, by then I was newly enrolled in a Ph.D. program where, at least for the first couple of years, there would be little time for such research. And, I was informed by electrical engineering professors that the equipment was outdated, and "modern" equipment likely would be too expensive for most graduate students (including me).

Shortly after graduating and returning to employment at the state hospital, I started a research project based on the Doman–Delacato neurological organization "patterning" procedures, but adding auditory, visual and tactile stimulation to the accompaniment of children songs. This was designated "rhythmic, simultaneous multiple sensory stimulation". To my chagrin, the hospital superintendent, fearing that such stimulation might induce seizures, permitted only a few of the most severely disabled residents to serve as subjects, and insisted that all be involved simultaneously in a behavior modification program. Needless to say, the limited positive findings were confusing at best! That same year of 1969 I contacted a Dr. Coons of NYU regarding a tangentially related matter, and was informed, " There is something which may be of interest to you. A man at the University of Chicago, named Joe Kamiya has discovered that some persons can control their EEG rhythms, and it seems to have positive effects on anxiety levels". Could it be that, with the help of electronic equipment, one could modify the rhythms of his or her central nervous system to facilitate healing? That notion, at the core of neurofeedback, has consumed much of my professional life since then.

After leaving state hospital employment, I became an assistant professor at a university. The interests in EEG biofeedback and healing potential of rhythmic stimulation persisted, but the combination of publishing pressures, lack of interest or encouragement of such topics by other faculty members and the high cost of quality equipment kept those interests in the background. Then, in the early 1990s high quality equipment sold by the Lexicor Corporation at reasonable cost, combined with the Internet information explosion changed the picture. Several of my graduate students became involved in QEEG and neurofeedback research, and I began doing related work in clinical practice.

The interest in external rhythmic stimulation as an aid to healing continued, but for many years was focused mainly on the evolving use of auditory-visual entrainment (AVE) equipment as an independent treatment modality, or as an adjunct to neurofeedback. I did, however, continue to seek more general information on rhythm and its effects on humans, and by the 1970s discovered an increasingly larger number of books on the topic. In 1986 I added to this list by coediting, with Manfred Clynes, the book Rhythm In Psychological, Linguistic and Musical Processes.

Although increasingly aware during more recent years that there were other emerging "stimulation" techniques purporting to be effective for neuromodulation and healing, I was surprised to discover so many which were new to me on display at the 2012 conference of the International Society for Neurofeedback and Research. It was at that time that I decided there was need for a book designed to describe those which appeared to have a strong theoretical rationale, and considerable scientific research and/or extensive clinical support. Realizing the complexity of many of these approaches and my limited knowledge of biology, music, and advanced mathematics, I was very pleased that Robert Turner, M.D., a neurologist with considerable musical training, agreed to coedit the book. Hopefully, our interests and skills, combined with the training and expertise of the individual chapter authors, have enabled a book which provides a clear overview of the rationale, procedures and to-date support for several of the more popular rhythmic stimulation techniques specifically targeting neuromodulation.

James R. Evans

Acknowledgments

My initial acknowledgment is difficult to put into words—perhaps of an ineffable "force", for lack of a more specific term, for whatever ignited and has continued for most of my life to motivate belief in the critical importance of rhythm and related phenomena, such as resonance, and entrainment. Perhaps a more "down to earth" acknowledgment would be of the influence of a beloved maternal grandmother, Ida V. Rogers. Possibly during preverbal and earliest verbal stages of development I experienced the rhythms and content of her language as she read to me from her Christian Science literature concerning healing and "mind over matter". Whatever the case, I move now to acknowledgment of persons with influences on this book of which I have consciousness awareness.

I would like to acknowledge the support and encouragement of a former supervisor, as well as a fellow psychologist during my days in the Psychology Department at Polk Center in Pennsylvania. Frank I. Varva arranged my job description to allow time for creative activities, such as designing a rationale for and conducting pilot research, on multiple simultaneous rhythmic sensory stimulation. And, fellow psychologist Joanne M. Borger provided much assistance, encouragement, and support for that project, as well as for subsequent research and writing on rhythm-related topics.

There are many others whose direct or indirect influence, I wish to acknowledge. For example, Scott Makeig, now a well known neuroscientist, shared a common interest in music, rhythm, and the brain during the 1970s, and introduced me to Manfred Clynes, with whom, I subsequently coedited an earlier book on rhythm. And, the recognition by Robert Thatcher and David Cantor, pioneers and present-day leaders in quantitative EEG and neurofeedback, of the potential value of my 1980s idea for a book on rhythm-related topics gave needed momentum to that project and to this.

There are several persons whose more recent influences are acknowledged. Robert Turner, neurologist, coeditor of this book and director of one of the largest medically supervised neurofeedback clinics in the world, shares a common interest in music and the brain. His acceptance of plans for this book was a strong motivating force. The continuing encouragement of my stepdaughter, Ann Guyer, served as a needed energizing force during the book's progress. My computer-savvy business associate, Mary Blair Dellinger, was of much help with designing the book cover, and preparing references. And, I acknowledge the encouragement, and continuing interest, patience, and support of Natalie Farra, Acquisition Editor, and Kathy Padilla and Julia Haynes, Project Managers, both with Elsevier.

The very large amount of time required for writing and editing, has been costly in regard to time together with my wife, Martha Ann Young-Evans. For her support and patience, I always will be especially grateful; thank you "MAYE".

James R. Evans

I would like to acknowledge, first and foremost, the privilege and honor it has been to be working with Dr James "Jim" Evans as coeditor of this volume. Dr Evans' wisdom and experience acquired during decades of clinical, academic, and research endeavors in the fields of neurofeedback and neuromodulation were my sole sources of inspiration in the conception of this volume, and seeing it to completion. In addition to his productive career, Dr Evans has been an unparalleled mentor, trusted friend, and utterly patient guide and counselor during this phase of my journey. He also has been seemingly untiring in his patience with my interruptions and postponements during the laborious process of bringing this book to fruition. He gave countless hours to reading, revising, and editing the chapters, making me even more grateful to be part of this publication.

My lifelong love and passion for music, including my continual quest to learn more of the universal principles of rhythm, synchronicity, and pattern, have brought our paths together. However, I now realize that my prior years of learning and experiences in musicology, neurophysiology, and clinical neurosciences—though establishing a wonderful foundation for coediting this textbook—were quite elemental from what I was to further explore and learn during the preparation and editing of "Rhythmic Stimulation Procedures in Neuromodulation."

I gratefully acknowledge the legacy of music and thirst for knowledge of all things human—spiritual, psychological, and physiological—given to me by my brilliant, compassionate, and musical parents—Dr. Robert W. Turner and Barbara Whitlock Turner. They gave my siblings and me the "gift of music" all of our lives, with classical music filling the house, encouraging me in the love of piano from the time I learned to walk, taking me to concerts, providing musical scores to "read" the music that was saturating my whole being while being a listener and performer.

The legacy of music and thrill of learning also came from both my maternal and paternal grandparents, who shared their journeys and experiences as professional pianists and violinists. I retain early memories, before and during elementary school, of playing the piano for them, knowing I would receive their love and praise, as well as constructive criticisms, ultimately enriching my musical life in eternal and inexpressible ways.

All of this family and cultural history cultivated an intuitive belief in the critical importance of the multicultural beauty and universal aspects of rhythmicity, resonance, and harmony—and how they are designed to bring us not only joy and beauty, but also health, wholeness, and healing.

I also would like to acknowledge the amazing piano instructors and mentors who patiently guided me from my beginnings with piano lessons at the age of 3 from Ms. Irene Most, all the way through to my postgraduate training with Mr. Clarke Mullen who mentored and refined my musical training which I pursued simultaneously during my training in Medical School.

Without the selfless dedication, teaching, and mentoring of Dr. Gordon Shaw and Dr. Mark Bodner of the MIND Research Institute in Irvine CA, who have been bringing research and application of music to a universal applicability, and the joyful teaching from Dr. Joyce Nicholson who inspired my love of neuroepidemiology, my career paradigm expansion into the noninvasive music neuromodulation would have never occurred. Their genius, coupled with passionate inspiration, gave me the vision and tools to pursue clinical research and evidence-based medicine in the dramatic, life-changing influences, which can occur in neural networks as a result of exposure to, and training with, music, rhythm, and neurostimulation.

My further acknowledgements include the infinitely valuable support and encouragement of my wife and children (and now grandchildren) who have tolerated countless hours listening to my piano practicing, or have permitted my absence from many family activities while I invested years in neuroscience study in preparation for an unbelievably beautiful career in integrative neurosciences, incorporating neurofeedback, neuromusicology, and neurostimulation to bring health, healing, and enrichment of life to those I am privileged to call my "patients".

Finally, I acknowledge the seemingly infinite patience and encouragement of Natalie Farra, Acquisition Editor, and Kathy Padilla, Editorial Project Manager, both with Elsevier.

Robert P. Turner, MD

Introduction

Following general acceptance of the reality of neural plasticity in the late 20th century, a plethora of potential methods of neuromodulation have been developed and attempted, and their numbers continue to proliferate. Many seem quite fanciful, and motivated more by expectation of monetary gain than scientific reasoning and support. Some, however, are being found effective and considered by many as evidence based, even if the evidence remains solely anecdotal in nature. A few others, such as neurotherapy and transcranial magnetic stimulation, have increasing scientific support, as from research demonstrating predicted, and often enduring, modification of specific brain electrical activity (EEG) during and following their application. See Herrmann et al. (2016) for a recent review of some such evidence.

The editors of this book recognized that a great many of these proliferating methods for neuromodulation are stimulation techniques, and that most, if not all, of the stimulation involves periodicity, that is, is frequency based, whether, for example, in regard to the electromagnetic, light, or sound frequencies involved, and/or the frequency with which the stimulus is applied. Furthermore, they noted that the frequencies, which are involved range from simple vibrations to complex combinations related to phenomena, such as overtones and harmony, thus seemingly making the term "rhythmic" most appropriate for these stimulation techniques. Given the increasing popularity and diversity of such techniques, a need was perceived for a source of information concerning their nature, theoretical mechanisms of efficacy, and present evidence for their value. Hence, this book was born.

Many of those who develop or use many of today's stimulation programs seem unaware of, and/or choose to ignore, the fact that the use of rhythmic stimulation for healing purposes has existed for many centuries. With Chapter 1 of this book the author attempts to rectify this situation by pointing out that heritage. Chapter 1 could appropriately be subtitled: "What is New is Old, and What is Old is New". The author mentions many historical phenomena ranging from stimulation via electric eels by ancient physicians to the common use of rhythmic dance, poetry, and music in the "hospitals" (aesculapia) of ancient Greece, to the early 20th century popularity of Rife equipment for stimulating diseased organs and their pathogens with specific frequencies, supposedly to cure a range of maladies, including cancer. In Chapter 1 the author notes that past and present stimulation techniques vary widely in regard to such matters as whether they are applied generally (e.g., dance, music), or target a specific symptom, organ, or location (e.g., transcranial magnetic stimulation targeting a specific brain region). While many, if not most, techniques specify or imply, that the theoretical mechanisms of efficacy involve entrainment (enhancement) or disentrainment (inhibition) of a targeted bodily frequency via principles of resonance, others emphasize stimulation's role in facilitating the body's own natural healing abilities. In this chapter the author suggests that a "common denominator" of all the methods is facilitation of rhythm, and that the many practitioners who use them might be referred to as "rhythmizers".

The author of Chapter 2 is a true pioneer in development, use and research on audio-visual entrainment devices for attention deficit and other disorders, and more recently has been heavily involved in development and use of electrocortical stimulation procedures for neuromodulation. In this chapter he describes similarities and differences among various electrocortioal stimulation methods and equipment, and provides supportive findings from his own related research.

In Chapter 3 the author of the second chapter joins with an electrical engineer, inventor, and practitioner famous in the area of neurofeedback/neuromodulation, to discuss auditory-visual entrainment (AVE) with which both authors have had extensive experience. The evolution of specifics of AVE procedures is covered, along with research findings supporting their efficacy. Although AVE effects often have been criticized as temporary, some research suggests otherwise. And, cited research suggests that the relatively low cost of AVE equipment can make it a very cost- and treatment-effective approach as compared to more expensive procedures which thus far have not been consistently demonstrated to be more effective. Finally, Chapter 3 includes discussion of possible reasons why the repetitive (or rhythmic) aspect of stimulation procedures, such as those involved in AVE may be a key to their effectiveness.

The authors of Chapter 4, a highly experienced clinical practitioner and developer of equipment, and a Harvard University-based researcher, discuss the rapidly growing use of stimulation procedures involving electromagnetic energy in visible and near infrared light wavelengths, for example, low level laser light therapy or near infrared light treatment. They note that there is increasingly accepted medical use of such stimulation, as for speeding the rate of wound healing, and treating/preventing such circadian rhythm disorders as seasonal affective disorder. They report on some of the known biochemical and electrophysiological effects of light stimulation, and note critical importance of variables, such as optimal wavelength, and stimulus delivery rate and amount. Speculation is provided on the potential of light stimulation to modify neural connectivity in the brain, with positive effects on cognitive, emotional and physical function, even in cases of neurodegenerative disorders, such as Alzheimer's Disease. Research findings of the authors and associates are presented which involved intranasal and transcranial stimulation with light frequencies, and incorporated QEEG findings, as well as other measures to demonstrate effectiveness.

Neuromodulation using EEG neurofeedback directly affects only cortical regions, yet subcortical structures may be contributing to a client's symptoms, and need targeted for modulation. The authors of Chapter 5, both highly experienced psychotherapists and neurotherapists, were well aware of this, and recognized the need for a noninvasive stimulation procedure which could address subcortical function. Reasoning that stimulation of the hypothalamus with its rich connectivity with other brain structures should prove especially effective, they began using the Nexalin procedure designed for such purposes. In this chapter they describe the procedure, and explain the rationale underlying its development. They also provide details of some of their many successful clinical applications of the procedure.

In Chapter 6, two experienced music therapists and music therapy researchers combine with a musically talented neurologist and an experienced neuropsychologist to consider music as an example of one of the oldest rhythmic stimulation procedures used by medical practitioners. Citing the many parallels between brain electrical activity and aspects of music (e.g., frequency/musical tone and phase relationships/musical harmony), and considering the richness of music with its complex rhythms, harmonics and other features, they suggest that music has potential to be the most effective neuromodulator of all. They provide a brief history of medical uses of music since antiquity, discuss the rise of modern music therapy, and describe details of, and supportive research concerning, some of today's music-based procedures for treating neurological disorders. Results of research by the main author and associates are presented using both QEEG and behavioral measures as dependent variables in studying effects of music-related rhythmic stimulation methods with ADHD. The authors note that use of the human voice as a musical instrument in toning, chanting and some forms of meditation may have been the earliest form of "musical" therapy, and remain in common use among some societies. Following a detailed discussion of methods using the voice as a musical instrument, some likely physiological mechanisms involved in its efficacy are mentioned, and supportive research findings presented. Results of authors' pilot research comparing effects on QEEG of three different music-related stimulation procedures also are presented.

Transcranial magnetic stimulation (TMS) may well be the most commonly mentioned, medically accepted and scientifically studied of the "new" non-invasive stimulation techniques for neuromodulation. However, many persons are unaware that the general term TMS subsumes a variety of specific techniques and underlying theories, each with variable research support for its enduring efficacy. Chapter 7, written by an internationally known expert on TMS and his associates helps to clarify this by describing details of some of the more common TMS approaches, and reviewing related scientific research.

In Chapter 8, two of the world's most experienced neurotherapists and trainers, and prolific writers on theories of neurofeedback efficacy emphasize the frequency basis of brain function, and the importance of considering frequency, as well as location during neurofeedback and other approaches to neuromodulation. They indicate that current findings of the importance of coordination of neural activity within and among brain regions imply a necessity for precise timing between and across specific frequencies, and that the frequency relationships involved extend far below the commonly considered 1–40 Hz range, that is, into infralow frequency (ILF) ranges. They note that individual neurotherapy clients have optimum response (or reward) frequencies which need to be determined and taken into account to attain best treatment results. In this chapter the authors also describe their observations that effective neurofeedback training does not necessarily involve operant conditioning as commonly believed, but that the brain can function as observer of the EEG signals (including ILF) reflecting its own activity, and, in its further role as agent, modify its own brain rhythms in positive directions. One might consider this as the brain serving as its own rhythmic stimulation agent. Among many other relevant topics, the authors emphasize their findings of the importance of attending to varying roles of the right and left cerebral hemispheres during neurofeedback treatment.

A primary emphasis in this book is upon the rhythmic aspect of the many stimulation procedures covered. Rhythm and related topics, such as entrainment and oscillation, however, are complex constructs, and can have subtle differences in meaning across cultures, including across professional groups. In Chapter 9, a distinguished cognitive ethnomusicologist provides a scholarly discussion of this fact. Related to that, he reports on research indicating that effects of rhythmic stimulation can be culture-specific. For example, the same rhythmic music which stimulates strongly altered states of consciousness in listeners during religious services in one culture may have no effect on listeners in another culture. Such findings indicate that cultural variables need consideration when one is searching for variables which may make any given rhythmic stimulation technique effective for one client, patient, or research subject, but not for another.

In Chapter 10, a neuropsychologist with long-standing clinical and research experience with QEEG and neurofeedback, and interests in rhythm, covers a potpourri of neuromodulation-related topics, including some rather free-wheeling musings. He questions why so many alternative/complementary medicine approaches "work", suggesting that the answer goes beyond placebo, and relates to their rhythmic stimulation components. Noting the extreme complexity and precise timing requirements of efficient brain function, he attempts to provide a rationale for why simple rhythmic stimulation often seems to suffice for correcting an aberrant "neural symphony". The primacy of rhythm is emphasized, with special reference to its role in neuromodulation and healing. Later in the chapter he speculates on such popular, but as yet scientifically unsupported, notions as person-to-person or person-to-external object energy transfer as hypothesized by Reiki practitioners, and believed by some to underlie the effectiveness of "natural healers". Discussion of matters, such as reasons for lack of general acceptance of many of the procedures discussed in this book, issues concerning professional ethics, and speculation on future directions of stimulation techniques complete this last chapter.

Much has been written from ancient times, and increasingly continues to be written, about the primordial status of vibrations, oscillations and rhythms, often said to characterize all "energy". Especially in literature for popular consumption, the latter term is used in a generalized sense, that is, not specified as thermal, electromagnetic, etc, energy, and often presumed to be ineffable and unmeasurable. Terms such as "life force", "chi", "prana", and "silent pulse" frequently are applied. However, this has not always been presented only in religious, esoteric or philosophical contexts. Nikola Tesla, the famous early 20th Serbian–American engineer, inventor and physicist, and compatriot of Thomas Edison, often is quoted as stating, " If you want to find the secret of the universe, think in terms of energy, frequency and vibration". The theorist Meerloo (1969) wrote of rhythm as "life itself", and described how World War II prisoners in Nazi Germany, suffering under unbearable conditions of deprivation, spontaneously vocalized rhythmic "emergency poetry" which "spoke for one's essential me". And, more recently, the physicist, Brian Greene, in describing string theory as the unified theory of everything stated, "everything at its most microscopic level consists of combinations of vibrating strands" (Greene, 2003). Given the importance increasingly attributed to rhythm-related phenomena and facilitated by Internet communication, it is not surprising that rapid proliferation of purportedly effective techniques is occurring for use with all manner of human needs and ills. The present book is only one of many, which deal with such themes, differing primarily in its emphasis upon application of specified vibrating, oscillating, rhythmic stimuli for neuromodulation purposes.

The field of stimulation techniques for neuromodulation is developing at a very rapid pace. Methods are evolving which were not being used or reported upon when some of the chapters of this book were being prepared. For example, methods for application of stimuli to parts of the ear or to the nose in the expectation of more directly accessing brain regions, and methods to stimulate the vagus nerve are being used. Neuroscientists increasingly are discovering the critical importance of neural timing within and among brain networks, and are researching variables, such as cross frequency coupling and phase reset duration, all of which, of course, involve (or are) rhythms of the brain. Stimulation techniques are being experimented with which raise many ethical questions as they go well beyond alleviating symptoms of disorders to what might be termed "steering" the mind toward improved self-confidence or stronger motivation or "positive thinking". And, there are considerations of relationships between rhythms of the brain and other bodily rhythms, suggesting that future researchers and practitioners will study not only symphonies within the brain, but an entire body symphony. One has to wonder however, if we ever will know who or what the conductor may be!

References

Greene, B. (2003). *The elegant universe*. New York: Norton.

Herrmann, C.S., Struber, D. Helfrich, R.F., & Engel, A.K. (2016). EEG oscillations: from correlation to causality. *International Journal of Psychophysiology, 103*, pp. 12–21.

Meerloo, J. (1969). The universal language of rhythm. In J. J. Leedy (Ed.), *Poetry therapy: the use of poetry in the treatment of emotional disorders*. Philadelphia: Lippencott.

James R. Evans

CHAPTER 1

Historical Overview of Rhythmic Stimulation Procedures in Health and Disease

James R. Evans

Sterlingworth Center, Greenville, SC, United States

INTRODUCTION

There are a great many different types of rhythm and stimuli; therefore, many definitions of "rhythmic stimulation" are possible. In this chapter, as in this book in general, "rhythm" basically will refer to a flow, or movement, of stimuli characterized by regular recurrence of features or elements alternating with different elements or patterns of elements. Since "different" may include presence–absence of an element, then even simpler vibrating (oscillating) stimuli are included. "Stimuli" refers to any agent that causes, or modulates, activity in an organism; and the term "rhythmic stimulation" as used herein will refer more specifically to procedures involving deliberate application of rhythmic stimuli known (or presumed) to incite changes in the functioning of bodily systems. Many familiar rhythms known to affect bodily systems are referred to as infradian, recurring (cycling) less than once a day, or circadian (c. 24 h). Examples include, seasons, planetary movements, and the day–night cycle. However, the rhythmic stimulation procedures discussed in this chapter deal exclusively with ultradian rhythms, that is, those with features which recur (cycle) with a frequency (periodicity) of more than once a day. Their recurrence can range from several times (cycles) a day, to several times per second, to millions or more times per second.

Readers may wonder why this chapter concerns "rhythmic" stimulation and not simply "stimulation". Some may cite findings or conclusions of physicists and others that all matter, and even apparently nonmaterial "energies", are vibratory in nature, thus making superfluous the adjective "rhythmic" (e.g., Green, 2003; Cole, 1985). They have a point, and a chapter, such as this might as accurately be entitled something like "Energy Procedures in Health and Disease" and follow a path more or less parallel to texts such as *Vibrational*

Rhythmic Stimulation Procedures in Neuromodulation. http://dx.doi.org/10.1016/B978-0-12-803726-3.00001-8

Medicine (Gerber, 2001), and *Energy Medicine* (Oschman, 2016). However, in this author's opinion it is not only the vibratory nature of matter (or nonmatter) that is of importance, but also the unique *patterning* of the stimulation such that simple rhythmic vibrations (oscillations) embedded, or "nested" in and dynamically interacting with a matrix of multiple other vibrations emerge as the complex rhythms likely to be of greatest importance in influencing health and disease. If such is the case, then relatively simple rhythmic and vibratory stimulation procedures, such as involved in earlier auditory–visual entrainment procedures (see Chapter 3), and many "energy medicine" approaches may not be as universally powerful as, for example, music-based approaches with their complex structures including melody, overtones, harmonics.

HISTORICAL HIGHLIGHTS

Antiquity

It is easy to speculate that the earliest humans were soothed by deliberately attending to the rhythmic flickering of campfire flames, or the sounds of primitive musical instruments. And, for millennia humans necessarily lived in synchrony with the rhythms of the seasons and the night–day cycle in regard to planting–harvesting, sleeping–working, and so on. Whether due to astrological beliefs or direct rhythmic influence, exposure to planetary movements in the smog-free atmosphere of ancient times also may have played a role in health and disease. However, it is largely from the written records of early Greek and Roman civilizations that we know of deliberate application of rhythmic stimulation in the treatment of disease. Information relevant to the latter is summarized in the following paragraphs, with its primary source being Chapter 3 of the book *Music and Medicine* (Schullian & Schoen, 1948).

The Greek and Roman God, Apollo is said to have had among his many functions the cleansing of the body and soul of disease, and, via his rhythmic movement through the skies, the production of harmony in the universe. He was seen as the God of healing, poetry, dancing, and music, and regarded by many as the founder of medicine. In Greek/Roman mythology Apollo's son, Aesculapius, subsequently became the God of healing and medicine, with his temples, or Aesculapia, serving also as "hospitals" of the time. Therapeutic procedures used are reported to have involved not only patient *belief* in the divine powers of the Gods Apollo and Aesculpius, but also "natural" remedies, such as nondistracting, healthful surroundings, gymnasia, theaters, thermal springs, and, of special significance here, singing, poetry, and dancing. Main components shared by the latter procedures included harmony, melody, and rhythm. Since there was belief that disease is a lack of harmony of elements composing one's "psychical" and physical nature, and that health requires that the elements be in harmony, it is not surprising that those music-related procedures were important aspects of what today might be termed holistic or psychophysiological therapy.

As cited in Schullian and Schoen (1948), both Plato and Aristotle were influenced by work of the philosopher/mathematician Pythagoras. The latter, believing that order, proportion, and measure were the essence of life, studied the physics of sound, and, using a monochord, developed the ratios of his "perfect musical consonances" (Schullian & Schoen, 1948, p. 56), the octave, fifth and fourth intervals which remain basic to our present tonal system. He is said to have held the view that music, poetry, and dance help insure emotional stability, and to have composed specific melodies to treat psychotic disorders and restore persons with physical ailments to former health. Similar views were expressed by the Greek philosophers Plato and Aristotle. Plato is quoted in Schullian and Schoen (1948), as, "Music was bestowed on man for the sake of effecting harmonious revolutions of the soul within us whenever its rhythmic motions are disturbed. Thus, when the soul has lost its harmony, melody, and rhythm, assist in restoring it to order and concord" (p. 57). Related to this, Aristotle is quoted as, "Anyone seems to have an inborn affinity with harmonious sounds and rhythms; therefore many wise men assert that the soul is harmony, whereas others say that it has harmony. And, since the soul rules the body and music is akin to the soul, appropriate melodies, harmonies, and instruments must affect both soul and body" (Schullian & Schoen, 1948, p. 59).

Especially if the term "mind" is substituted for "soul", ancient Greek and Roman attitudes, such as those cited previously are very similar to beliefs underlying modern music, dance, and poetry therapy, as well as a great many of today's alternative and complementary medicines and "stimulation" approaches to neuromodulatuion.

Nineteenth and Early Twentieth Centuries
Mesmerism and Hypnosis
During the late 1700's and throughout much of the 19th century there was great interest in the purported healing powers of magnetism, perhaps culminating in the concept of "animal magnetism" as espoused by Mesmer and other practitioners of what today would fall under the heading of hypnosis. According to many practitioners of animal magnetism (or mesmerism), a magnetic fluid exists, which emanates from the "magnetizer" and is willfully communicated to the magnetized person, usually via the hands and eyes. Most practitioners eventually came to realize that "trances", and passes of hands over client's bodies, and so on, were not necessary for results, and that mental factors, such as suggestion, belief, or sustained attention sufficed for desired results. The latter, of course, is a conclusion still made by many critics of today's alternative medicine procedures.

From an historical perspective the practice of animal magnetism was a "stimulation procedure", albeit based on faulty theorizing, and not necessarily rhythmic in nature. However, hypnosis as practiced throughout the 19th century (and later) remained a powerful therapeutic technique, and often did involve rhythmic stimulation in the tone and cadence of the hypnotist's voice, rhythmic movement of his or her hands and arms, and so on, sometimes deliberately aimed at "getting in synchrony with" a client's rhythms or moods to help effect desired change. Thus, many hypnotic techniques could have been, and still can be, classified as rhythmic stimulation procedures.

HOMEOPATHY

The early 1800s saw the rapid rise in popularity of homeopathy as a means of treating disease. Founded by a German physician, Samuel Hahnemann, in the late 1700s, this approach was based largely on the notion that administering tiny doses of substances known to produce symptoms of certain diseases in healthy persons could cure persons actually suffering from those diseases, that is, "like cures like". The substances administered often are diluted to the point where none of their original molecules remain, leaving no active chemical ingredients. Thus, the treatment seems incompatible with modern biology, chemistry, and physics, and has been condemned by mainstream medicine almost from its beginning. It has become a favorite target of skeptics of alternative medicine in general, with accusations, such as "utter nonsense", "quackery," and "pseudoscience". Despite such criticism, and the fact that decades of scientific research have failed to support its efficacy, it remains a popular treatment approach, especially in parts of Europe. It is nonregulated in many places, thus allowing persons with minimal or no medical training to practice. On the other hand, some persons with MD degrees here and abroad are graduates of 3 year homeopathic training programs and regularly use its procedures, often as a complementary approach to their otherwise conventional medical treatments.

What may enable homeopathic treatment to qualify as a rhythmic stimulation procedure follows from the view of some of its defenders that, although no chemical molecules may remain in an extremely diluted homeopathic substance, nevertheless an "essence" or "imprint" remains, involving an "electromagnetic signature" of a vibratory/frequency nature. As such, treatment would appear to some to involve rhythmic stimulation, not via stimulation of the auditory, visual or kinesthetic senses, but via ingestion of specific frequencies, which then may interact with and neutralize the effects of vibration-related activity causing the disease symptoms.

Some defenders of the efficacy of homeopathy have been encouraged by findings of research on so-called "water memory"—the notion that water can retain a "memory" of substances once dissolved in it. This was suggested, for example, in the work of Benveniste (1988), and supported later in a study by Louis Rey published in the scientifically reputable journal *Physica* (Rey, 2003). The latter found that a pattern of hydrogen bonds could survive multiple dilutions of substances, thus giving hope to those searching for scientific support of some basic beliefs of homeopathy. A relevant subsequent study supported by the US Department of Defense (Jonas et al., 2006) investigated whether biological signals, such as electromagnetic ones occurring in molecule-receptor interaction, could be digitally mimicked and produce specific biological effects. If found to be true, it could give some support to the claim of homeopathy that an "electromagnetic signature" remains in diluted substances to become a mechanism of therapeutic efficacy. Results of this thoroughly controlled study, found no significant differences in the biological effects of digitally mimicked electromagnetic signals and effects of two control conditions. Interestingly, however, strong biological effects of the digitized electromagnetic signal inexplicably were found when experiments were being conducted by one specific member of the research team. Authors concluded that, "experimenter factors, such as the influence of chemical residues, energetic emanations, or intentionality from individual experimenters could be an explanation for these findings" (Jonas et al., 2006, p. 27). They note that future research needs to attempt to control for such factors. Parenthetically, this chapter author recalls a similar statement having been made by one of the pioneers in the field of neurofeedback who confided that he was perplexed by his observation that some research participants controlled their EEG activity significantly more effectively when certain experimenters were present. Perhaps the term "personalized" medicine has especially great significance when electromagnetic force dynamics are critical to success.

RADIONICS

During the early 1900s Albert Abrams, a neurologist and one-time director of clinical medicine at Stanford University, invented electronic equipment, which he claimed could detect radiations from a field of energy emitted from tissues of living organisms, including humans. His subsequent diagnostic and treatment techniques became known as radionics. He found that radiations of diseased tissues differed from those of healthy ones, thus making it possible to diagnose disease even in early stages, often before any symptoms became obvious. His treatment approach involved stimulating the "energy field" where he believed disease originated, neutralizing abnormal radiations via use of vibrations and resonance to raise the tissues to a "normal standard of frequency" (MacIvor & LaForest, 1979). Radionics attracted a large following, including some prominent physicians of the time, as well as several practitioners here and abroad who, modified Abrams' original equipment and his theories of efficacy. Eventually claims were made that radionic diagnoses could be made by some practitioners at a distance (as by phone), or from only a photograph of the client, and without necessarily using electronic equipment. Theories were put forth of involvement of subtle energy fields which are not within the electromagnetic spectrum, such as "thought fields" (T-fields), which can be generated by, but exist independently of, the brain. Such theories almost certainly helped push radionics even further from scientific acceptance.

The practice of radiology is and always has been a favorite target of skeptics, and rejected by mainstream medicine. For example, according to Wilfried and Nowak (2007), in 1925 the AMA described Abrams as the "Dean of 20th Century Charlatans" (p.21). However, as with homeopathy, it continues to this day to be used in various forms, and can be considered as one among the many and rapidly increasing "energy medicines". Many claims of miracle cures and glowing testimonials exist.

COLOR THERAPY

Stimulating diseased or dysregulated body organs with specific colors as a therapeutic procedure has had a long history. It was a commonly used treatment throughout the 19th and early 20th centuries, and, in various forms, still is practiced today. Also sometimes referred to as chromotherapy, it may be considered a subset of more modern light stimulation therapies, such as covered in detail in Chapter 4. Hypotheses concerning the therapeutic efficacy of color stimulation often involve a color's specific frequency within the color spectrum interacting with abnormal vibrations of an energy field surrounding the body, or emanating from a diseased organ or body system, in a manner which restores balance. For example, if an excess of some frequency exists in an organ (apparently as evidenced by disease symptoms) a specific opposite or "affinity" color would be chosen of a frequency which, via resonance, would neutralize the excess and bring balance to the organ's functioning.

One of the leading color therapists of the early 20th century, Dinshah Ghadiali, proposed a type of chemical imbalance theory of the mechanisms of his Spectro-Chrome approach to healing. Based on the discovery that every chemical element has a characteristic color, and the fact that our bodies are composed of chemical elements, he concluded that bodies also contain colors (with their corresponding frequencies). If there was an imbalance of body chemistry resulting in disease, there also should be a corresponding frequency imbalance. Irradiation of the body with colors (frequencies) associated with any deficient chemical elements then should in effect, introduce the needed elements into the body and restore balance (MacIvor & LaForest, 1979).

Ghadiali is said to have had a "forty year running battle" with the American Medical Association (AMA), and later with the Food and Drug Administration

(FDA), but, convinced that he had found a very effective healing technique, managed to continue his practice and study for years until his death at the age of 92. Many therapists continue to use color therapy. However, as true for most of the long-standing rhythmic stimulation procedures, mainstream medicine still considers color therapy to be a pseudoscience with no credible scientific research supporting its therapeutic efficacy.

THE WORK OF ROYAL RAYMOND RIFE

Overlapping in time and theoretical orientation with the frequency healing methods of Radionics and color therapy was the work of Dr. Royal Rife. This American inventor developed the "Universal Microscope" which, in the 1930–1940 period, apparently was the first to enable viewing the individual structure and chemical composition of microorganisms, including that of a virus found in cancer cells. Details of this microscope and its mode of functioning are given in the 1944 Annual Report of the Board of Regents of the Smithsonian Institution (1994) in an article entitled, "Rife's Microscope". Basically, it involved using crystal quartz prisms to create polarized light of varying (and adjustable) frequencies passed through a specimen (microorganism) from a diseased tissue to "illuminate" the specimen. When the light was adjusted to the point that its frequency and that of the specimen resonated (vibrated) in "exact accord", a definite and characteristic spectrum (perceived as a specific color) was emitted by the specimen. This was reported to have been independently replicated by various other medical scientists of the day. It was theorized that it is not bacteria themselves which cause disease, but, rather their chemical constituents acting upon "unbalanced cell metabolism of the human body". Dr. Rife further claimed to have discovered that by directing "rays" of certain specific frequencies onto an organism it could be deactivated or destroyed via principals of resonance. The frequencies of the "rays" would be precisely adjusted such that, when beamed upon the frequencies unique to and emitted from the chemical constituents of an organism, the latter would be selectively destroyed. There were claims that use of his Universal Microscope uniquely enabled actual observation of the destruction or deactivation of live pathogens, which was impossible with later more powerful electron microscopes that necessarily killed them prior to observation.

According to one source (Foye, 2002), Rife and associates demonstrated the significance of his approach to the treatment of cancer by creating tumors in rats and then consistently controlling their growth with his resonant frequency technology. Subsequently, in 1934 a cancer clinic trial of Rife technology was conducted in a Pasadena, California hospital involving 16 cancer patients, who had been medically diagnosed as terminal. After 3–6 months all were pronounced apparently cured by a University of Southern California medical research committee. Over the next 3 years, two additional cancer clinics reportedly used Rife's procedures, and replicated his cure rate (Walter, 1998).

During the earlier days of research, development, and demonstrations of efficacy of his procedures, Rife had support from many prominent physicians, medical clinics, and private investors. Although some support continued through the 1940's and 1950's and beyond, multiple sources (Walter, 1998; Valone, 2003) note, that beginning about 1938, his procedures and claims came under attack from the AMA, which indicted him for fraudulent medical practices. Although he went to court and won, much has been written about how the AMA (even during Rife's trial), ordered all physicians to cease using frequency instruments or lose their licenses, and continued to harass physicians who advocated a "germ theory" of cancer, and "frequency" treatment. Following this, many either quit using this technology or did so surreptitiously. Many variants of "Rife machines" remain available today, but may be illegal in some areas. And, both Rife's germ theory of cancer and his notion of frequency healing remain rejected by mainstream medicine, while being favorite targets of Quack Watch and other groups supposedly "policing" alternative medicine practitioners who would fleece the public for financial gain.

The Rife story is replete with conspiracy claims, and many charges and counter charges, such as those of retribution by an AMA leader, backstabbing and instrument tampering by Rife associates hoping to "cash-in" on his discoveries, a mysterious death of an associate, suspected arson, and the protection of vested interests by mainstream medicine, the National Cancer Institute and the FDA. Although written in a manner which to most readers would appear extremely biased, *Rife's World of Electromedicine* (Lynes, 2009), provides some interesting points concerning Rife's professional and personal trials and tribulations.

ELECTRODIAGNOSIS, DIAGNOSCOPY, AND DISCOVERY OF THE HUMAN EEG

An interesting article by Borck (2001), discusses a diagnostic/treatment procedure from the late 1920's period in Germany involving electrical stimulation. Although rarely mentioned today, it would seem to have special historical relevance to present-day quantified EEG (QEEG)-guided neurofeedback treatment.

In line with newly popular fascination with radio and the power of electricity in the early 20th century, entrepreneurs had successfully marketed many electrotherapeutic devices designed to "recharge the batteries of life" and treat symptoms of a myriad disorders. In the same era scientists and neurosurgeons were placing electrodes on exposed human brains and applying stimulation to help localize functional centers. It seems that it was such fascination, combined with early discoveries of functional anatomy which led to a sort of "electrophrenology"-type procedure known as diagnoscopy. Popularized by a physician from the Ukraine named Zachar Bissky, and originally used only for diagnosis, this procedure once was touted as a superior and more cost-effective method than psychological testing for vocational guidance and professional counseling. The procedure involved electrically stimulating, in a sequential manner, about 50 specific "reaction zones" on the surface of the head, each time eliciting a tone. Volumes of the tones were said to reflect the client's relative levels of development of the "psychical quality" represented by each zone (and the presumed specialized functions of underlying brain regions). Scores were assigned based on the tones and recorded to form a graphic profile depicting a client's status regarding personality traits such as idleness, selfishness, and violent temper. Eventually Bissky also began writing of the treatment potential of his method, claiming that touching an electrode to specific areas on the head would, for example, modify speech, elicit anger, induce calmness, or alleviate symptoms of migraines.

Borck notes that Bissky initially found strong public support for diagnoscopy, and even support from "scientific psychology" with some researchers reporting equivalent or even greater, diagnostic efficacy than aptitude tests and interviews. However, many others were alarmed by what they saw as a revival of Franz Gall's 18th century practice of phrenology in which personality characteristics were determined from study of head shape and skull protuberances. A commission established in 1926 to further evaluate validity of Bissky's method reported entirely negative results, and diagnoscopy soon more or less faded from use. It has been suggested that its demise also was due in part to the discovery of the electroencephalogram by Hans Berger during that time period. This presented researchers with a more direct and objective measure of brain activity, and eventually led to today's QEEG measures which are widely used in diagnosis of brain function/dysfunction, and as a means of evaluating effectiveness of neuromodulation procedures, such as many of those discussed in this book.

1950 to Present

During the 1950's and 1960's little scientific literature was available concerning effects of rhythmic stimuli on human behavior or brain function. The role of rhythm in the voices of charismatic religious figures to induce altered states of consciousness, and the suspicions of some parents that the rhythm and tempo of rock and roll music would corrupt the younger generation's morals occasionally was discussed in newspaper and magazine articles. However, even music therapists of those days with whom this author spoke did not seem to have considered the basic role or effects of rhythm on brain function. This probably was due in large part to a widespread belief prior to the 1980's or 1990's that neural dynamics were immutable.

The 1970's were years when there was to be major development of interest in the nature and effects of rhythm, and several books on the topic were published. The "hippies" of the era even spoke of "being on the same wave length" rather than "having good chemistry" with others. However, it was during and after President H. W. Bush designated the 1990–1999 period as the "decade of the brain", and the National Institutes of Health (NIH) and other groups introduced members of Congress and the general public to results of scientific brain research, when concepts of brain plasticity and neuromodulation became widely accepted. One might even say "wildly accepted", since during the last 15 or so years there has been an "explosion" of interest in the topic. Newly minted "neuro"—(psychologists, chiropractors, counselors, life coaches, etc.) appear with increasing regularity, and with highly variable training, promising to balance, exercise, heal, regulate, train, and so on, the brain, and alleviate symptoms of ADHD, depression, PTSD, anxiety, sleep disorder, substance abuse, and many other disorders using a plethora of (often patented) procedures. A great many, if not all, of the procedures involve rhythmic stimulation. Most, however, have little or no high quality research to back their claims. Several of those which the editors of this book consider currently to be ones with most scientific or clinical support, and/or have solid theoretical bases are covered separately in other chapters. Brief overviews of those, along with somewhat more detailed descriptions of several others constitute the contents of this section. The majority can be categorized as involving primarily electrical, magnetic, electromagnetic, sound, or light energy, although some procedures use some combination of these. Others, however, claim to involve hypothetical "subtle" energies, that is, other forms of energy not yet understood or measureable in a consistent, scientifically acceptable manner.

ELECTROCORTICAL STIMULATION PROCEDURES

As mentioned by Oschman (2016) in the book *Energy Medicine*, there are records from nearly 5000 years ago indicating use of shocks from electric eels in treatment of the sick. However, it is likely that most persons in the United States, especially if over the age or 40 or so, associate healing via electricity with electroconvulsive therapy (ECT). This treatment, which was developed in 1938, by the 1940s had become a favorite treatment for clinical depression, and other major mental disorders. It involved placing electrodes on a patient's head and passing an electric current through the brain, causing a seizure to occur. Because of major side effects, such as memory loss and broken bones associated with the seizure, it nearly disappeared from use after development of effective psychotropic medications by the 1970s. While some psychiatrists continue to use ECT, especially in cases of severe depression which has not responded to medication, it is said to be of a much safer type than earlier, with far fewer side effects. Its mechanisms of action basically remain unknown, but "causing multiple changes in brain chemistry" is the rather vague explanation often mentioned.

There are additional brain stimulation therapies involving rather strong electrical or electromagnetic pulses in use today. These include vagus nerve stimulation, magnetic seizure therapy, and deep brain stimulation. All involve surgery or seizure generation, and have potentially serious side effects. They must be used by or under the supervision of a physician. Due to space limitations (and the limited qualifications of this author) they will not be discussed here.

In recent years evidence has been presented that electrical stimulation involving far less voltage is effective for many disorders. Some cite the so-called "butterfly wing" effect where weather scientists had discovered that the exact paths of hurricanes may be determined by extremely minor weather perturbations occurring weeks earlier (perhaps even the flapping of a butterfly's wings). In other words, what appears to be an insignificant action can have major effects, and "less can be more" (perhaps similar to the homeopathic notion of the power of extremely diluted substances). Unlike ECT, many of these electrical stimulation procedures are of such low power that clients often cannot perceive the stimulation, and they cause no tissue damage or obvious harm. Many are FDA approved and are being used legally by non-physicians. A few are described in some detail in Chapter 2 of this book by David Siever, a pioneer in that field.

Two examples of more commonly used cranial electrotherapy stimulation (CES) procedures are Alpha-Stim and the Fisher-Wallace Stimulator. Each is described briefly later, based on information included in the sales literature of the companies selling the devices, which is readily available on the internet.

Alpha-Stim devices are reported to have been developed by Dr. Daniel Kirsch and consultants, Kirsch being a pioneer in the field of electromedicine since 1972. The two versions of the device are described as using microcurrent electrical therapy to treat pain, anxiety, insomnia, and depression. One version (Alpha-Stim-M) stimulates through probes applied to the body for pain treatment, and through ear clips for treatment of the other disorders. The actual stimulation is described thusly, "Alpha-Stim's proprietary waveform works by moving electrons through the body and brain at a variety of frequencies, collectively known as harmonic resonance". It is stated that this normalizes electrical activity, as can be measured by an EEG.

Potential customers and others often are puzzled by the name "Alpha-Stim", which could lead one to expect that the EEG alpha frequency is being trained or affected in some manner. However, the author could find no evidence that this is so other than a sentence stating that "maintenance of a relaxed, yet alert state is generally achieved with treatment three times per week" —a state commonly reported by persons during "alpha" neurofeedback training. The sales literature claims that "over 3 decades of 55 research studies all support the safety and efficacy of Alpha-Stim devices". Outcome data from 2500 Alpha-Stim user self-reports, and from a postmarketing physician survey of 500 of their patients using these devices were provided, indicating >25% improvement in 79% (insomnia), to 95% (anxiety/depression) of cases. Readers were advised to consult the book, *The Science behind Cranial Electrotherapy*, (Kirsch, 1999) for more detailed information.

The Fisher-Wallace stimulator is advertised online as "cleared by the FDA to treat anxiety, depression, insomnia, as well as chronic pain (such as fibromyalgia) when used on the body". It is said to function through use of patented waveforms to "gently stimulate the brain to use serotonin and other neurochemicals responsible for healthy mood and sleep", thereby improving the brain's ability to regulate the limbic system. It is claimed to have been developed by world class engineers and doctors, and to have been proven in multiple published studies.

ELECTROMAGNETIC STIMULATION

There are some, for example, Meyers (2014), who claim an advantage of electromagnetic stimulation over electrotherapy and many other forms of "energy medicine", noting that the human body is "transparent to magnetic fields" supposedly making magnetic stimulation uniquely capable of penetrating deeper to "energize cells at the core level". Transcranial magnetic stimulation (TMS) and pulsed electromagnetic field therapy (pEMF) are two current categories of such stimulation receiving much attention by mainstream medicine, as well as by alternative and "energy" medicine practitioners. Several forms of such stimulation procedures are described in considerable detail in Chapters 5 and 7.

In contrast to many other rhythmic stimulation procedures mentioned in this chapter, many, if not most, TMS and some pEMF approaches "target" specific brain regions commonly found to be dysfunctional in certain disorders rather than simply "exposing" the entire brain to one or more frequencies. Physician as well as non-physician use of TMS for treatment of depression is increasing quite rapidly despite limited research support for its enduring effectiveness.

In regard to pEMF, Meyers (2014) touts it as "the crown jewel of energy medicine" (p. xiv), one of five elements (along with fresh air, sunlight, water, and food), necessary for health and vitality. He references a 2011 "Dr. Oz" television episode in which the latter cited pEMF machines as "one of the most important breakthroughs in pain management that he's ever discovered" (p. xi).

Meyers states that, "PEMF therapy devices use a computer or control unit to administer precise natural pulsating magnetic fields to the body (with specific frequencies and intensities) through various applicators, such as a full body mat, localized pad applicators, and pinpointed probe or pen "spot" applicators" (Meyers, 2014, p. xi.)

Some of the pEMF treatments now available claim to energize or regulate bodily systems or promote natural healing in a generalized manner, as by lying on a full body mat, which emits an electromagnetic field. Other pEMF procedures "target" specific body parts or regions (now including brain areas) with specific, sometimes "personalized" frequencies. The latter are reminiscent of the work of Rife, although pEMF therapy most often purports to energize cells, not kill microorganisms. While most of the increasingly large numbers of pEMF procedures advertised by persons of widely varying training and making extreme claims for efficacy are rejected or looked upon with much skepticism by mainstream medicine, there are some specific uses of this type stimulation which have been accepted for years, for example, for speeding healing of wounds, and for fusion of broken bones.

What this author believes to be a unique and especially promising approach to use of such stimulation is the "Neurofield" system designed by Nichols Dogris and his engineering consultant, Brad Wiitala (Dogris, 2009). A psychologist and neurofeedback practitioner, Dr. Dogris became interested in "energy" medicine to help his son, who was born prematurely and suffered brain trauma associated with anoxia. Drawing upon the writings of persons, such as McTaggart (2002) in *The Field*, he came to adopt the concept of a "biofield" composed of the summation of the energy emitted from every cell in the body, and having a "neurofield" subdivision into which energy could be introduced to promote healing. To that end he and his consultant worked to develop a system which could emit different frequencies of variable amplitude, duration and wave shape, for example, sine, square, and potentially deliver energy to areas of the body, which match the resonant frequency of the energy being delivered. To their surprise they found positive effects without the need for placing electrodes directly on the body, and concluded that the frequencies delivered must be interacting with a client's unique biofield (or regions thereof) rather than going directly to cells of the body. Importantly, Neurofield is sold only to licensed health care practitioners, its evolution is guided by on-going research, and it is being integrated with neurofeedback by many practitioners. Neurofield stimulation not only is being reported by many to facilitate neurofeedback progress in clients who had been responding slowly or not at all, but, additionally, the gathering of quantified EEG (QEEG) measures used in conjunction with neurofeedback is allowing objective measurement of Neurofield effects. Recently, a TMS capability has been added to the Neurofield system.

SOUND THERAPIES

As noted earlier, the use of sound as a therapeutic tool dates to ancient times. And, of course, we all are familiar with the relaxing effects many experience from listening to babbling brooks and ocean tides. Thus, it may not be surprising that a very large number of formal rhythmic stimulation therapies use sound as the stimulus, often alone, but sometimes in conjunction with other type sensory stimuli. In fact, a search of the internet regarding the various stimulation procedures to be discussed in his chapter revealed more involving sound therapy than any other. There are professional organizations for such therapies, such as Sound Healer's Association and British Academy of Sound Therapy. Many music-based procedures, including music therapy of course, can be considered a subset of sound therapy. The latter are the topics of Chapter 6.

As with many of the electrical and electromagnetic procedures discussed previously, most of the multitude of sound therapies advertised today seem to be practiced by persons with minimal or no formal training in relevant fields (e.g., audiology or physics of sound in this case). It is common for developers and sales representative to refer to the primacy of sound in the human experience, sometimes even citing a Biblical passage, such as "In the beginning there was the word, and the word was God", as support that the universe began with a tone. They frequently state that hearing is the first sense to develop, and may suggest that a fetus' first sensory experience is that of sounds in the womb and the frequencies and other rhythms of the Mother's body and voice. Some suggest that the latter "sets the stage", for better or worse, in regard to language development. Generally, the advocates of sound therapies note that sound has been used as a tool for healing for millennia, and has been presented in many ways, for example, by drumming, tuning forks, Peruvian singing bowls and the human voice (as with chanting, "toning", poetry, and the mantras involved in some forms of meditation).

A very commonly given explanation of the mechanisms of efficacy of these techniques is that the frequencies and rhythms involved entrain abnormal, "out of synch" or "unbalanced" body rhythms, thus bringing about improved health, alleviation of symptoms, and so on. Whether based on clinical observation, anecdotal reports, or research (or to gain a larger share of the market), the various providers of sound healing tout the unique value of their specific procedure by emphasizing, for example, that theirs involves harmonics, uses stereophonic sound, uses both ear and bone conduction, uses ultrasound (sound outside the normal hearing range, i.e., subliminal), stimulates acupuncture pressure points, balances the two cerebral hemispheres, modifies brain waves, and/or uses binaural beat frequencies (the latter being frequencies elicited when a frequency presented to one ear differs from that presented to the other, such that the listener hears a frequency which is equal to the difference between the two presented frequencies). As with most of the procedures discussed here, a very wide range of positive treatment effects are claimed, including improved concentration, deep relaxation, better sleep, less pain, better stress management, and alleviation of symptoms of dyslexia, anxiety, and depression. Rarely is controlled research cited, with most support coming from client testimonials, many of which are glowing, that is, "This procedure was life-saving or live-changing for me (or my child)".

Many of the specifics of present sound therapies are based on ideas first popularized during the mid- to late 20th century by Alfred Tomatis, a French ear, nose, and throat specialist who devoted most of his professional life to study of the many functions of the ear and listening. He taught that many persons hear passively, but do not actively listen, and that deficiencies in the latter impede spoken and written language development, social skills, and self-confidence. He developed specialized techniques and equipment (e.g., the Electronic Ear) to help correct such deficiencies which led to many success stories and positive research support over several decades. Alleviation of symptoms of dyslexia and of autism were often reported. He is reported to have believed that sound coming through the ears and/or via bone conduction can be considered a needed nutrient which can "charge" one's nervous system, and that many persons do not receive adequate exposure to sounds rich in higher frequencies. He presented clients with frequency filtered sounds of Gregorian chants, their Mother's voice as may have been heard in the womb, and/or Mozart music. Regarding the latter it is reported that Tomatis believed Mozart's music had been dictated to him (Mozart) by divine inspiration, capturing the rhythms of the universe and enabling a listener's body to resonate with those in a manner facilitative of health (Sollier, 2005).

As with most, if not all, innovative, alternative approaches to medical practice, Tomatis' work was subjected to, much criticism from French Medical authorities. However, unlike most of the very large number of "sound healing" approaches being marketed today, the Tomatis method has been the subject of many scientific research studies in Europe, Canada, and the United States, much of it involving autism or various types of specific learning disabilities in children. In his autobiography, Tomatis (1991) mentions that, although the group comparison designs used in much of the research probably was inappropriate, some statistical support for his method has been found. Given the heterogeneity of symptoms in persons with learning disabilities and autism, individual case study designs may have been more appropriate than those involving group comparisons.

Despite extensive criticism and limited research support, the Tomatis method is widely used today in conjunction with special education services in many schools; and aspects of it are incorporated into various other "sound therapy" programs. Two if the latter, which are marketed quite widely, are the integrated listening system (ILS) and The Listening Program. ILS claims to improve many neurological conditions and produce enduring restorative effects on the brain through activities designed to coordinate function of the auditory, visual, and vestibular systems. Based on the notion that different frequency bands of sound affect different abilities, ILS activities centered around Tomatis listening training are integrated with visual-motor and balance activities. Thus, it might accurately be labeled a multiple simultaneous rhythm stimulation approach. There appears to be very little research on the effectiveness of the program, but it is touted by marketers, and supported by many anecdotal reports, as effective for remediating disorders of attention, learning, reading, and sensory integration.

The listening program (TLP) by Advanced Brain Technologies is another "sound-based" approach which, as with the Tomatis approach, involves listening to acoustically modified instrumental music via both air and bone conduction. Their website claims "global effects on the brain" which support changes in such areas as executive function, auditory processing, motor coordination, social and emotional function, and creativity. In a manner similar to most, if not all, other sound therapies its effects apparently are believed to be due largely to particular frequencies affecting particular brain functions. TLP claims to offer "personalized adaptive sound brain fitness programs", including use of a patent-pending artificial intelligence system which can "target specific areas of brain performance" and can lead to better health, increased productivity and more self-confidence. Claims are made that considerable research supports TLP's efficiency, and that its therapies now are covered by at least one major insurance provider. However, a now somewhat dated review of TLP by the Macquarie University Special Education Centre (Stephenson & Wheldall, 2010) concluded that the evidence base for efficacy of TLP was slight, and composed largely of unpublished pilot studies, or single subject case reports. They added that was true as well for claims of various other music-listening interventions. This situation may have changed over the past 6 years, but a quick search by this author did not find evidence that it has.

LIGHT THERAPIES

Who could doubt the value of sunlight to nearly all living matter? From ancient peoples with their Sun Gods, to today's "Sun worshippers" crowding the sunny beaches of the world, that natural source of full spectrum light remains much sought after. Records indicate that many ancient civilizations practiced "heliotherapy" or the use of sunlight for treating both physical and mental disorders, and it should not be especially surprising that this continues to be true today. Since light has wave/frequency features, it could be considered a source of rhythmic stimulation as defined in this chapter. Some current, related procedures are covered in detail in Chapter 4.

One subtype of light therapy briefly mentioned earlier in this chapter is color therapy. Although varying somewhat in theoretical rationale, a basic notion seems to parallel one commonly presented in sound therapies, that is, that different frequencies (light frequencies in this case, and perceived by humans as separate colors) have differential effects on specific organs, regions of organs, and/or mental or physical functions. Some of today's light therapies extend the frequency ranges outside the normal range of human perception to infrared or ultraviolet, while others make use of "coherent" light or lasers. Some procedures appear to involve global (systemic) effects, while others purport to target functioning of specific organs or regions, for example, a specific lobe of the cortex of the brain. Often-mentioned advantages of light stimulation is that it does not have to be transmitted via only the eyes, but can "bathe" an entire body, and can penetrate deeply.

Some uses of light therapy are well accepted by mainstream medicine, for example, use of ultraviolet light to prevent jaundice in premature infants, exposure to full spectrum lighting in cases of seasonal affective disorder. However, as with other stimulation techniques mentioned herein, mainstream medicine considers most to be unproven and "experimental" at best and, more often, as dangerous, outright quackery.

There may well be other potentially therapeutic phenomena involving light which usually are not considered light therapies. Art therapy, for example, has a history of its own and involves dynamics such as enabling cathartic, mentally healthful, nonverbal expression of emotion, internal conflict or repressed needs. Conceivably, however, producing and/or observing patterns of colors (i.e., of light frequencies) or the ordered spatial patterns (visual rhythms) of environmental objects or scenes serves to facilitate rhythmicity in, and, hence, regulation of, various bodily functions. Perhaps the special patterning of a variety of colors in some "abstract" art, or the contemplation of a mandala (geometric pattern involved in some forms of meditation), or being exposed to actual or symbolic examples of various symmetries and patterns in nature, such as mountain ranges also could be considered examples of rhythmic stimulation procedures involving light.

AUDITORY-VISUAL STIMULATION

Simultaneous stimulation with pulses of both light and sound to induce altered states of consciousness, promote neuromodulation or facilitate neuromodulation in conjunction with other procedures such as neurofeedback has been in use for at least the last fifty years. Its methods and examples of present-day use are described in Chapter 3 of this text.

As a brief overview, several variations will be mentioned here. In one of its earlier and simpler forms light emitting diodes (LEDS) (white or red) embedded in goggles flash at a single rate (frequency) within the range of 1–30 cycles per second (Hz.), while simultaneously a tone is presented at the same rate via headphones. It was demonstrated that for many, but not all, persons the frequency used elicited, via a process of entrainment, the same frequency in one's EEG activity. This often is referred to as an example of "frequency following". Frequencies involved may be preset by the AVE manufacturer, and supposedly chosen because they correspond to EEG frequencies commonly associated with specific states, for example, 4–7 Hz (theta) with "dreamy-like" altered states, 8–12 Hz (alpha) with states of relaxed alertness, >15 Hz (beta) with being alert and actively engaged. Some equipment allows users or therapists to customize frequencies, which might help accommodate variability among individuals in regard to frequency-state relationships.

Over the years various modifications have been made to AVE procedures and equipment. These have included, (1) coordinating stimulus presentation with ongoing EEG activity. For example, if attempts were being made to train a neurofeedback client to produce 12–15 Hz activity a greater percentage of the time, but he/she seemed "stuck" at 8 Hz, some AVE equipment could detect this and automatically and sequentially adjust the light and sound frequency to 8.5 Hz, then 9 Hz, then 10 Hz, and so on, to entrain the client's EEG toward higher frequencies (or, stated alternatively, disentrain it from 8 Hz); (2) adding tactual stimulation in the form of rhythmic taps or vibrations to create auditory–visual–tactual stimulation; (3) permitting separate adjustment of each sensory frequency such that the stimulus frequencies (pulsing rates) were independent; (4) permitting adjustment not of the actual pulse frequency but of the specific color frequency of each light pulse or flash, and/or of the frequency (and/or harmonics of) each of the tone pulses; (5) arranging two or more light bulbs in each side of the goggles, and programing the equipment so that pulses of light may be presented alternately to left and right sides (or to left and right visual fields), and sound pulses presented alternately to left and right ears.

Another variation, which has seen much acceptance and use, is the ROSHI equipment designed by Chuck Davis (Ibric & Davis, 2007). There have been several versions developed over the years, and interested readers can consult roshicorp@roshi.com for information on the latest iterations. The ROSHI approach was unique in that it involved not only EEG-guided light stimulation but also made use of low level electromagnetic stimulation to adjust EEG frequencies and amplitudes, as well as phase relationships. Modification of phase relationships to promote region to region synchrony within an individual's EEG was made possible by programming specific phase relationships between the right- and left-side LEDs.

AVE stimulation in various configurations remains in use, with frequent reports of effectiveness for a variety of symptoms alone or in combination with other neuromodulation procedures. Frequent criticisms have been that effects of AVE stimulation alone are not as long-lasting as those with procedures such as neurofeedback which involve operant learning. They also have been criticized as being too general and having "whole-brain" effects, rather than targeting specific dysregulated brain sites presumed to have causal relationships with symptoms. (Much of such criticism is countered by research cited in Chapter 3.)

THERAPIES INVOLVING RHYTHMIC PROPRIOCEPTIVE STIMULATION

As a young psychologist in the 1960s, the writer became familiar with what for awhile was considered by many to be a revolutionary and effective treatment for brain injury being used at the Institutes for the Achievement of Human Potential in Philadelphia. Developed using ideas of a neurologist, Temple Fay, and promoted by a physical therapist Glen Doman and a reading specialist named Carl Delacato, the program often came to be referred to as the Doman–Delacato approach. Despite much skepticism by mainstream medicine, it was accepted as effective by some cerebral palsy associations, and training programs were instituted in many hospitals and special education programs in the United States. The writer recalls attending a seminar given by Glen Doman in the Pittsburgh area where professors from a local university questioned the lack of research support for the program. They were countered by someone from a local industry criticizing them for "interrupting this man's presentation", claiming that he was there to consider giving financial support, and asking if they were "not aware that this method will make it possible to close the state hospitals for the mentally retarded." Such was the program's acceptance for awhile; however, the program rather abruptly lost its popularity by the early 1970s. Later the writer had occasion to speak with a physical therapist who had worked for some time with the program leaders in Philadelphia. He confided that, in fact, he personally had seen cases where clients diagnosed as severely mentally retarded had developed average level intelligence and shown other types of "near miracle" gains. However, he also stated that there was poor prediction of which clients would show major gains and which would not.

The Doman–Delacato treatments consisted of various components, including visual, auditory, tactual and olfactory stimulation. A major component was referred to as "patterning". This involved having a client lie on a cot or similar surface while one therapist rhythmically moved the arm and leg of one side, and another moved the arm and leg on the other side in a cross pattern fashion (as seen in creeping patterns of infants). Meanwhile a third therapist moved the client's head from one side to the other, coordinating neck and head movement with the rhythm of the arm/leg movements. Obviously, rhythmic proprioceptive and vestibular stimulation was being imposed upon the client. Similar stimulation techniques have been, and still are, used by some occupational therapists.

Various other activities, some used as therapy and others during recreation, include rhythmic movement either applied by a therapist or trainer or intentionally self-generated by the individual, and involve proprioceptive stimulation.

A currently popular treatment technique is the interactive metronome (IM) requiring intentional generation of increasingly precise timing of movement. Basically, this involves individuals practicing rhythmic movements so that they become increasingly synchronized with the beat of an electronic metronome. Motor planning and sequencing ability are involved, along with auditory-motor integration, This is not an entirely new idea—50 years or so ago the Keystone View Company was advertising metronomes for use with improving timing skills of learning disabled children, especially those with dyslexia. Developmental vision professionals were among the major users in the days when most cases of dyslexia were believed due primarily to visual disorders rather than to auditory-related disorders involving phoneme awareness as subsequently has been supported by extensive research. However, modern neuroscience research strongly supports the critical importance of timing in all sensory modalities (as well as for intersensory integration) such that "learning disabilities" in general may as accurately be labeled "dyschronias", (see, for example, Llinas, 1993; Condon, 1986), and hence perhaps amenable to treatment via procedures such as IM specifically designed to facilitate timing ability and sensory-motor synchrony. And, with present metronomes capable of millisecond, or greater interval timing, results with IM could be expected to yield even better results than those claimed a half-century ago.

In contrast to many other strongly advertised procedures claiming effectiveness for a large number of learning, psychological and/or medical conditions, the IM website currently cites twenty-nine instances of published research supporting its efficacy for a variety of conditions, including various learning disabilities, cerebral palsy, ADHD, blast-related brain injury, and improvement of golf and soccer skills. Although only seventeen of the citations are from journals, with several of those being pilot or single subject studies, and the 12 other citations being "white papers" or conference presentations, the IM producers do show definite interest in providing scientific support for their methods. The writer has noticed that various groups, when advertising their programs as being effective for training or balancing client's brain functions, include IM among their other treatment procedures (such as diet, right–left side integration, or sound-based activity). Given the increasing evidence for the critical importance of neural timing in brain function, one may wonder if IM is not their most therapeutically effective ingredient.

Manipulations by massage therapists very often involve rhythmic patterns of proprioceptive and tactual stimulation. One type is, in fact, termed "rhythmical massage therapy" or RMT. That particular approach often is described in a manner which inevitably brings criticism from those who tend to reject most if not all alternative medicine as "quackery", and lacking in scientific support. For example, a recent Web description (Rhythmical Massage Therapy Association of North America, 3-28-2016) refers to the hand movements involved as based on "archetypal forms of line, circle and lemniscates", and "rounded and rhythmical like living water", and having a "musical breathing quality". It was further described as "a warm, nurturing, enveloping massage therapy which enlivens and balances the body, while calling forth the client's own healing forces", and "addresses the relationship between the body, soul, and spirit by working with the life supporting forces of warmth, rhythm and breathing as well as the interplay of levity and gravity, center and periphery". No research support was provided in this reference for its claims of usefulness with behavioral health disorders, "frail constitution", hormonal imbalances and various other physical disorders.

Although usually guided by a trainer or therapist, dance therapy is an example of intentional rhythmic movement which can facilitate organization within individuals and groups, promote healing and modify affect. Thus, it also can be considered a rhythmic neuromodulation stimulation technique. Because it usually is accompanied by music, it could be seen as another multiple simultaneous rhythm stimulation procedure, that is, involving primarily rhythmic sound and proprioceptive stimuli. The American Dance Therapy Association, founded in 1966, stresses body/mind/spirit interconnectivity, and defines dance/movement therapy as, "the psychotherapeutic use of movement to further the emotional, cognitive, physical, and social integration of the individual". Berrol (1992) discusses theoretical and empirical evidence for neurophysiological factors, such as neural connectivity and neurotransmitters, being involved in effects of dance therapy. She cites the role of rhythm and movement in neurophysiological response, and their ability to organize individuals, promote healing and alter affect. Of course, salutary effects of dance/movement therapy are not limited to those relating to rhythmicity as such; cardiovascular and metabolic benefits of the aerobic exercise involved, and the potential social benefits of dance group, or dance partner, interaction must be considered. An overview of this field is provided in an edited book titled *Dance Movement Therapy: Theory and Practice* (Payne, 1992).

It seems likely that, aside from their positive effects on health in general, running and other intentional, often recreational, activities involving rhythmic movement may entrain and thus facilitate organization of neural activity. The often reported "runner's high" or athletes' experiences of being "in the zone" may be due to such neuromodulation. Related to this it even could be speculated that the positive effects of "equine therapy" are not due solely to variables such as development of self-efficacy associated with control of, or empathy with a large living creature (the horse being ridden), but also to the proprioceptive and other stimulation provided by the rhythmic gait of the animal.

AROMATHERAPY

Until rather recently, this writer had not considered aromatherapy as falling within the category of rhythmic sensory stimulation. However, while preparing for this chapter, the topic of the "vibration theory of olfaction" was encountered. This still controversial theory claims that smell receptors are tuned to different frequencies, and serve as "switches". That is, when a molecule of odor with a distinctive vibration pattern binds to a smell receptor with a matching vibration pattern the "switch" is thrown enabling flow of electrons directly to certain brain regions. Since the receptor-to-brain route is direct, the blood–brain barrier is by-passed, thus potentially making odors more efficient than most medications which cannot pass through the barrier (Keville & Green, 2012). This theory stands in contrast to what seems at this time to be the more popular "lock and key" theory which states that each odor has a specific shape which fits only into specific receptors. However, if accurate, the vibration theory would make plausible such notions as specific vibrations (or rhythmic *patterns* of vibrations) of an odor resonating with certain brain electrical activity frequencies to entrain, disentrain or otherwise modulate the latter.

Aromatherapy involves use of essential oils defined as "volatile liquids that are distilled or pressed from aromatic or fragrant parts of plants, such as seeds, bark, leaves, stems, roots, flowers, and fruits" (Oschman, 2016, p. 314). Such oils are among the oldest of medicines, with records of their use dating back thousands of years, and continue to be among the most used and researched of the alternative medicines today. Oschman notes 188 references to essential oils in the Bible, as well as over 12,000 articles on such oils listed in the database of the US National Library of Medicine, PubMed.

Given a direct smell receptor-to-brain route reported to initially impact subcortical brain structures, such as those of the limbic system, one could expect significant effects on the amygdala and hippocampus, and their roles in emotion and long term memory. This is commensurate with the commonly recognized ability of certain odors to stimulate memories and emotions. Concerning specific ways with which aromatherapy may achieve desired neuromodulation results, Keville and Green (2012), note that "euphoric" odors may stimulate the thalamus to secrete neurochemicals, such as enkephalin with natural pain-killing properties, and a sedative odor, such as marjoram, may facilitate secretion of serotonin which aids sleep. Regardless of whether it is the shape or the vibrational patterns of odor molecules, or whether the neuromodulation effects are via entraining/disentraining neural rhythms or via stimulation of neurochemicals, there appears to be growing clinical and scientific evidence supporting therapeutic use of aromas.

OTHER RHYTHM-RELATED THERAPIES

Rivaling aromatherapy in terms of longevity and popularity are various techniques claimed to involve "subtle energy" emanating from a healer's body, and interacting with those emanating from the body, or body part, of another. Techniques such as Reiki and therapeutic touch fall into this category. Although some advocates of these approaches claim valid measures of such energies, to the writer's knowledge there are no objective measures available which meet scientific standards of validity and reliability. Many believe that these subtle energies are vibratory in nature, and therefore exert their healing effects through some sort of harmonic resonance between healer and patient, whether involving specific "targeted" body parts or regions, or more generally via neuromodulation. If true, they could be considered rhythmic stimulation procedures with potentially major promise for neuromodulation. However, since evidence for the actual existence of these subtle energies and their rhythmical nature remains much more circumstantial than scientific at this point in time, they are not covered in any detail in this book.

Some human bodily processes which may not readily be perceived as frequency-based, however, may be periodic and thus rhythm-related, and become the focus of specific treatment procedures. One such process is described as involving periodic changes in cerebrospinal fluid production, causing rhythmic rise and fall of *pressure* and movement within the craniosacral system composed of bones of the skull, face, mouth, sacrum, brain, spinal cord, connective tissues, and cerebrospinal fluid (Wedel, 2006). A related treatment procedure, referred to as craniosacral therapy is used by some osteopathic physicians, chiropractors, massage therapists, and others. It consists of detecting and then decreasing or eliminating membrane tension patterns related to dysrhythmias of pressures within this system through a hands-on technique during which a trained therapist palpates a patient's head and other regions of the craniosacral system.

The therapist first seeks to detect subtle pressure dysrhythmias believed due to "restrictions" or "energy" blocks caused by actual physical trauma and/or by psychological stress or trauma. As with certain other techniques, often referred to as "body-work", craniosacral therapists generally accept the "tissue memory" theory that injury, neglect, various types of childhood abuse, etc can become "frozen" in the body and exert chronic negative effects on physical and mental health. The following description of the actual therapeutic process is based on the authors' interpretation of contents of a book chapter by Wedel (2006).

Once the location of a restriction is determined by detection of an asymmetry of rhythm, the therapist palpates the relevant region, applying no more than 5 g of pressure of "gentle traction compression or lifting" (Wedel, 2006, p.158) while simultaneously monitoring membrane tissue patterns, fluid wave patterns and subtle energy for signs of release of tension and adjusting palpation accordingly. It is deemed critical to "follow where the patient's body takes you" (Wedel, 2006, p.155) since the body knows how to heal itself. The therapist's light touch is said to "assist hydraulic forces inherent in the body to improve its own natural internal environment". It was emphasized that the therapist needs to blend with, "be in tune with" and work with the patient. Deep breathing by the patient was noted as likely to assist the process. It is claimed that successful craniosacral therapy often results in detailed recall and cathartic reliving of traumatic events. Indications for use of this type therapy are touted as including a very large number of disorders including traumatic brain injury, chronic back pain, migraine headache, chronic fatigue, post traumatic stress disorder, fibromyalgia, and autism.

A somewhat dated review of craniosacral therapy and its clinical effectiveness is provided by Green, Martin, Bassett, and Kazanjian (1999). There is much skepticism about the plausibility of the theory behind the method, as well as the reliability of its diagnostic methods. For example, it supposedly now is known that craniosacral therapists (or any others) cannot move skull bones to a degree which would affect cerebrospinal fluid movement; and multiple studies have shown very poor diagnostic agreement among therapists. Even many osteopaths, chiropractors and others in related fields question the effectiveness of the method, or at least the stated rationale for its effects.

Another therapy, which very obviously involves rhythm is poetry therapy. It sometimes is defined as a form of bibliotherapy, but unique in its use of rhythm, metaphor and imagery for healing and personal growth. A history of the development of poetry therapy is provided in the book *Poetry as Healer: Mending the Troubled Mind* (Leedy, 1985). As mentioned earlier in this chapter, it commonly accompanied music in the ancient Greek approaches to healing. According to historical notes provided by the National Association for Poetry Therapy (NAPT) (2016), the Greek God Apollo was considered the god of both poetry and medicine. The same NAPT source named a first century AD Roman physician, Soranus, as the first poetry therapist on record. Poem writing was said to have been an activity used with patients by the "Father of American Psychiatry", Benjamin Rush in hospitals of the 19th century; and Sigmund Freud was quoted as saying, "Not I, but the poet discovered the unconscious" (pp.1–2). Developments in more recent times include a burgeoning increase in interest in the 1960s and 70s as reflected, for example, in publication of several poetry therapy-related books, and the establishment of NAPT and the Journal of Poetry Therapy in the 1980s.

Poetry therapy is form of stimulation which can be self-generated (composing/reading poetry), and/or received from an external source (the human voice, or via written language). Both music and poetry are strongly associated with emotion, and appear to have special value in facilitating access to and cathartic expression of repressed emotions and related memories. It is of interest to note that metaphor, imagery and rhythm are features not only of poetry therapy, but also of some forms of therapeutic hypnosis, guided imagery approaches used by music therapists (see Chapter 6), and other procedures sometimes referred to as "royal roads to the unconscious". Due to their rhythmicity, both music and poetry therapy also may help bring organization to chaotic thought processes and emotions.

Obvious rhythmic body processes, such as breathing and heart rate have long been targeted for self-regulation via methods such as autogenic training and, more recently, biofeedback with therapeutic goals in mind. While these methods involve regulation of rhythmic activity, they generally do not qualify as rhythmic stimulation procedures as such. An exception would be where breathing using prescribed timing patterns is practiced intentionally to modulate neural function, especially of the autonomic nervous system. "Deep breathing" is a very commonly used method for control of anxiety; and specific breathing patterns are essential aspects of most methods of heart rate variability (HRV) biofeedback training—an increasingly popular treatment among therapists who emphasize mind–body connections.

A currently very popular therapeutic technique being used for various disorders, and especially for posttraumatic stress disorder (PTSD) and other trauma-related disorder is eye movement desensitization and reprocessing (EMDR). It is reported to have been discovered and developed serendipitously in the late 1980's by a psychologist, Francine Shapiro, while she was exercising. It basically involves a therapist moving one of his or her fingers back and forth in front of a client's face with the latter following those hand motions visually while simultaneously recalling a disturbing event, including associated emotions and bodily sensations. In early phases of the treatment this is said to facilitate desensitization to the trauma and related memories. In later phases the client is to replace the negative memories, thoughts and sensations with positive ones while again following the hand motions in order to facilitate "reprocessing".

It is an increasingly accepted therapeutic technique for PTSD, covered by some insurance companies and considered useful by the American Psychiatric Association and the Department of Defense. Various theories exist regarding mechanisms of its efficacy. One considers that recall of traumatic events, while simultaneously diverting attention away toward the finger movements makes it a form of desensitization or exposure therapy. Another, more neurologically based theory, is that the forced left-to-right-to-left, movements alternately activate the left and right hemispheres of the brain, thus facilitating integration of their specialized functions, such as processing of emotional memories and processing of conscious (verbalizable) access to episodic memories.

An alternative theory compatible with this chapter is that the rhythmic nature of the finger movements of the EMDR therapist entrain, or otherwise restore, a normal rhythm to eye movements (saccades). Although it is a bodily rhythm not commonly considered other than by vision specialists or neurologists, saccades occur normally with a periodicity of 3–4 times per second. Dysregulation of this rhythm can occur, and most often may be considered due to cerebellum, brainstem or other subcortical damage or malfunction (Ramat, Leigh, Zee, & Optican, 2007). If the rhythmicity of finger movements and the client's correlated eye movements during EMDR function as neuromodulators, this would be similar to the various other procedures covered in this chapter (as, e.g., where audio–visual stimulation entrains the rhythm of brain electrical activity). It is conceivable that if EMDR affects and regulates cerebellar and other subcortical function, it might prove to be an excellent supplement to procedures such as EEG neurofeedback (aka neurotherapy) which *directly* affect only cortical function.

Various rhythmic stimulation procedures frequently have been and continue to be used in conjunction with neurofeedback (NFB) for facilitating neuro-modulation. As mentioned earlier, AVE equipment has been used in cases where, for example, a training goal for a client may have been to raise 12 Hz. EEG amplitude, but he or she seemed "stuck" at 8 Hz with excessive power at that frequency, and failing to move higher. Exposure to light/sound stimuli at 9, then 10, then 11 and finally 12 Hz often entrained the client's EEG activity to the higher frequencies, eventually allowing successful training of the desired 12 Hz amplitude. Such AV stimulation still it used, with tactile stimulation occasionally added. Today, however, it is becoming increasingly common to supplement NFB training with transcranial magnetic, electrocortical, or pEMF stimulation, targeting not just specific frequencies, but specific brain regions as well. Presently, the Neurofield equipment aforementioned, is, in this author's opinion, the most plausibly designed, and promising example of this.

NFB itself, is rarely considered a stimulation procedure, but a form of operant conditioning wherein a client is rewarded when changes deemed desirable by the therapist occur in her or his brain electrical activity (EEG). "Rewards" vary widely, ranging from audio tones, to numbers on a scoreboard to being allowed to successfully play a video game or maintain an ongoing movie. While there is great variation among NFB practitioners concerning what constitutes "desirable" EEG change and how to accomplish it, most, but not all, believe that operant conditioning is the basic and essential mechanism of its efficacy. An exception to that belief, or at least a supplementary view, is presented in Chapter 8 of this book by Siegfried and Susan Othmer who provide insights from their decades of NFB experience indicating that the brain, given sufficient information concerning its electrical activity, essentially can regulate that activity with or without external "reward". In their theories and NFB practice they place strong emphasis upon individual frequency preferences, and not just brain regions. Given the central role of frequency in this theory, this writer finds it tempting to speculate that the brain itself functions as a rhythmic stimulator perhaps regulating its own functioning via production of its own unique "preferred" resonant frequency. If so, we each own built-in rhythmic stimulation equipment.

No survey of stimulation procedures for neuromodulation would be complete without at least some mention of what might appropriately be called, "the most popular and effective non-stimulation stimulation technique" in the field of NFB. Here I refer to the Low Energy Neurofeedback System (or LENS) developed in the early 1990's by psychologist Len Ochs, Ph.D. It is a very popular system, highly praised by many, including leaders in the field of NFB, and commonly considered to involve stimulation. An advertisement by one vendor described it as involving radio frequencies which "gently stimulate sluggish brain activity". In a personal communication, however, Ochs (2014) quite vehemently denied that it is a stimulation technique. LENS does not involve any of the typical rhythmic stimulation procedures to induce neuromodulation, that is, those based on a priori decisions regarding type, frequency, and patterning of stimuli. Although noting the possibility that some very low voltage electrical current may be transmitted to a client's scalp via amplifier-to-scalp wires, he emphasized that this is true for any EEG equipment, and not unique to that used with the LENS.

Ochs admitted that neither he nor anyone else knows the exact mechanisms of LENS effectiveness, but shared his belief that the "paradoxical" feedback provided by the LENS disrupts chronic, maladaptive cortico-thalamic timing, phase settings and associated defensive disconnections which the brain has been using. He considers the feedback "paradoxical" because it is reflecting both a person's dominant frequency at the specific electrode site involved (actual EEG feedback), plus a variable "offset" frequency (not actual EEG feedback). The formula is (feedback = dominant frequency + offset). He speculates that such feedback confuses the brain in a fashion reminiscent of the "confusional communication" some hypnotists use to induce trance, and leads to disruption of chronic maladaptive phase setting, and emergence of adaptive neural connectivity, which then continues in a self-sustaining fashion. The feedback received by the person is described as "profoundly weak" (less than 1/1000th of a cell phone's power), and transmitted via the same wires that connect the scalp electrodes to the EEG amplifier.

In a later communication (Ochs, 2016) it was stated that there are other considerations in the LENS which involve personalizing treatment to match each client's unique "neurological fingerprint". He noted that the form of mapping provided with LENS may differ in significant ways from others in the field, and play a significant role in its effectiveness, for example, "the particular sequence in which each site is treated is a function of the EEG standard deviation of the sites being examined, regardless of the putative neuropsychological functions determined at the site". He also stated his belief that vasodilation under the EEG electrodes plays a significant role in the symptom-reduction effects of neurofeedback, as evidenced by, for example, sensations of vascular activity reported by patients. However, at the heart of the system is "very low energy, very precisely timed".

CONCLUSIONS

What conclusions may be reached based on this overview? Later, are several which seem to the author to be warranted:

- Rhythmic stimulation as broadly defined in this chapter has been incorporated into healing practices since earliest recorded history whether in the form of music, chanting, dancing, drumming, touching of electric eels, or (assuming validity of the vibration theory of olfaction) exposure to aromas of essential oils. In more recent times discoveries of electricity and magnetism led to both ECT and the far gentler stimulation of today's electrocortical methods, for example, TMS and pEMF.
- In most if not all of the procedures reviewed concepts of frequency, vibration, oscillation, resonance, harmony, entrainment, and timing either were implied, or used in attempts to explain mechanisms of action.
- The majority of treatment approaches now classified as alternative (or complementary) medicine involve some type rhythmic stimulation, and/or deliberate attempts to modify body rhythms, such as heart rate or EEG.
- Throughout much of this history advocates often presented these procedures and their reported effects in esoteric, quasi-religious or mystical terms. While less true in recent years, proponents of many of the procedures discussed in this chapter appear to accept their apparent healing efficacy with little concern for details of how they may function, and actually may shun rigorous scientific study of mechanisms involved.
- Advocates of many of these procedures claim effects to be due to their stimulating, releasing, allowing or otherwise facilitating the body's (or brain's) own natural regulating or healing powers, rather than directly due to the procedure itself.

■ The variety of procedures covered in this chapter can be separated into relatively distinct categories; for example: (1) those which do not deliberately target any specific brain region or symptom, and may have generalized, systemic effects, for example, cranio-sacral, rhythmic massage, music and some applications of electrocranial and nonspecific pEMF stimulation. In some cases, advocates seem to imply that these are not unlike needs for a balanced diet, and appropriate amounts of water intake, sleep, and exercise, for example, implying that we all need daily doses of pEMF, high frequency sound or music, and so on; (2) those which do not target a specific brain area, but are claimed to address specific symptoms, for example, Tomatis method for central auditory processing; selected frequencies of AVE to alleviate symptoms of anxiety or ADHD; (3) those targeting specific brain regions with a goal of alleviating symptoms believed mediated by those regions, for example, TMS for depression, Neurofield using pEMF of specific frequencies for alleviating anxiety.

■ Despite lack of support for most of the procedures based on the current "gold standard" scientific method (double blind, randomized, placebo-controlled studies), rapidly increasing numbers of rhythm- (vibration-; frequency-) related procedures are being discovered, rediscovered or advertised as "new and improved". Reasons for this include: (1) disillusionment with effectiveness of standard medical practice for some chronic disorders; (2) the many negative side effects of medication; (3) decreasing personalized care by physicians under the current domination by pharmaceutical and insurance companies; (4) increasingly sophisticated advertising; (5) the internet with its ready access to relevant information; and (6) probably not least of all, some of the procedures have proven very effective with certain persons.

■ Just as surely as there will continue to be growth in use of these type procedures, there will continue to be "knee-jerk" opposition to most of them on the part of the majority of scientists and main-stream medical professionals. Reasons for this can include: (1) competition for clients among the many different medical professionals; (2) a mind-set that most of these procedures appeal only to "new-age" dreamers of questionable intellectual ability, or to those who are jealous of the financial status and respect enjoyed by practitioners of mainstream medicine; (3) the lack of exposure to such methods in current medical school curricula; and (4) honest beliefs that these procedures have very little or no treatment value, and simply reflect wishful thinking by "wanna-be healers" at best, and willful "fleecing" of unsuspecting clients at worst.

■ There are many documented and undocumented stories of near-miraculous healings or remissions of symptoms using each of these procedures. Well-designed scientific research on some, while often finding very limited or no support using groups statistics, had individual cases showing expected, or predicted, positive effects. Critics, of course, tend to attribute such to belief, expectation or placebo, and not to the treatment as such. Possibly true, but, if so, it would seem that the placebo mechanism needs to become a major focus of medical research and training. And, if not so, it behooves present skeptics to open their minds to the notion that, for a truly personalized medicine approach, rhythmic stimulation procedures cannot all be dismissed. Some may be very effective for some persons aside from placebo, even though exact reasons remain unknown at this time.

■ Given that rhythm and closely related concepts such as harmony are basic to the many procedures summarized in this chapter, perhaps we should refer to all as "rhythmizing techniques", and to their practitioners as "rhythmizers".

References

Benveniste, J. (1988). Dr. Jacques Benveniste replies. *Nature, 334*, 291.

Berrol, C. F. (1992). The neurophysiologic basis of the mind-body connection in dance/movement therapy. *American Journal of Dance Therapy, 14*(1), 19–29.

Borck, C. (2001). Electricity as a medium of psychic life: Electrotechnological adventures into psychodiagnosis in Weimar Germany. *Science in Context, 14*(04), 565–590.

Cole, K. C. (1985). *Sympathetic vibrations: Reflections on physics as a way of life.* New York, New York: William Morrow & Company, Inc.

Condon, W. S. (1986). Communication: Rhythm and structure. In J. Evans, & M. Clynes (Eds.), *Rhythm in psychological linguistic and musical processes* (pp. 55–77). Springfield, Illinois: Charles C Thomas Publisher.

Dogris, N. (2009). Neurofield. *Neuroconnections.* International Society for Neurofeedback and Research, pp. 21–24.

Foye, G. F. (2002). *Royal R. Rife humanitarian: Betrayed and persecuted.* Spring Valley, California: R. T. Plasma Publishing.

Gerber, R. (2001). *Vibrational medicine: The #1 handbook of subtle-energy therapies* (3rd ed.). Rochester, Vermont: Bear & Company.

Green, B. R. (2003). *The elegant universe.* New York, New York: W.W. Norton & Company, Inc.

Green, C., Martin, C. W., Bassett, K., & Kazanjian, A. (1999). A systematic review of craniosacral therapy: biological plausibility, assessment reliability and clinical effectiveness. *Complementary Therapies in Medicine, 7*(4), 201–207.

Ibric, V. L., & Davis, C. J. (2007). The ROSHI in neurofeedback. In J. Evans (Ed.), *Handbook of neurofeedback: Dynamics and clinical applications* (pp. 185–211). New York, New York: The Haworth Medical Press, Inc.

Jonas, W. B., Ives, J. A., Rollwagen, F., Denman, D. W., Hintz, K., Hammer, M., Crawford, C., & Henry, K. (2006). Can specific biological signals be digitized? *The FASEB Journal, 20*(1), 23–28.

Keville, K., & Green, M. (2012). *Aromatherapy: A complete guide to the healing art.* Hoboken, N.J: Potter/Ten Speed/Harmony.

Kirsch, D. L. (1999). *The science behind cranial electrotherapy stimulation: A complete annotated bibliography of 106 human and 20 experimental animal studies plus reviews and meta-analyses, a current density model of CED, side effects and follow-up tables, all indexed and cross-referenced.* Edmonton, Alberta, Canada: Medical Scope Pub.

Leedy, J. J. (Ed.). (1985). *Poetry as healer: Mending the troubled mind.* New York, New York: Vanguard Press, Inc.

Llinas, R. (1993). Is dyslexia a dyschronia? *Annals of the New York Academy of Sciences, 682*(1), 48–56.

Lynes, B. (2009). *Rife's world of electromedicine: The story, the corruption and the promise.* South Lake Tahoe, CA: BioMed Pub. Group.

MacIvor, V., & LaForest, S. (1979). *Vibrations: Healing through color, homoeopathy, and radionics.* York Beach, Maine: Samuel Weiser, Inc.

Meyers, B. A. (2014). *PEMF: The 5th element of health.* Bloomington, Indiana: Balboa Press.

McTaggart, L. (2002). *The field: The quest for the secret force of the universe.* New York, New York: HarperCollins Publishers Inc.

National Association of Poetry Therapy (2016). Available from: http://www.poetrytherapy.org/history.html

Ochs, L. (2016). Personal Communication., 07-20-2016.

Ochs, L. (2014). Personal Communication, 06-23-2014.

Oschman, J. L. (2016). *Energy medicine: The scientific basis* (2nd ed.). Elsevier Health Sciences.

Payne, H. (Ed.). (1992). *Dance movement therapy: Theory and practice.* New York, New York: Routledge.

Ramat, S., Leigh, R., Zee, D., & Optican, L. (2007). What clinical disorders tell us about the neural control of saccadic eye movements. *Brain, 130*(1), 10–35.

Rey, L. (2003). Thermoluminescence of ultra-high dilutions of lithium chloride and sodium chloride. *Physica.A: Statistical mechanics and its applications, 323*, 67–74.

Schullian, D. M., & Schoen, M. (Eds.). (1948). *Music and medicine.* New York, New York: Henry Schuman, Inc.

Smithsonian Institution. (1944). *Rife's microscope.* The Smithsonian Report. From the Annual Report of the Board of Regents of the Smithsonian Institution.

Sollier, P. (2005). *Listening for wellness: An introduction to the Tomatis method.* Walnut Creek, California: The Mozart Center Press.

Stephenson, J., & Wheldall, K. (2010). *The listening program.* Macquarie University Special Education Centre MUSEC Briefings, Issue 23.

Tomatis, A. A. (1991). *The conscious ear: My life of transformation through listening.* Barrytown, New York: Station Hill Press, Inc.

Valone, T. F. (2003). *Bioelectromagnetic healing: A rationale for its use.* Washington, DC: Integrity Research Institute.

Walter, A. (1998). *Royal Raymond Rife, a timeline.* Available from: http://www.aceguru.com/Health/rife.htm

Wedel, A. (2006). Craniosacral therapy for traumatic brain injury clients with neurobehavioral disorders. In G. Murrey (Ed.), *Alternate therapies in the treatment of brain injury and neurobehavioral disorders: A practical guide* (pp. 149–180). New York, New York: The Haworth Press, Inc.

Wilfried, A., & Nowak, H. (2007). *Magnetism in medicine: A handbook.* NewYork: Wiley.

CHAPTER 2

Cranio-electro Stimulation: An Effective yet Simple Technique for Calming the Mind

Dave Siever

Mind Alive Inc., Edmonton, AB, Canada

INTRODUCTION

Cranio-electro Stimulation (CES) is primarily a brain-calming technique, which delivers small pulses of electrical current through the brain. One advantage of CES over other brain modulation techniques is in its simplicity, which allows it to be used even by those with mild cognitive impairments. CES is a subset of Transcutaneous Electro-neural Stimulation (TENS). TENS refers to any form of electrical stimulation that is delivered through the skin. Typically, TENS consists of a fairly strong pulse and is used to contract muscles. CES, on the other hand, involves a much weaker pulse and typically is applied bilaterally across the cranium via the placement of two small electrodes, one on each side of the head. Most studies have used electrode placements on the mastoid process, the ear lobes or temporal lobes and sometimes on a shoulder or arm. CES devices employ alternating currents (AC) in audio frequencies typically from 0.5 to 100 Hz, and one device uses 15 KHz. There are various theories as to how CES affects the brain. The most popular are that a direct action is enacted on the brain via the brain stem; the limbic system; the reticular activating system; and/or the hypothalamus, and this in turn affects neurotransmitter production and possibly the default-mode network. Some research suggests that CES affects the brain via the reticular activating system and/or the hypothalamus (Gibson & O'Hair, 1987; Brotman, 1989).

Rhythmic Stimulation Procedures in Neuromodulation. http://dx.doi.org/10.1016/B978-0-12-803726-3.00002-X

HISTORY

The roots of simple brain-electrostimulation span back as far as 129 AD, when the philosopher and physician named Galen used eels to provide electric shocks to treat a variety of ailments including melancholia, depression, and epilepsy (Kneeland & Warren, 1994). In 43 AD, Scribonius Largus, a Roman physician, used torpedo fish (also a type of electric eel), to treat various ailments including headache and gout (Kneeland & Warren, 1994). Modern CES research was begun by Leduc and Rouxeau (1903) in France (McClintic, 1978). CES was initially studied for insomnia and was called electrosleep therapy (Gilula & Kirsch, 2005). In 1949, the Soviet Union expanded research on CES to include the treatment of anxiety, as well as sleeping disorders.

To date, approximately 40 different CES devices have been marketed in the USA, Canada, and Europe. About 200 studies have been published that cover a wide variety of clinical applications and findings, including, for example, improved drug abstinence and cognitive functioning in recovering alcoholics and drug users.

On the regulatory side, the Food and Drug Administration (FDA) recognizes CES as a treatment for anxiety, depression, and sleep (serotonin effect), but, surprisingly, not for pain (endorphin effect) even though the endorphin effect and efficacy for pain treatment have been well documented.

ABOUT ELECTRICITY AND NERVE PHYSIOLOGY

To understand how electrical stimulation works, it's important to know some basics of nerve and neuron physiology. The first thing to consider is that when a nerve or neuron is in its normal resting state, also called a polarized state, it is positively charged on its outer membrane and negatively charged on the inside by a difference of 65 millivolts. Therefore, if a negative stimulus is applied to the membrane, it will put a negative charge on the membrane, thus reversing the nerve/neuron's polarity; it will flip the polarity and cause it to depolarize and activate. This action happens very quickly and given that CES is AC, the stimulating polarity rapidly flips and then flips back for another activation.

The second important consideration is to realize that it is only the amperage (known as current) that can produce stimulation. Voltage does not. However, being that the skin is resistive (it impedes the flow of current), a voltage is needed to push the electrons (amperage) through the resistance (in this case—the body) in order to produce a current. When using ear clips or small electrodes, the AC electrical resistance [called impedance (Z)], through the skin under one electrode, through the body, and out through the skin under the

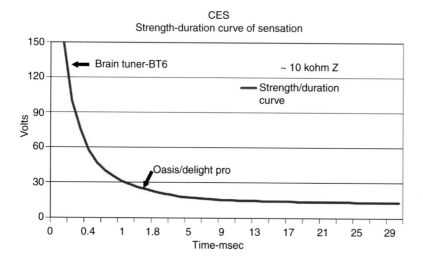

FIGURE 2.1 Strength-duration curve for nerve stimulation at an impedance of approximately 10k Ohms.

other electrode, ranges from 10,000 to 40,000 ohms. This electrical resistance varies widely, as it depends on thickness and dryness of the skin, mineral content in the water, and/or if whether or not conductive gel is used.

It is fairly common for CES devices to stimulate at 1–3 milli-amps (ma). One ma is 1/1000 of an amp—a rather small number. To put this in perspective, a typical wristwatch uses a few microamps (1/1,000,000) of current and a typical LED indicator light found on many small electronic items uses 2–5 ma of current. The little lights on a CES device might use more power from the battery than from the actual stimulation.

The third thing to note is that nerve excitation is a result of both the current flowing by it and also the *length of time* that the current has been flowing. In basic terms, it is the area of the pulse, and this area has been plotted into what is known as the strength-duration curve (McClintic, 1978). A strong, short pulse will activate nerves and neurons just as well as a weaker pulse of longer duration. By exciting nerves or neurons with varying pulse widths and intensities, scientists have developed a strength-duration curve for firing of the nerves and neurons. Fig. 2.1 shows that as long as the strength-duration of the stimulus is above the curve, nerve and neuron stimulation will occur. Therefore, a very long pulse can exert an effect with very low voltage (and low current). This, in part, is why a 9-volt battery tingles the tongue, because of its good, wet connection, and because the current flows indefinitely (direct current or DC).

Some devices can deliver pulses above 150 volts with a very high current, but with a very short period of time, such as less than 0.1 milliseconds (ms).

This approach has a tendency to utilize more of the capacitance of the body. Capacitance is the ability of any object, and in this case, a living body, to store a charge. As a result, this can reduce the risk of blistering an earlobe. However, high voltages can push current through high impedances and might even accidentally trigger a heart contraction if the electrodes fell onto the user's chest.

The alternate approach is to use a low voltage and current, but for a longer period of time, typically a few ms. The plus side to this approach is that it is relatively safe. The down side is that there is a larger flow of electrons and it can burn and/or blister the skin if the intensity is turned up excessively. Most CES devices use lower voltage-current and longer duration stimulation for the sake of safety. The Delight Pro and Oasis Pro stimulate in the 1.5–2 ms range with a maximum of 30–40 volts and a maximum current of about 4 ma, whereas the Brain Tuner (BT6) produces very short pulses of 0.1 ms, but with voltages of 150 volts and currents of 15–20 ma.

HOUSEHOLD POWER

To further our understanding of electricity and stimulation, let's consider household power and what's in a standard electrical plug. Our household power alternates from positive to negative, back to positive and back to negative, 60 times per second, or 60 Hz AC. If you should accidentally grab a "hot" power source, you will get a much more severe shock than from an electrostimulator with the same voltage. This is because at 60 Hz, one cycle is 16.7 ms. The "stimulus" is strong on the positive side about 6 ms (a very long time), then flips polarity for another strong burst of 6 ms (strong stimulation for about 12 ms/cycle × 60 cycles/s = 720 ms of stimulation/s) and so a great deal of current can pass through a person from household power, as shown in Fig. 2.2. Due to the long stimulus time, a 60 Hz line shock can really excite nerves and make muscles contract harder than anyone ever could of their own accord. As a result, torn muscles, heart attacks, and burnt tissue are possible, depending on where the hazardous live wire contacts the body and how well the connection is. All electricians must, by law, use non-metal ladders, such as fiberglass to prevent electric shock from a current flowing through their bodies and into a grounded ladder on wet ground or any grounded metal, in case they accidentally grab a hot wire. Table 2.1 shows the shock hazard at varying currents of 60 AC.

So, it's not the voltage that does harm, but rather the amperage plus the length of time that the amperage is present. With a low voltage, a better connection is needed to get the amperage through. A high voltage, however, can push an excessive current through the body even if the connection is poor.

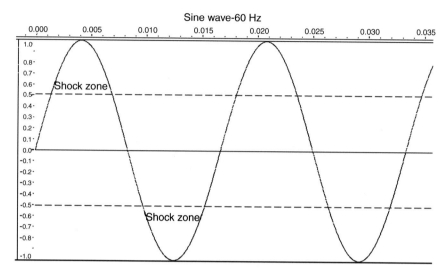

FIGURE 2.2 Electrical current and stimulus time of household power.

Table 2.1 60 Hz AC Currents and Human Experience of Them	
1 ma	Perception level.
5 ma	Mild shock felt, not painful but startling.
6–16 ma	Painful shock, but able to let go if through the hand.
20 ma	Paralysis of respiratory muscles and no longer able to let go.
50–100 ma	Extreme pain, unable to breathe, severe muscle contractions.
100 ma	Mild skin and internal tissue burns. Heart goes into ventricle fibrillation (heart flutters but does not pump blood).
2000 ma and up	Cardiac arrest and internal organ damage.

(Fish, R., & Geddes, L. (2009). Conduction of electrical current to and through the human body: a review. Open-access Journal of Plastic Surgery, 9, 407–421.)

THE ELECTRONICS BEHIND CES

CES exerts its effects on the brain by presenting short negative (cathodic) electrical pulses into the cranium. The pulses alternate from side to side making the current flow back and forth. This back and forth motion constitutes AC in a similar fashion to how current alternates back and forth in a common power plug. As the outside of a neuron is positively charged, a negative pulse will flip its resting state and cause an action potential to occur. This is not to be confused with transcranial DC stimulation (tDCS), in which a direct current is applied through a group of neurons. The principles governing DC stimulation and AC stimulation are not the same. With tDCS, it's the anode (positive) that enhances neuronal activity. With CES and TENS, it's

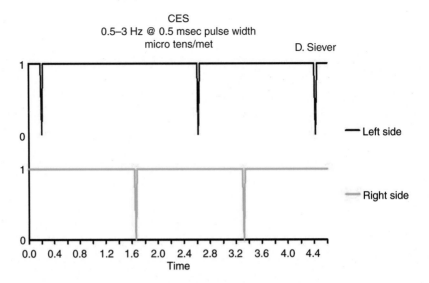

FIGURE 2.3 Pulses of short duration at any given frequency.

the negative going (cathodic) pulse that triggers action. With anodal tDCS, the voltage gradient becomes more negative as the current flows across our six layers of neurons. Given that the axons of our neurons lie away from the scalp, the axons will be more negatively charged than the soma, or neuron body. This reduces the overall membrane resting state from −65 millivolts to possibly −55 to −60 millivolts. So, whereas CES flips the resting state of a neuron, tDCS increases the probability that a neuron will fire when stimulated by adjacent axons (from other neurons) onto its dendrites (incoming branches).

In Fig. 2.3, we can see that when a stimulus is presented (a negative pulse), the other electrode provides the return path by being relatively positive and vice versa. The polarity alternates back and forth as the electrodes take turns acting as a complete circuit: one electrode providing stimulation (negative pulse), while the other provides the return path (positive) and vice versa. So long as both pulses are not negative or positive at the same time, there will always be a current. In Fig. 2.3, both sides are positive much of the time, so current flows during a downward pulse because the other circuit is positive. Fig. 2.4 shows the same idea, but with a very long pulse. Some devices make pulses at 50% duty cycle, meaning that each electrode is negative or positive 50% of the time. This enhances the *cup of wine effect*, that fuzzy-headed sensation from consuming alcohol, which some people strive to experience during CES therapy. Some people respond very well to this feeling, while others experience nausea and must stop the stimulation. Regardless of the subjective

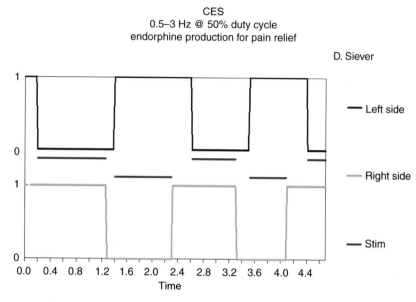

FIGURE 2.4 Pulses at a 50% duty cycle.

experience, the pulses generating this effect are shown in Fig. 2.4. In this case, the pulse has been randomized somewhat as with Fig. 2.3. In Fig. 2.4, the current flows only when the electrodes *are of opposing polarity*. This current may be flowing very little or very much of the time, depending on how the opposing pulses are timed. Here we see that current flows only when the red line is present, as there are opposing polarities on the electrodes. When the polarities are the same (both positive or both negative), there is no current flow and no stimulation (no line).

CES PHYSIOLOGICAL STUDIES

Methodology

The best scientific studies are double-blind, meaning neither the participant nor the researcher know when the real or the placebo treatment is being applied. To accomplish this using CES, the real stimulation must be low enough that the participant cannot feel it (subthreshold), and therefore cannot differentiate it from the placebo stimulation. Double-blind CES studies show excellent clinical results even when the person cannot feel the stimulation at all. However, for actual clinical work, it is suggested that the user feel the stimulation mildly.

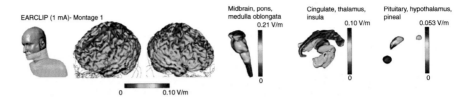

FIGURE 2.5 Peak electric field distribution from CES from 3–100 Hz.
(From Datta, A. et al. (2013). Neuroimage, pp. 280–287.)

Many clinicians experience confusion about the various frequencies and wave forms used with CES because of buzz words created for the sake of marketing. Most CES devices employ frequencies somewhere in the 0.1–40 Hz range, one device uses 78 Hz and another uses 15 KHz, modulated with 15 and 500 Hz. However, it is mainly the electrical nature of CES itself that produces the benefits. Frequency has some influence, but a minor amount. As a general guideline, it is believed that CES increases neurotransmitters at any frequency. However, 100 Hz has been promoted to increase serotonin more so than slow frequencies, and thought to be more effective for reducing anxiety and depression and improving sleep. Subdelta (in the 0.5–3 Hz range) is generally believed to increase endorphins, and therefore is used to reduce pain and promote feelings of well-being.

A quantitative EEG (qEEG) study by Kennerly (2004) of 72 participants receiving 0.5 Hz (n = 38) and another group receiving 100 Hz (n = 34) stimulation found that both groups showed increases in alpha with decreased delta and beta activity. However, the 0.5 Hz group showed decreases in a wider range of delta EEG, whereas the 100 Hz group showed decreases in a wider range of beta EEG (serotonin effect). This is consistent with the literature that CES at 100 Hz is mildly more effective at increasing relaxation while reducing anxiety than stimulation at subdelta frequencies.

A study by Datta, Dmochowski, Guleyupoglu, Bikson, and Fregni (2013), used computer-based, high-resolution modeling to simulate the current flowing through the brain during CES. The modeling placed some of the current flowing through the temporal lobes (which makes sense as the temporal lobes are close to the ears where the electrodes are placed). It predicted that the bulk of the current path went directly through the brain stem, in particular the medulla and the posterior aspect of the hypothalamus (areas in red) as shown in Fig. 2.5. This correlates with the liberation of neurotransmitters, such as serotonin, norepinephrine, dopamine, and acetylcholine, which are produced within the brain stem and modulated by the hypothalamus as shown by Silverthorn (2003), in Fig. 2.6. Endorphins are thought to be liberated via the hypothalamus, as well as some other brain areas.

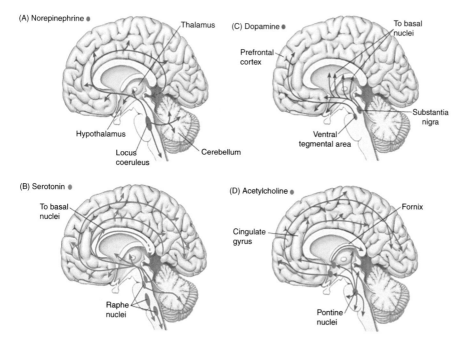

FIGURE 2.6 Neurotransmitter pathways originating from the brain stem. *[From Silverthorn, D.U. (2003). Human physiology: an integrated approach with interactive physiology (3rd ed.). Chapter 9. As illustrated by Howard Booth, Eastern Michigan University. Pearson Education Inc., publishing as Benjamin Cummings, (2004)]*

Neurotransmitter Effects of CES

As can be deduced from the studies aforementioned, CES primarily modulates the brain via neurotransmitter production, and many of the effects of CES involve the reduction of arousal (presumably overarousal). Shealy, Cady, Culver-Veehoff, Cox, and Liss (1988) performed a study of the Liss Pain Suppressor ($n = 5$) comparing blood-serum measures with that of cerebral spinal fluid (CSF) (drawn from lumbar subarachnoid space in the spine). Their study was intended to prove that CSF measurements were considerably more accurate than blood serum measures, but in the process, they found that average increases of beta endorphin and serotonin production following CES were quite dramatic, as shown in Fig. 2.7.

Another study (Shealy et al., 1989) using the same device with depressed ($n = 11$) and chronic pain ($n = 23$) subjects showed less impressive results. However, the largest increases involved the neurotransmitters that it was believed the participants needed most (Table 2.2). For depressives, serotonin and norepinephrine are needed most, and these show the largest increases. Those with chronic pain showed the largest increases in endorphin, exactly what they likely needed most.

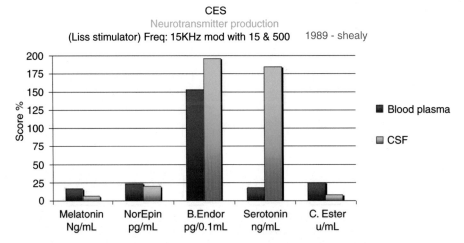

FIGURE 2.7 Blood serum versus CSF measures of neurotransmitters following CES.

Table 2.2 Neurotransmitter Changes Following CES Among Those With Depression or Chronic Pain

	Depressed (n = 11)			Chronic Pain (n = 23)		
	Pre CES	Post CES	% Change	Pre CES	Post CES	% Change
Serotonin	28	44	57	40	42	5
B. Endorphin	10.5	7.5	−29	8.9	10.2	15
Norepinephrine	212	239	13	214	224	4.6

Clinical Studies

As of 2002, there have been approximately 200 clinical studies using CES, spanning 50 years and over 25 devices (Kirsch, 2002). Smith (2006) completed an extensive analysis of the more credible studies of CES and cross-referenced them into *effect-size*, a unique type of statistical analysis, so they could be compared with other treatment modalities, as shown in Table 2.3.

Drug Use vs. CES for the Treatment of Depression

Depression is a very debilitating condition. Functional neuroanatomy studies of depression have shown a direct relationship with hyperactivation of the ventromedial prefrontal cortex (vmPFC) and hypoactivation of the left dorsal lateral prefrontal cortex (dlPFC). As depression subsides, the vmPFC becomes less active while the left dlPFC becomes more active (Koenigs & Grafman, 2009). QEEGs that the author has performed in his office usually reflect excessive alpha activity at F3, and to a lesser extent at FP1 and FP2, in conjunction with depression.

Table 2.3 Studies of CES by 2005						
Syndrome	Total Studies	Double Blind	Single Blind	Other	Total Participants	Average Improve (Effect-Size) (%)
Insomnia	18	7	3	8	648	67
Depression	18	7	2	9	853	47
Anxiety	38	21	1	16	1495	58
Drug Abstinence	15	7	4	4	535	60
Cognitive Dysfunction	13	3	4	6	648	44

There is evidence for a nutritional connection to depression. Shealy et al. (1992) studied blood-serum levels of five neurochemicals (melatonin, norepinephrine, beta-endorphin, serotonin, and cholinesterase) in depressed persons. He found that 92% had abnormal levels in at least one of the five neurochemicals tested and 60% showed three or more abnormalities. In over half, he found either elevated or low levels of norepinephrine/cholinesterase ratios. He also found magnesium deficiencies in 80% of depressed patients while 100% were deficient in taurine.

Neurotransmitter levels also play a role in depression. Depletions in serotonin, norepinephrine and dopamine are well documented with major depressive disorder (Nutt, 2008).

These widely varying factors involved in depression may well contribute in part, to the general failure of drug treatment. Given the direct neurotransmitter effects of CES and the absence of negative side-effects, CES studies of over 1000 depressed persons have shown relative success across the board. A meta-analysis study by Gilula and Kirsch (2005) of 290 subjects made a direct comparison of CES against various depression medications. Fig. 2.8 shows the treatment-effect improvement in depression over placebo obtained from freedom-of-information data as provided to the FDA from pharmaceutical companies when seeking FDA approval. The CES data came from eight studies submitted to the FDA from Electromedical Products International, Inc., to reclassify CES from class III to class II for the treatment of depression, anxiety, and insomnia. The CES studies had no reported negative side-effects, whereas the drug studies indicated side-effects ranging from interruption of liver metabolism to reactions with other medications and an increase in thoughts and behaviors related to suicide. This graph shows CES reduces depression better than placebo. Note that Prozac was found to be only 11% better than placebo.

Several studies point to the success of using CES to help with cognitive dysfunction. One study by Smith (1998) examined the emotional and IQ outcomes of

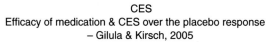

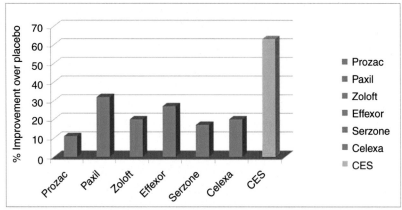

FIGURE 2.8 Drug treatment versus CES in the treatment of depression.

23 children diagnosed with ADD/ADHD following three weeks of 45 minutes of CES stimulation five times per week, and again at an 18-month follow-up with no treatment for 18 months (although some participants admitted to occasional use during the 18-month interval, which may have influenced the outcome). Smith compared three different CES devices (CES Labs, Liss Stimulator, and Alpha Stim), all with different stimulation parameters, such as frequency, pulse-width and intensity, and found no statistical differences in treatment efficacy. Fig. 2.9 shows the reduction in both state and trait anxiety and depression as measured on the IPAT Depression index and the State-Trait Anxiety Inventory (STAI). Fig. 2.10 shows the improvements in IQ as indicated by pre- to post- Wechsler Adult Intelligence Scale (WAIS-R) and Wechsler Intelligence Scale for Children (WISC-R) scores.

SOME EXTRAORDINARY APPLICATIONS OF CES

A study by Engelberg and Bauer (1985) used CES as auricular acupuncture at 13 sites (15–20 minutes total time) around the ear for the treatment of tinnitus. Improvement was perceived by 20 subjects in 27 of 33 ears (82%) in which tinnitus was perceived. The permanence of the improvement ranged from 20 minutes to at least six months (final assessment).

A small study ($n = 6$) by Smith (2002) of multiple sclerosis patients showed improvements after using CES at 0.5 Hz for approximately four weeks. The

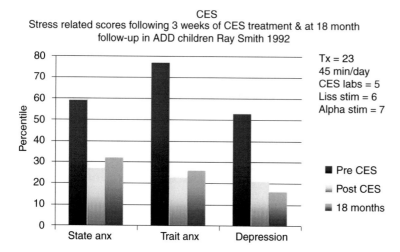

FIGURE 2.9 Behavioral outcomes following CES treatment at 3 weeks and at 18 months.

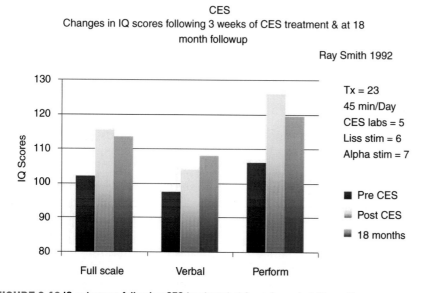

FIGURE 2.10 IQ outcomes following CES treatment at 3 weeks and at 18 months.

improvements spanned improved mobility and gait, improved motor control in the hands, improved vision, reduced pain, muscle spasticity and sensory problems, such as numbness and tingling.

Maldan and Charash (1985) published a double-blind study of 20 children suffering from Cerebral Palsy relating to birth anoxia. The study assessed the degree to which CES would reduce primitive spastic reflexes as measured on the Malden Gross Motor Rating Scale. In six weeks, most children showed improvements in gross motor ability with reductions in spasticity that exceeded those of 18 months to three years of physiotherapy.

MODERN DEVELOPMENTS IN CES TECHNOLOGY

Frequency

Recent research involving CES frequency has resulted in the discovery that CES may entrain brain waves (Zaehle, Rach, & Herrmann, 2010). Although brain wave entrainment research has not been officially done under the label of CES, it has been done with transcranial AC stimulation (tACS). TACS, in the brain wave frequency range, appears to be nothing more than relabeled CES. It may be that the developers of tACS were not aware that CES already existed. However, CES has been around for over 50 years, and it becomes confusing to call it something completely new. As with tACS, CES has been used in the brain wave frequency range and with temporal and frontal placements. And the protocol for adjusting the intensity of the stimulus with both tACS and CES is to increase the stimulus until a mild sensation is perceived. Given the identical nature of tACS and CES, it is likely that CES in the brain wave frequency range also has entraining properties.

Waveforms

Some manufacturers promote that they use a unique "special" waveform, which is better than what is being used by other manufacturers. Most CES research has used the simplest waveform to generate the pulse. This is a square wave (on and off). Although simple and inexpensive to generate, the down side to a square wave is that the stimulus can sting the ears, especially with a poor connection on the ears. People with naturally thicker and/or drier skin may experience more pain directly from the stimulus than a person with thinner and/or moister skin. The author has found that rounding the front end of the pulse, so it does not turn on so quickly (Siever, 2013), reduces the stinging/burning sensation by about 80% without loss of effectiveness. This increases user compliance, as most people will not use something that hurts. Fig. 2.11 shows an oscilloscope tracing of a rounded waveform versus a traditional square wave.

Another aspect of clinical effectiveness is to randomize the stimulus frequency as suggested by Heffernan (1997), who demonstrated EEG smoothing and pain reduction following CES treatment. Fig. 2.12 is an example of randomized frequency/time stimulation (Siever, 2012a, 2012b).

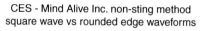

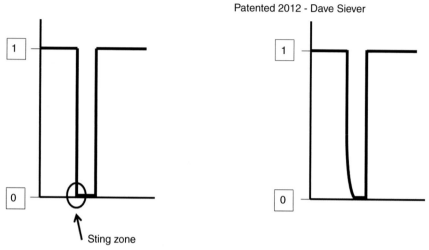

FIGURE 2.11 Oscilloscope tracing of a square waveform versus a rounded wave.

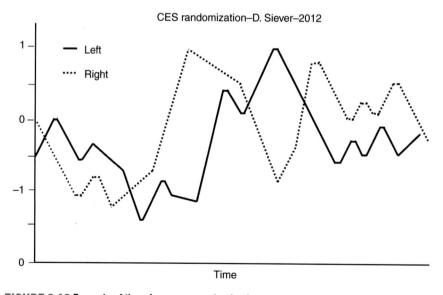

FIGURE 2.12 Example of time-frequency randomization.

CONCLUSIONS

There is ample physiological and clinical evidence showing the effects of CES on the brain and improved quality of daily living, as well as its having a widely varying number of specific clinical applications. Although CES achieves much of its effect through the relaxation response, studies show that some brain activation also may be involved. Perhaps the best features of CES are that it is easy to use, its efficacy is reasonable to good, and the equipment is fairly inexpensive and may be competently operated even by persons with mild cognitive impairments. In the author's opinion, CES is well proven and should be part of the therapeutic "tool chest" of all clinicians working with generalized cognitive challenges, affective disorders, and pain, including prophylactic, acute and chronic applications.

References

Brotman, P. (1989). Low-intensity transcranial stimulation improves the efficacy of thermal biofeedback and quieting reflex training in the treatment of classical migraine headache. *American Journal of Electromedicine, 6*(5), 120–123.

Datta, A., Dmochowski, J., Guleyupoglu, B., Bikson, M., & Fregni, F. (2013). Cranial electrotherapy stimulation and transcranial pulsed current stimulation: a computer based high-resolution modeling study. *Neuroimage, 65,* 280–287.

Engelberg, M., & Bauer, W. (1985). Transcutaneous electrical stimulation for tinnitus. *Laryngoscope, 95*(10), 1167–1173 Presented at the Meeting of the Southern Section of the American Laryngological, Rhinological and Otological Society, New Orleans, Louisiana, January, 1985.

Gibson, T., & O'Hair, D. (1987). Cranial application of low level transcranial electrotherapy vs. relaxation instruction in anxious patients. *American Journal of Electromedicine, 4*(1), 18–21.

Gilula, M., & Kirsch, D. (2005). Cranial electrotherapy stimulation review: a safer alternative to psychopharmaceuticals in the treatment of depression. *Journal of Neurotherapy, 9*(2), 7–26.

Heffernan, M. (1997). The effect of variable microcurrents on EEG spectrum and pain control. *Canadian Journal of Clinical Medicine, 4*(10), 2–8.

Kennerly, R. (2004). QEEG analysis of cranial electrotherapy: a pilot study. *Journal of Neurotherapy, 8*(2), 112–113.

Kirsch, D (2002). *The science behind cranial electrotherapy stimulation—second edition.* Edmonton, Alberta, Canada: Medical Scope Publishing.

Kneeland, T., & Warren, C. (1994). *Pushbutton Psychiatry: A history of electroshock in America.* Westport, CT: Praeger.

Koenigs, M., & Grafman, J. (2009). The functional neuroanatomy of depression: distinct roles for ventromedial and dorsolateral prefrontal cortex. *Behavioural Brain Research, 201,* 239–243.

Leduc, S., & Rouxeau, A. (1903). Influence du rythme et de la period sur la production de l'inhibition par les courants intermittents de basse tension. *Comptes Rendus des Séances de la Société de biologie, 55,* 899–901, VII–X.

Maldan, J., & Charash, L. (1985). Transcranial stimulation for the inhibition of primitive reflexes in children with cerebral palsy. *Neurology Report, 9*(2), 33–38.

McClintic, J. (1978). *Physiology of the Human Body.* NY, NY: John Wiley & Sons Inc, p103.

Nutt, D. (2008). Relationship of neurotransmitters to the symptoms of major depressive disorder. *Journal of Clinical Psychiatry, 69*(Suppl E1), 4–7.

Shealy, N., Cady, R., Culver-Veehoff, D., Cox, R., & Liss, S. (1988). Cerebrospinal fluid and plasma neurochemicals: response to cranial electrical stimulation. *Journal of Orthopedic Medical Surgery, 18*, 94–97.

Shealy, N., Cady, R., Wilkie, R., Cox, R., Liss, S., & Clossen, W. (1989). Depression: a diagnostic and neurochemical profile & therapy with cranial electrotherapy stimulation (CES). *Journal of Neurological & Orthopaedic Medicine & Surgery, 10*(4), 319–321.

Shealy, N., Cady, R., Veehoff, D., Houston, R., Burnetti, M., Cox, R., & Closson, W. (1992). The neurochemistry of depression. *American Journal of Pain Management, 2*(1), 13–16.

Siever, D. (2012a). Patent USA 8,265,761. Improved Cranio-Electro Stimulator (Randomization at standard frequencies (0.5–40 Hz) and at 100 Hz).

Siever, D. (2012b). Patent USA 8,265,761. Improved Cranio-Electro Stimulator (Phase-shift randomization at 100 Hz.).

Siever, D. (2013). Patent US 8,612,007B2. Cranio-Electro Stimulator (Rounded pulse and 3 Hz randomization.).

Silverthorn, D.U. (2003). Human Physiology: An Integrated Approach with Interactive Physiology, 3rd ed., Chapter 9. As illustrated by Howard Booth, Eastern Michigan University. Pearson Education Inc., publishing as Benjamin Cummings, (2004).

Smith, R. (1998). Cranial electrotherapy in the treatment of stress related cognitive dysfunction with an eithteen month follow-up. *Journal of Cognitive Rehabilitation, 17*(6), 14–19.

Smith, R. (2002). The use of cranial electrotherapy stimulation in the treatment of multiple sclerosis. *The Original Internist*, September, 25–28.

Smith, R. (2006). Cranial Electrotherapy Stimulation; Its First Fifty Years, Plus Three—A Monograph. Self published.

Zaehle, T., Rach, S., & Herrmann, C. (2010). Transcranial alternating current stimulation enhances individual alpha activity in human EEG. *Plos One, 5*(11), 1–7.

Further Reading

Siever, D. (2013). Transcranial DC Stimulation. *Neuroconnections*, Spring Issue, pp. 33–40. Avavilable from: http://www.mindalive.com/1_0/article%2010.pdf

Siever, D. (2013a). Patent Canada 2,707,351. Improved Cranio-Electro Stimulator (Phase-shift randomization at 100 Hz.).

CHAPTER 3

Audio–Visual Entrainment: Physiological Mechanisms and Clinical Outcomes

Dave Siever*, Tom Collura**

*Mind Alive Inc., Edmonton, AB, Canada; **Brain Enrichment Center, Bedford, OH, United States

INTRODUCTION

Audio–visual entrainment (AVE) is a technique in which lights flash into the eyes while tones are pulsed into the ears. The frequencies of the lights and tones are in the most common brain wave frequencies, typically ranging from 1 to 40 Hz. AVE is one of the most intriguing stimulation technologies. AVE devices have been shown to have a myriad of influences and effects on brain activity, which often far exceed that of a simple frequency following response (where the frequency of brain waves matches the frequency of the light and sound stimulation). Meditation and hypnotic induction, autonomic calming, cerebral blow flow, neurotransmitter production, and neuronal/network activation occur almost simultaneously during AVE.

When considering that the frequency aspect of entrainment is such a small portion of its many effects, one could ponder if we should simply rename the technique audio–visual stimulation (AVS). Our senses are constantly bombarded by random AVS as when watching TV or sitting on a street corner watching traffic go by. These activities comprise abundant AVS, yet they do not appear to have much specific impact on the brain. For instance, when AVE is randomized at ±1 Hz (for instance, 10 Hz varying randomly between 9 and 11 Hz), entrainment is reported to provide a significant clinical impact, while at ±2 Hz, the clinical effect is poor, and at ±3 Hz, the clinical effect is lost. So, it appears that the myriad of effects observed with AVE only occur when the stimulation is kept fairly consistent and rhythmic, and therefore, entraining, whether or not an obvious frequency effect appears within the brain. Hence, AVE appears to be a unique and clinically important subset of AVS, so AVE will be the term used within this chapter. While we often think of AVE as only entraining

Rhythmic Stimulation Procedures in Neuromodulation. http://dx.doi.org/10.1016/B978-0-12-803726-3.00003-1

its frequencies into the brain, as mentioned earlier, a host of complementary effects are occurring simultaneously. The effects of AVE include:

- brain wave driving
- dissociation/hypnotic induction
- autonomic nervous system calming and heart rate variability
- increased cerebral blood flow
- increased neurotransmitters
- unexpected effects of light and sound stimulation

BRAIN WAVE DRIVING

The main mechanism of action attributed to AVE, and to some degree all stimulation technologies, is related to frequency driving of brain wave activity. By definition, entrainment occurs when a biological rhythm reflects the rhythm of the stimuli to which it is exposed. For example, brain waves observed via EEG reflect the dominant brain wave frequency duplicating the frequency of an auditory, visual, or tactile stimuli. Photic driving of brain waves was first discovered by Adrian and Matthews (1934), while auditory entrainment was first demonstrated by Chatrian, Petersen, and Lazarte (1959). Photic entrainment occurs best near one's own natural alpha frequency (Toman, 1941; Kinney, McKay, Mensch, & Luria, 1973). AVE above 10 Hz tends to inhibit the half frequency of stimulation. Therefore, when treating ADHD, 14 Hz stimulation can both boost the low sensorimotor rhythm (calming hyperactivity) and suppress excessive frontal theta at 7 Hz thus increasing attention and emotional control (Collura & Siever, 2009). Beta at 20 Hz is used to inhibit excess left-frontal alpha at 10 Hz.

AVE utilizing square-wave photic stimulation should be avoided in those with photosensitive epilepsy, where flashing lights of certain frequencies may trigger a seizure (Ruuskanen-Uoti & Salmi, 1994; Erba, 2001; Trenité, Guerrini, Binnie, & Genton, 2001), while sine-wave stimulation does not produce harmonics and may well be safer for epileptics (Donker, Njio, Storm Van Leewan, & Wieneke, 1978; Regan, 1966; Townsend, 1973; Van der Tweel & Lunel, 1965). Fig. 3.1 shows the EEG effects of square-wave (xenon flash, checkerboard pattern) photic entrainment at a variety of frequencies.

AVE effects are primarily associated with frontal, parietal, occipital brain regions and near the vertex (Frederick, Lubar, Rasey, Brim, & Blackburn, 1999). For example, Fig. 3.2 shows a qEEG (quantitative electroencephalograph), or "brain map" from the Sterman-Kaiser Imaging Labs (SKIL) database, in 1 Hz bins showing the frequency distribution of AVE at 8 Hz. The area within the circle at 8 Hz shows maximal effects of AVE in central, frontal, and parietal regions (at 10 µv in this case) as referenced with the oval area on the legend. It is through associated influences on frontal brain regions that AVE

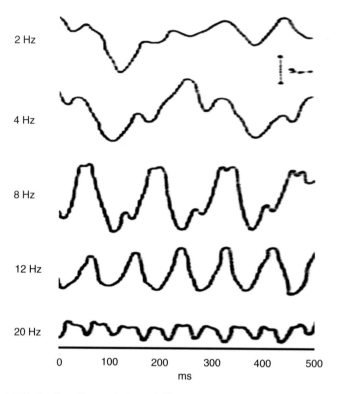

FIGURE 3.1 EEG showing photic entrainment (Kinney et al., 1973).

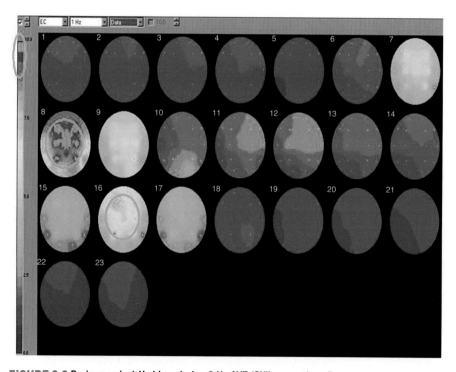

FIGURE 3.2 Brain map in 1 Hz bins, during 8 Hz AVE (SKIL-eyes closed).

has been shown effective in reducing depression, anxiety, and attentional disorders. For the more advanced reader, a harmonic is also present at 16 Hz (the circled image), which is typical of semi-sine wave (part sine/part square wave) stimulation.

DISSOCIATION/HYPNOTIC INDUCTION

Several related studies have been completed since the 1950s on hypnotic induction and dissociation (Walter, 1956; Lewerenz, 1963; Sadove, 1963; Margolis, 1966; Leonard et al, 1999 & 2000), and on altered states of consciousness (Hear, 1971; Lipowsky, 1975; Glicksohn, 1986-1987). The first study on dissociation induced via entrainment involved hypnotic induction, which found that photic stimulation at alpha frequencies could easily put subjects into hypnotic trances (Kroger & Schneider, 1959) as shown in Fig. 3.3. Notice that nearly 80% of the participants in the study were in a hypnotic trance within 6 minutes of photic entrainment.

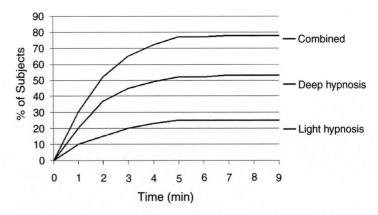

FIGURE 3.3 Photic stimulation induction of hypnotic trance (Kroger & Schneider, 1959).

AUTONOMIC NERVOUS SYSTEM CALMING

Assisting clients with trauma histories to dissociate (in a constructive and meditative way), during the course of treatment, can be very important. AVE has been shown to dramatically reduce sympathetic and autonomic activity. This rapid return to homeostasis or restabilization has been given the term *dissociation and restabilization* (DAR) (Siever, 2003a). For example, Fig. 3.4 shows a typical reduction in forearm Electromyogram (EMG) and Fig. 3.5 shows a typical increase in finger temperature during AVE (Hawes, 2000). Notice that restabilization begins after approximately 6 minutes of AVE, when the user begins dissociating and the autonomic nervous system calms down.

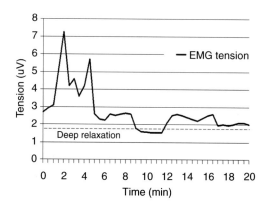

FIGURE 3.4 Forearm EMG levels during AVE (Hawes, 2000).

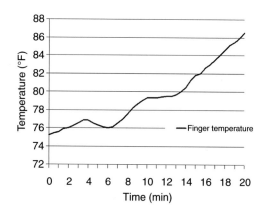

FIGURE 3.5 Peripheral temperature levels during AVE (Hawes, 2000).

HEART RATE VARIABILITY (MEASURE OF AUTONOMIC AROUSAL)

Heart rate variability (HRV) is a measure of variance of heart rate swing during inspiration and expiration. It also measures consistency in heart rhythm over time. It is an excellent measure of autonomic arousal and commonly is used with biofeedback for learning meditative breathing (Gervitz, 2000). Fig. 3.6 is a profile taken from the emWave PC (www.heartmath.com). The upper-left window indicates the actual heart rate throughout the recording. The bottom-left window shows a score as determined by an algorithm which looks at the high versus low heart rate, roundedness of the waves and consistency over time. The lower-right window indicates the average heart rate and a score as to how well the heart followed a breathing rhythm of 6 beats per minute (bpm) (10-second breath cycles). The upper-right display shows a spectral analysis of the breathing. Given a 10-second breathing cycle, there should be a peak at 0.1 Hz and the spectral display would look much like a witch's hat. Sympathetic activity is indicated at low frequencies (below 0.1 Hz) while compensatory parasympathetic activity is indicated at frequencies above 0.1 Hz.

Fig. 3.6 is the profile of a 30-year-old mother of two, struggling with post traumatic stress disorder following a disturbing event with her recently separated husband. She reported that her mind is constantly bombarded with distressing thoughts that she simply cannot stop. Although she is trying to breathe at 6 breaths per minute, she shows a "spike and clamp" (and unhealthy) heart rate frequency, which is typical with fear and anxiety. Her score is zero, her coherence ratio is 100% in the low category and her average heart rate is 99 bpm. Her spectral display looks like a mountain range with all her autonomic activation.

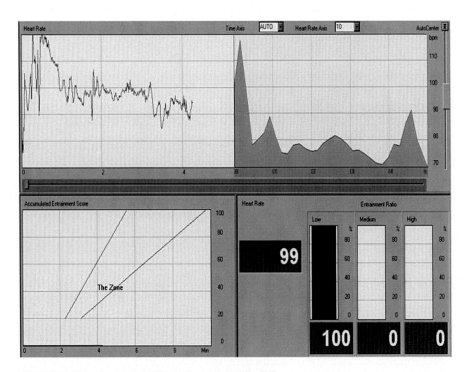

FIGURE 3.6 Heart rate variability just prior to 7.8 Hz AVE.

A few minutes after the data shown in Fig. 3.6 were gathered, this client was stimulated with AVE at 7.8 Hz and immediately began to calm down (as can be seen in Fig. 3.7). Recording began after 5 minutes of AVE, at which time her heart rate had already fallen by 22 bpm. This recording shows the effects of dissociation and restabilization. Her breathing and heart rate quickly became stable. Aside from a few thoughts that crept in just past the 2-minute mark, she commented that she was completely relaxed.

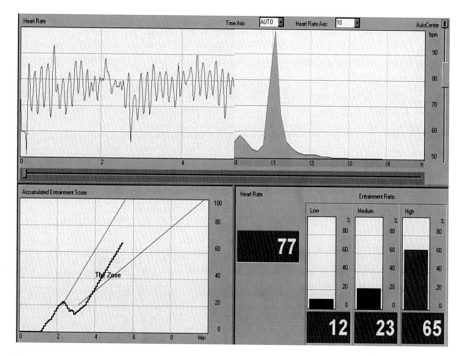

FIGURE 3.7 Heart rate variability during 7.8 Hz AVE.

When an HRV study is done with AVE, the preferred device is a DAVID AVE unit from Mind Alive Inc. The specific procedure used with each participant will be described here. First, it is powered on with a meditation protocol selected. The eyesets (containing the entraining lights) and the headphones then are placed on the participant, but the entraining lights and tones are shut off at this time. The DAVID devices have a heartbeat sound that is played through headphones at 24 bpm. The participant is instructed to breathe in for two heartbeats and exhale for two heartbeats. This makes a four-beat breathing cycle. Consider 24 bpm divided by 4 beats/breath = 6 bpm. After about 10 minutes of breathing, the session is halted. There is a short break of a few minutes and the session is repeated, but now with the entraining lights and tones turned on. Notice how quickly the autonomic nervous system quietens down, by observing the drop in heart rate and smoothing of the spectral graph.

INCREASED CEREBRAL BLOOD FLOW

Normal cerebral blood flow (CBF) is essential for good mental health and function. Measures of CBF show that hypoperfusion of CBF is associated with many forms of mental disorders including anxiety, depression, attentional, and behavioral disorders (Teicher et al., 2000), and impaired cognitive function (Amen, 1998; Meyer et al., 1994; Meier et al., 2015). Some of the purported beneficial effects of AVE have been attributed to increases in frontal region CBF (Fox & Raichle, 1985; Fox, Raichle, Mintun, & Dence, 1988; Sappey-Marinier et al., 1992). For example, Fig. 3.8 shows an increase of 28% in CBF within the striate cortex, a primary visual processing area within the occiput. As an interesting side note, maximal increases in CBF have been shown to occur with stimulation around 7.8 Hz, which is known as the "Schumann resonance" frequency of the electromagnetic propagation around Earth (Balser & Wagner, 1960).

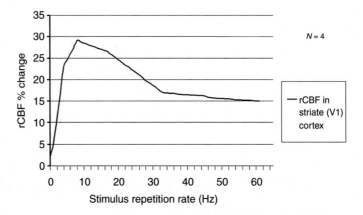

FIGURE 3.8 Cerebral blood flow at various photic entrainment repetition rates (Fox & Raichle, 1985).

INCREASED NEUROTRANSMITTERS

Seasonal affective disorder (SAD) involves reduced levels of melatonin, a neurotransmitter that is related to hibernation in animals and is associated with slowing of brain waves and reduction of CBF (Murphy et al., 1993). Endorphins are essential for blocking pain and are the basis of analgesic medication. Low levels of serotonin are part of most every psychiatric disorder. Moderate levels of norepinephrine (the "brain's adrenaline") are involved in mental vigilance. Norepinephrine is increased by caffeine (which is why we enjoy a cup of coffee in the morning). The coffee effect is much like stimulant medications like Ritalin. It wakes up our frontal lobes, increases mental sharpness, and stabilizes emotions. Many psychiatric disorders, including anxiety, depression, obsessive-compulsive disorder, schizophrenia, and memory and cognitive disorders have been linked to brain neurotransmitter action (Emmons, 2010; Arco & Mora, 2009). Of special relevance to this book chapter, there is evidence that cerebral-spinal fluid levels of melatonin fall, while serotonin, endorphin, and norepinephrine levels rise considerably following 10 Hz, white light AVE (Shealy et al., 1989) as shown in Fig. 3.9. Increases in endorphins are associated with increased relaxation, while increased norepinephrine, along with a reduction in winter daytime levels of melatonin, typically increases alertness.

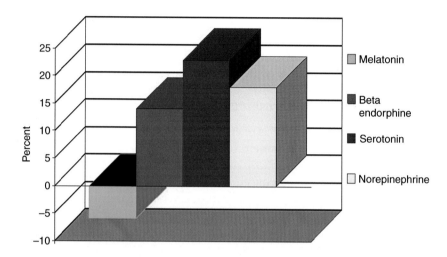

FIGURE 3.9 Neurotransmitter levels following AVE (Shealy et al., 1989).

UNEXPECTED EFFECTS OF LIGHT AND SOUND STIMULATION

Both nonentraining aspects of light and sound stimulation and unexpected side effects of entrainment are emerging within the literature. For instance, for treating auditory processing disorder (APD), often comorbid with autism spectrum disorder, the Tomatis Method, a three-part technique in which a variety of filtered prenatal sounds, then filtered sounds of Mozart music and Gregorian Chants, followed by an active phase, are presented via headphones to the patient. Marked improvements in auditory processing and various forms of memory and cognitive improvements have been observed (Ross-Swain, 2007).

Intranasal light therapy (ILT) is a technique involving the shining of near-infrared light into the nasal cavity from a light-emitting diode (LED) mounted on a small clip. Ample evidence of ILT has shown effectiveness in treating a multitude of neurological disorders including insomnia, mild cognitive impairment, Alzheimer's disease, Parkinson's disease, schizophrenia, migraine, and stroke. The current theory of function has been based on optimizing the results for blood irradiation, which may be holding back the brain's potential for better neurological outcomes (Lim, 2013). Although the main premise of entrained light is that the frequency of the light pulses are affecting the brain, this does not preclude the fact that entrained light shines through the skull and into the brain directly, as does light therapy.

Many studies on near-infrared (NIR) light therapy for a variety of brain issues ranging from depression to concussion have been emerging in recent years (Doidge, 2015). NIR light has been shown to improve mitochondrial function in hypoxic cells, increase adenosine triphosphate (ATP), important for cellular metabolism because it releases local nitric oxide, which in turn increases regional CBF. A recent study utilizing an array of head mounted transcranial near-infrared LEDs on people with mild traumatic brain injury, showed significant improvements in cognitive performance, social, interpersonal, and occupational functions (Naeser et al., 2014).

One of the most intriguing effects of entrained light involves the application of 40 Hz (gamma) photic entrainment for 1 hour into the eyes of mice with the mouse equivalent of Alzheimer's disease. This study showed that, within a few hours, amyloid plaques were reduced in the range of 50%–70% and microglia cell body diameter increased by 165.8% from consuming the amyloid deposits (Iaccarino et al., 2016). As has been demonstrated throughout this chapter is the notion that the actual brain frequency driving aspect of entrainment is just a fraction of the overall effects.

RESEARCH

Over the past several decades, AVE has been associated with several types of beneficial outcomes in a wide variety of clinical studies. It has been found effective for SAD (Berg & Siever, 2009); for improving concentration and memory in college students (Budzynski & Tang, 1998; Budzynski, Jordy, Budzynski, Tang, & Claypoole, 1999; Budzynski, Budzynski, & Tang, 2007; Wuchrer, 2009; Siever, 2003c); for reducing worry in college students (Wolitzky-Taylor & Telch, 2010); for the treatment of ADHD and behavior disorders; (Carter & Russell, 1993; Joyce & Siever, 2000; Micheletti, 1999); for regaining motor control in post aneurysm (Russell, 1996); for treating depression and risk of falling in seniors (Berg & Siever, 2004); for treating Alzheimer's (Budzynski, Budzynski, & Sherlin, 2002); for improving brain function and memory in seniors (Williams, Ramaswamy, & Oulhaj, 2006; Palmquist, 2014); for reducing chronic pain from occupational injury (Gagnon & Boersma, 1992); for reducing symptoms of fibromyalgia (Berg et al., 1999); for reducing symptoms of temporomandibular dysfunction (Manns, Miralles, & Adrian, 1981; Morse & Chow, 1993; Thomas & Siever, 1989); for reducing anxiety during dental procedures (Morse & Chow, 1993; Siever, 2003b); for treating PTSD in war veterans (Trudeau, 1999); and for improving sleep (Tang, Riegel, McCurry, & Vitiello, 2016). Whereas there is not space in this chapter to describe all the studies cited, Table 3.1 lists a variety of clinical conditions, along with the number of studies, the combined size of study populations, and the type of demographic.

Table 3.1 Clinical Studies Involving AVE

Condition	Studies (N)	Demographic
Attention Deficit Disorder (ADD)	4 (359)	School children
Academic performance in college students	3 (134)	College students
Reduced worry/anxiety in college students	3 (163)	College students
Drug rehabilitation	1 (44)	General population
Depression and anxiety	3 (93)	General population
Improved cognitive performance in seniors	1 (40)	From seniors' homes
Reduced falling and depression in seniors	1 (80)	From seniors' homes
Memory in seniors	1 (40)	Seniors - drop in
Dental (during dental procedures)	3 (>50)	Patients
Temporomandibular dysfunction (TMD)	3 (76)	General population
Seasonal affective disorder (SAD)	1 (74)	General population
Headache (migraine and tension)	2 (35)	General population
Neurophysiology	>100 (>3000)	General population
Heart rate variability and hypertension	5 (148)	General Population
Hypnosis/dissociation/meditation	10 (>2000)	General population
Pain and fibromyalgia	5 (178)	General population
Insomnia	1 (10)	General population
Post Traumatic Stress Disorder (PTSD)	1 (15) ~600 cases	Public, military, police
Literature reviews	5	Diverse population
Premenstrual syndrome (PMS)	2 (23)	General population
Diffuse axonal injuries (interruptions)	1 (5)	Closed-head injuries

THE ENIGMA OF RESULTS OF AUDIO–VISUAL ENTRAINMENT

Within the biofeedback and neurofeedback community, there has been a focus on the frequency following or entrainment effects of AVE. This is due primarily to neurofeedback practitioners using frequency parameters as their primary form of treatment, and because of this, they see AVE mainly through the same lens, that is, primarily as a brain wave driving technique. Although entrainment is a well-documented part of AVE, there are mysterious and quite frequently, amazing nuances to AVE that often are not considered by the biofeedback community and therefore missed. In the following pages, we will present a few case examples which indicate that the dissociative, neurotransmitter, and cerebral blood flow aspects of AVE may be just as, or even more, important than the frequency effects of brain wave driving. For example, as with some ADHD children, it is common for clients to fall into deep sleep with dual frequency AVE stimulation of an alpha-beta or SMR-beta nature. This makes AVE a valuable tool for assessing and treating sleep issues, such as poor sleep-onset and sleep apnea. Another example relates to successful treatment of cases involving diffuse axonal injury.

With diffuse axonal injury (where alpha magnitude especially is attenuated), it seems logical to provide entrainment at an alpha frequency. This technique has shown modest improvements, but has not been overly successful. What came as a surprise were clinical findings that dual frequency stimulation, where left visual fields and left audio were presented at a different frequency than on the right side, was highly successful. This technique provided right-hemisphere entrainment at SMR (12–15 Hz) and left-hemisphere entrainment at beta (20 Hz) frequencies. We are not aware of a physiological theory as to why it should work; only that it does clinically.

CASE STUDIES

In the following section, several case reports involving successful use of AVE are presented.

Definition of Some EEG and Quantitative EEG (qEEG) Terms

qEEG: First an electroencephalogram (EEG) is collected from the brain and then it is quantified, meaning that the EEG data over a period of time (usually several minutes) is averaged and processed. Most quantitative EEG (qEEG) software also includes a database, usually of normal persons (although some are of clinical conditions). Several measures are analyzed and a few of these measures are shown later. The qEEG software used to create the images below is from the SKIL EEG processing software (Sterman & Kaiser, 1999).

The electrodes are placed on the scalp at various positions as defined by the 10–20 electrode placement montage (Trans Cranial Technologies, 2012). These positions include regions, such as prefrontal (FP), frontal (F), temporal (T), central (C), midline (Z), parietal (P), and occipital (O) positions.

For the purpose of brevity, only qEEG measures of greatest concern in the following case examples are defined.

Absolute Magnitude: The average voltage amplitudes averaged over time, shown in microvolts (µv) for each frequency band.

Database Magnitude: The absolute magnitude in comparison to a database with an average healthy population.

Comodulation: EEG spindles and wavelets generally appear with some synchrony across and between brain regions. This is a measure of the degree to which the spindles and wavelets are in alignment. Comodulation is compared against a normative database.

Delta Phase: As used here, this is considered a measure of the integrity of the white matter extending from neurons to other regions of the brain.

Case Report 1: Boosting Cerebral Blood Flow

Many people struggling with ADD and cortisol damage from trauma show abundant slow alpha rhythms, which leaves them feeling lethargic, unable to have focus and attention and are foggy-headed. AVE stimulation at the exact same frequency of 7.8 Hz has been shown to inhibit this excessive slow wave activity.

Client information:		
Condition: ADD and Fibromyalgia	Age: 20	Gender: Female
Handedness: Right	Medications: None	

History

This college student has been struggling with low energy, poor focus, poor comprehension, and mild pain throughout her body. She had been diagnosed, by a medical doctor, with ADD and fibromyalgia. As she struggles to read more than half a page of her textbooks at a time, she rereads and rereads as she attempts to learn the study material and as a result, she often stays up until 3 am. Fig. 3.10 shows her linked-ears, magnitude on the SKIL database in standard deviations. Here we see that she is flooded with 7–8 Hz dominant and slowed alpha. She is only able to read two to three paragraphs before "fogging" out and quitting.

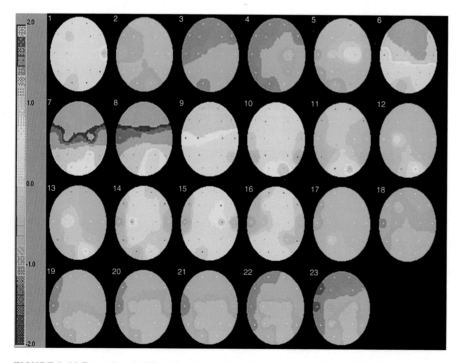

FIGURE 3.10 Eyes-closed alpha using linked-ears, Z-scored analysis (SKIL database).

During 30 minutes of squared, sine wave type of AVE at 7.8 Hz, her 7–8 Hz activity did increase as would be expected with AVE. However, 30 minutes following AVE, her abnormally high theta activity at 7 and 8 Hz normalized. This, seemingly counterintuitive effect is believed by the first author to occur from the increase in CBF, one of the benefits of AVE (Fox & Raichle, 1985). Fig. 3.11 shows her normalized brain wave activity 20 minutes following the cessation of AVE. This student was now able to read 10 pages before losing attention.

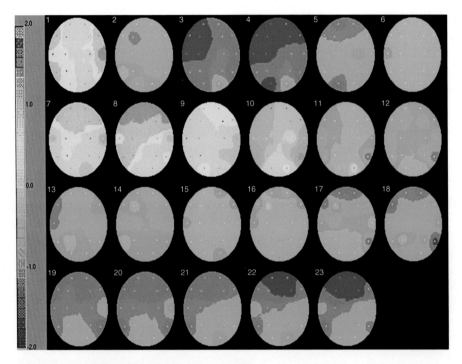

FIGURE 3.11 QEEG analysis 20 min post 7.8 Hz AVE.

Case Report 2: Calming Beta-Type of Anxiety Using Beta AVE

Client Information:

Condition: ADHD and generalized anxiety	Age: 40	Gender: Male
Handedness: Right	Medications: None	

History

It is well known that beta AVE often effectively puts children diagnosed with ADHD into a deep sleep, whereas alpha, theta, and delta AVE usually will not. The mechanism behind this counterintuitive finding may be that of dissociation. Or perhaps beta AVE suppresses slow frequencies and thus "puts on the brain's brakes," which in turn allows the child to relax. This is a case of a 40-year-old male diagnosed with ADHD, but also complaining of intense generalized anxiety for "no good reason" as commented by him. Fig. 3.12 shows his excessive activation as reflected in excessive randomized beta activity while he was experiencing anxiety.

To see if beta frequency AVE at 17 Hz would reduce his anxiety in the manner that often would put an ADHD child to sleep, he was placed on a DAVID Delight AVE unit using combined sine-square wave photic stimulation and isochronic auditory tones. Within 4 minutes of AVE, his excessive beta activity vanished along with his reported anxiety, as shown in Fig. 3.13. This effect lasted for about 2 days. He began a program using either beta or beta/SMR AVE every morning upon wakening so that he could go about his day with a calm demeanor.

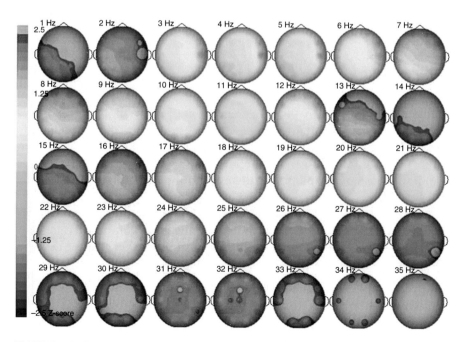

FIGURE 3.12 QEEG of a 40-year-old male with ADHD and severe anxiety.

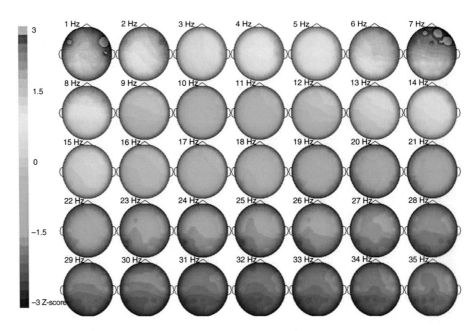

FIGURE 3.13 QEEG of a 40-year-old male at 4 min of 17 Hz AVE.

Thalamo-Cortical Issues/Closed-Head Brain Injury of the Diffuse-Axonal Type

Diffuse axonal injury (DAI) is a frequent result of traumatic acceleration/deceleration or rotational injuries. Depending on severity, it ranges from mildly noticeable to causing persistent vegetative state in patients. DAI is most commonly the result of motor vehicle accidents and sports injuries and the most significant cause of disability in patients with traumatic brain injuries. It is called diffuse because, unlike some other brain injuries that are in a specific area, DAI is widespread, sometimes affecting a large area. It is typically comprised of several small tears or distortions at the gray–white matter junction and within the corpus callosum causing electrical signals between the neurons of the cortex and the thalamus to be interrupted (as shown in Fig. 3.14).

The alpha rhythm is mediated by billions of synchronized thalamo-cortical circuits. This back and forth electrical volley takes close to 100 milliseconds (ms) and therefore healthy alpha is close to 10 Hz (Schreckenberger et al., 2004). When there is an electrical blockage to either axons or dendrites in the loop between the cortex and the thalamus, two things occur: (1) the alpha rhythm is attenuated or eliminated and (2) because they are no longer receiving a

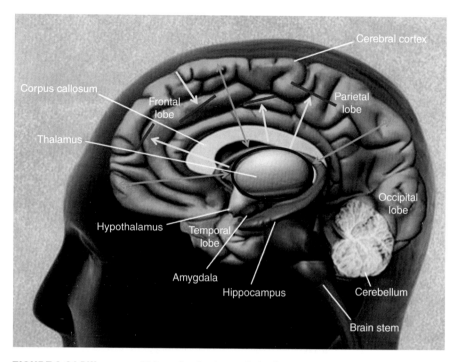

FIGURE 3.14 Diffuse axonal injury showing transmission breaks (Frewin et al., 2012).

synchronization pulse from the thalamus, they begin randomly oscillating on their own in the 1–2 Hz range (low delta) (Thatcher, Walker, Gerson, & Geisler, 1989; Steriade, Nuñez, & Amzica, 1993). As they are now oscillating without a sync pulse, delta phase is asynchronous and disconnected in relation to a database of a healthy brain (Thatcher et al., 1989).

Given that DAIs primarily affect the alpha rhythm and delta phase, these measures were most apparent clinically. Therefore, magnitude in 1 Hz bins and delta phase will be the focus of this next case.

Case Report 3: Jumpstarting the Thalamo-Cortical Loop Using Beta/SMR AVE

Client Information:		
Condition: Concussion from motor vehicle accident	Age: 43	Gender: Female
Handedness: Right	Medication: None	

History

This woman was in a motor vehicle accident in 2011. She was in immediate pain and judging from her self-assessment, developed the following neurological and behavioral issues:

1. *Frontal lobe slowing*: It is typical to see such brain wave slowing somewhat like would be observed in ADHD children or seniors with dementia. Here are some quotes from this woman:
   a. "I would do an activity one day and it would not hurt while I was doing it but I would be very sore and exhausted a day or two later and it would often take days or weeks to recover."
   b. "I would have a conversation with someone and my mind would go blank talking to them. Then a day or two later, out of the blue, I would think of what I wished I had said to them."
2. *Loss of spatial ability*: She was bumping into and dropping things frequently.
3. She has not been able to work and has had trouble organizing her thoughts.

Assessments and Work Done:

1. Symptom Checklist-90-Revised (SCL-90-R) was administered. Her overall state of mind improved from a score of 70 to a score of 10 following 3 weeks of AVE. She had a score of 5 following 15 weeks of AVE treatment, reflecting exceptional improvement.
2. QEEG was completed.
3. AVE was provided in the form of SMR/beta left-side stimulation (right hemisphere) of 14 Hz and right-side stimulation (left hemisphere) at 20 Hz, using split-field eyesets and headphones with the intention of inhibiting her slow brain waves. EEG was recorded during AVE treatment.
4. She was given a DAVID Delight Pro device, which provides a combination of AVE and cranioelectro stimulation (CES). The addition of CES helps promote calmness and reduce pain. CES presents a small electrical pulse across the cranium (see Chapter 2). The resulting stimulus flows across the brain stem and appears primarily to stimulate serotonin and endorphin production making it beneficial for calming the mind and reducing pain.
5. One month later, she was placed on several applications of transcranial DC Stimulation (tDCS) over T5 (10–20 montage) to help improve language comprehension and name recall. TDCS, as used here, is the application of a small direct current (1–2 mA) used to excite neuronal activity and performance (Siever, 2013).
6. Follow-up SCL-90-R questionnaires at 3 weeks and at 15 weeks.

EEG/qEEG Analysis

Raw Waveforms Condition

Referring to Fig. 3.15, we see multiple indications of eye twitches. Note the ticks at FP1 and FP2, F7 and F8. This typically reflects anxiety. In this snapshot, both slowed alpha waves and excess beta waves are present. Both are typically accompanied with subjective feelings of anxiety. Fast (beta) waves often are considered a "side effect" of too many slow waves and by "speeding up" the brain, the fast waves often diminish. Electrode sites FZ and CZ show excessive slowed alpha. All channels are high in beta and beta spindle activity.

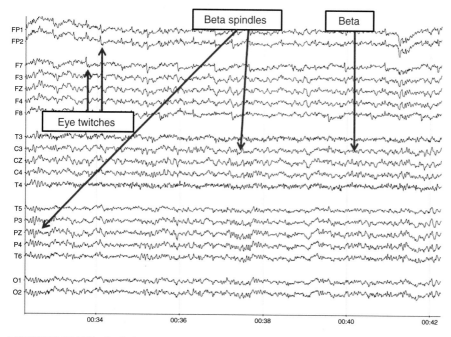

FIGURE 3.15 EEG prior to treatment with SMR/beta AVE.

The database indicates that the primary alpha frequency (9–11 Hz) is of quite low magnitude (Fig. 3.16). Loss of alpha is a classic sign of a DAI and generally an indicator of anxiety.

In Fig. 3.17, we can see desynchronization (areas in blue) between several areas commonly affecting reasoning, attention, verbal expression, emotional expression, cognitive processing, and impulse control.

In Fig. 3.18, we see several phase issues relating to a lack of connectivity between regions of the left and right hemispheres, as well as some front/back issues, an indicator of DAI.

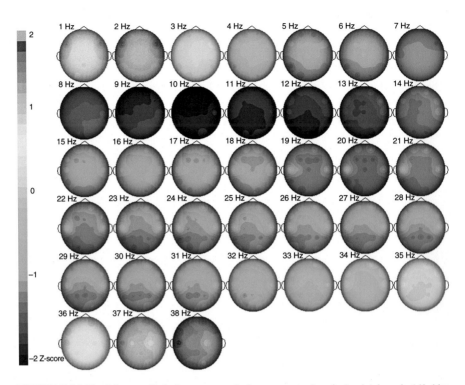

FIGURE 3.16 Absolute magnitude frequency analysis as compared against a database in 1 Hz bins.

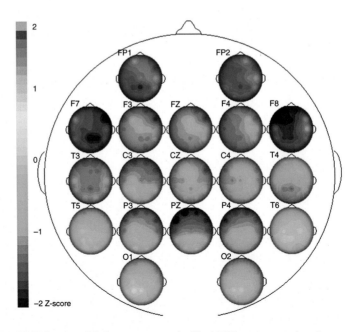

FIGURE 3.17 Alpha comodulation as compared with a database—eyes closed.

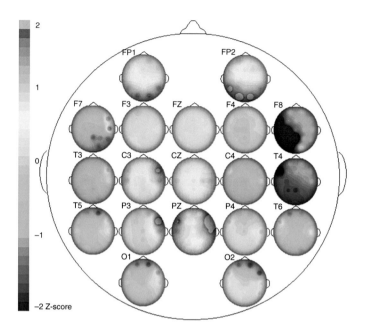

FIGURE 3.18 Delta phase as compared with a database.

EEG Assessment During Beta/SMR AVE

At roughly 20 minutes of SMR/Beta AVE, alpha spindles began forming. Notice the alpha spindles in Fig. 3.19. When the session ended at 30 minutes, the client exclaimed "Wow! My head is clear and I can think!"

Looking at absolute magnitude (Fig. 3.20), we see greatest EEG alpha at 10–11 Hz, which is where it should be (occipital and parietal regions).

Fig. 3.21 shows a normal brain map as compared with the database. The only difference is the heightened 20–21 Hz activity, which is considered a result of the AVE and not of concern.

Fig. 3.22 shows that comodulation is close to normal and not considered clinical.

Given that the neurons presumably now have a thalamic synchronization pulse, the phase has been restored (Fig. 3.23). It is near normal and not considered clinical.

While it's always good to treat the qEEG, what matters is that the client feels the effects of the treatment. Table 3.2 shows the client's subjective experience of the treatment.

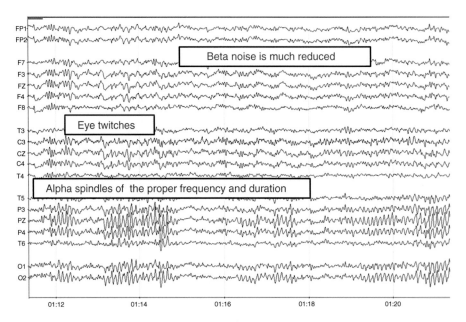

FIGURE 3.19 Raw waveforms following 20 min of left side 12–14 Hz and right side 19–21 Hz AVE.

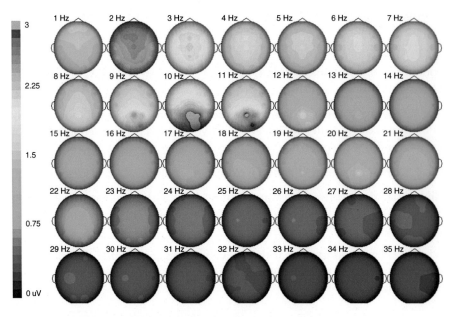

FIGURE 3.20 Absolute magnitude frequency analysis at 3 μV scale.

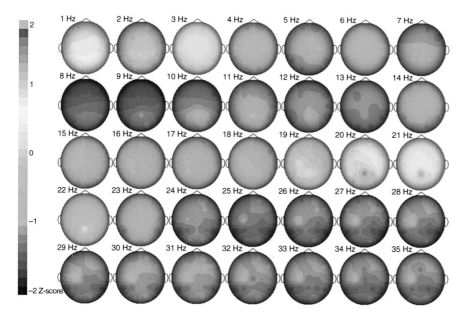

FIGURE 3.21 Absolute magnitude frequency analysis as compared against a database.

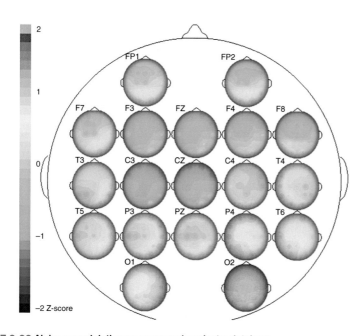

FIGURE 3.22 Alpha comodulation as compared against a database.

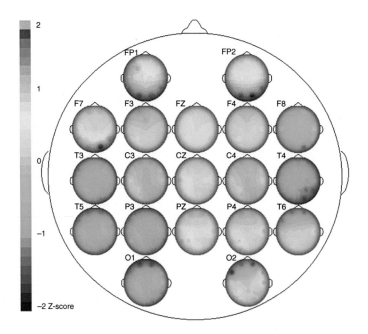

FIGURE 3.23 Delta phase as compared against a database.

Measure	Raw Scores Pre AVE	Raw Scores 3 Weeks Post AVE
Table 3.2 SCL-90-R Score Pretreatment and Post 3 Weeks of Treatment		
Somatization	1.09	0.09
Obsessive-compulsive	2.50	0.40
Interpersonal sensitivity	0.67	0.11
Depression	0.77	0.08
Anxiety	0.70	0.10
Hostility	0.33	0.17
Phobic anxiety	0.00	0.00
Paranoid ideation	0.33	0.00
Psychoticism	0.20	0.00
Grand total raw score:	72	10
Global severity index:	0.80	0.11
Positive symptom distress index:	1.31	0.18
Vegetative depression	1.2	0.0
Suicide ideation	0.9	0.3
Dysfunctional sleep	1.3	0.0
Chronic fatigue	2.4	0.2
Decreased cognitive functioning	2.3	0.5
Physical pain	0.8	0.3
Pain versus somatization index	0.7	0.0

DYNAMICS OF BRAIN RESPONSE TO REPETITIVE STIMULATION

In this section, we turn attention away from the traditional frequency driving effects of AVE to more general mechanisms and principles that may underlie the brain's particular responses to repetitive stimuli. Thus, the following paragraphs have relevance to all of the chapters of this book.

In particular, we are interested in how network properties and nonlinear effects may produce results *beyond* simple changes in arousal, such as activation or relaxation. On the surface, a repetitive stimulus is nothing more than the combination of single stimuli, in a time-dependent manner. However, because the function of the brain is acutely conditioned on factors, such as synchrony, network connectivity, and self-organization, it is likely that complex processes can occur that produce much more than simple changes in neural processes. Whereas physical processes, such as blood flow, metabolic activity, and related physiology might be expected to simply combine repetitive stimuli in a monotonic fashion, the way in which the brain *as a system* reacts to repetition at various frequencies and via different modalities could be expected to be much more complex, even unpredictable.

Neuroplasticity

Neuroplastic effects of repetitive stimulation include those that involve changes in neuronal responsivity and tendency to activate, even after the cessation of the stimulation. Hebb's law (1961) that "neurons that fire together, wire together" further suggests that connectivity changes as a result of the stimulated neuronal firing may also occur. Underlying mechanisms may include changes in receptor distribution and response, increase or decrease in the number of synaptic connections, alterations in genetic expression and protein synthesis, metabolic or even anatomic changes.

Neuroplasticity has its origin in response characteristics that are not simply linear, but can change. In particular, we are interested in response characteristics pertinent specifically to repetitive stimuli, hence being frequency specific. There are several potential mechanisms that relate to nonlinear frequency-specific responses. One of these is the presence of a refractory period after each stimulus, which includes the time that the system requires to recover completely. If another stimulus is presented within the refractory period, the response to that stimulus will be affected by the presence of the response mechanisms, so that the response is now different.

When looking at EEG evoked responses, if stimuli are presented slower than about 2 per second, there is no evident change to the individual responses. Even at higher rates, up to and exceeding 10 per second, minimal change due to successive stimuli can be found, even as the evoked potentials overlap in time. This is because of the "pipeline" effect, in that the brain passes information sequentially from primary to secondary and further processing locations, thus freeing up the primary sites to process subsequent stimuli. As the stimulus rate increases, several additional mechanisms come into play. One is the presence of refractory effects, in which the response to subsequent stimuli is modified, because stimuli fall within the time range of the aftereffects of prior stimuli. In addition to this, resonance effects that are reflected in a network's intrinsic natural frequencies, including nonlinear effects, put an entirely new dimension on the total brain response to the stimulation. The entire brain can be expected to go into alternate modes, which may be distinguished by particular sets of activation and connectivity, which can participate in learning processes.

Thus, responses to repetitive stimuli are much more than simply the concatenation of responses to individual stimuli. Grossberg (2013) articulated the principles of adaptive resonance in neuron-like networks. His theory is that the brain is able to carry out fast, incremental, and stable unsupervised and supervised learning in response to a changing world. This, then, may be a key foundation in the ability of the brain to respond to repetitive stimuli in a manner of learning and adaptation. In the real world, the brain consists of thousands of interrelated networks, each operating at its own dynamically changing set of frequencies and resonances. Different networks have different roles, so that operational modification of any network is reflected in the duties that it undertakes. As such networks need not be only sensory, but can involve secondary and tertiary activation of attentional, emotional, and decision-making networks, the effects of repetitive stimulation can be unpredictable and wide ranging.

Physioplasticity

One class of effects includes physiological changes incorporating metabolism, respiration, nutrition, and the interrelations of these mechanisms. Repetitive stimulation, particularly when it produces metabolically significant activity, such as action potentials, can have both immediate and lasting effects on the brain. These effects would primarily be due to the physical consequences of the increased activity of neurotransmitters, transmembrane transport mechanisms, and the associated vascular, respiratory, and recovery mechanisms. Vascular changes would include alterations in capillary structure and physiology, and changes in blood flow. Respiratory activity at the cellular level involves the intake of oxygen and fuel, and the consequent production of carbon dioxide and water as products. This activity occurs in mitochondria and involves the Krebs cycle (http://www.chemistrylearning.com/krebs-cycle/), which in turn operates by virtue of the cytochrome chains which organize the enzymatic activity, so that cells can produce and use energy. The associated recovery mechanisms take care of the removal of metabolic by-products, as well as the reuptake and transport of neurotransmitters. It is therefore reasonable to expect that repetitive stimulation can have effects, not only on neurons, but also on associated structures including the glia, vascular structures, and even supportive tissues.

Attractors and Random Stimulation Concepts

The dynamic stability of the brain is evident in the time-series behavior of the EEG (and qEEG) in both the endogenous (unstimulated) state and various stimulated states. Given the variability and complexity of the population dynamics, energetics, and communication/control throughout the brain, we would expect the systems networks of the brain to respond in a complex, non-linear fashion, rather than as the chiming of a bell, for example. Therefore, it is reasonable to ask whether specific spatial (and temporal) distributions of activation should be expected, in response to simple repetitive stimulation.

Following each stimulus, the system goes through a series of transitions (microstates) that define an attractor for that system. If the system is re-stimulated before that attractor has run its course, the system will respond differently to that stimulus, based on the nonlinear dynamics of that particular system and response. Based on the work of Basar (2013), we can view the brain as a chaotic system that exhibits attractors and limits cycles that are highly rate dependent.

A great emphasis has been placed on the specific pathways inherent in processing stimuli, such as the visual system incorporating the retina, optic nerve, thalamus, optic radiation, and cortical projections in Brodmann areas 17 and 18. Similarly, for auditory stimuli, a pathway involving cochlear nuclei, the inferior colliculus, the temporal lobes, and associated auditory processing areas are implicated. However, given the hyperconnectionism in the brain, it is not unreasonable to expect that activation pathways and changes may instantaneously involve alternate, nonspecific pathways, thus involving networks or hubs, rather than overtly mapped areas of sensory cortex.

Subjective Aspects

The pervasive emotional content of rhythm is clearly evident. It is, for example, a foundational aspect of the experience of many sports events, dance clubs, and drumming circles. The "Jaws" theme demonstrates the profound feeling elicited by a simple repetitive beat. Why does repetitive stimulation feel the way it does? Why is there something festive about seeing flashing lights or hearing the beat of drums? There is something deeply rooted in human experience that relates to the presence of repetitive stimuli, that goes beyond the simple mechanistic consequences. Why is there something emotional about the speed of a stimulus, in and of itself? Without appealing to content or learned experience, a subject may report "I like that," "that feels good," or "I don't like that, that's disturbing," simply as the rate of a repetitive stimulus is adjusted.

Huang and Charyton (2008) conducted a review of psychological effects of repetitive stimuli in 20 peer-reviewed experimental studies selected from the literature. The outcome measures included standardized reading tests, Wechsler Intelligence Scale for Children (WISC), Test of Variables of Attention (TOVA), Stroop Test, various continuous performance tests (CPT), state anxiety, and academic tests. They identified a wide range of positive outcomes including cognitive function, stress/anxiety, pain, headaches, mood, and behavior. Carter and Russell (1993) and Joyce and Siever (2000) observed behavioral and cognitive improvements in ADHD children, while Wolitzky-Taylor and Telch (2010) noticed a reduction in worry amongst college students. Their general conclusion was that this approach has therapeutic effectiveness for conditions involving these measured variables and that additional study should explore alternate protocols.

Rhythmic Stimulation of a Hyper-Connected, Allostatic System

Given the complexity and purposeful design of the brain, it is unlikely that it would respond to a repetitive stimulus in a direct, linear manner. The brain has been configured to respond in a myriad of adaptive and goal-directed manners to external events. Further, it is comprised of a collection of interacting, specialized subsystems that operate in a delicate balance of control and communication. Based on a superficial analysis, one might posit that repetitive stimulation produces an induced synchronization of brain locations that have access to the stimulus timing information. Primary sensory areas might be expected to have a primary or phase-locked response to the stimulus, while secondary and subsequent processing areas would be expected to have lesser amounts of induced synchrony, depending on the degree of connection (or, conversely, autonomy) with primary sensory areas.

Fundamentally, it seems that we are providing information to the system that is reflected in relative changes in local and global synchrony, with concomitant changes in neuronal activation, communication, and control. The observed outcomes indicate that the effects on the brain are far from simple or linear. They involve complex interactions of networks and systems that are tuned for specific frequency-related response behavior. Specific tuning aspects are further localized, so that regions of functional response characteristics emerge, and may change in time. Considering emotional responses alone, it is relevant to consider how a simple repeating event can invoke affective changes, modulating such factors as relaxation, arousal, and positive and negative moods.

The goal of a homeostatic system is to recover the initial state in the face of perturbations. Examples of homeostatic systems are thermostats and related control mechanisms. Often, the perturbations are anticipated, such as variations in the weather or the opening or closing of a door. The hallmarks of homeostasis are the speed and precision with which the system reacts, in order to maintain, as well as possible the stated goal state. The operating range of a homeostatic system is generally "designed in" and involves well-defined, fixed, and rigid parameters. The goal of an allostatic system, however, is different. It is to react to perturbations in such a way as to alter behavior toward a goal. Often, the perturbations are not anticipated, may be unpredictable, and may threaten the function of the organism. The operating range of an allostatic system may vary with time and may adapt itself to demands. One hallmark of such a system is its ability to spontaneously explore and widen its own operating range, in order to determine the possible range of healthy functioning. Biofeedback and neurofeedback modalities may make use of this concept. For example, heart rate variability training seeks to optimize the range and variability of heart rate, producing benefits in adaptability and resilience.

In the final analysis, what is being put forth is nothing more than information in a particular form. From an information point of view, a single flash of light or a single click contains the minimum amount of information that can be conveyed to the sensory system. The effects of repetitive stimulation can be viewed in terms of information and providing a system with the ability to explore alternative attractors and limit cycles.

Bertolero, Yeo, and D'Esposito (2015) described the dynamic organization of up to two dozen connector hubs, which are alternately recruited, then released as the brain adapted to tasks. According to their view, based on a review of 9208 studies using 77 different cognitive tasks, instantaneous brain activity is characterized by coordination and integration of information for specific purposes. Using graph theory, they identified 14 networks and 25 hubs that dynamically shared the processing burden according to immediate needs, as the brain handled problem-solving tasks.

De Ridder, Verplaetse, and Vanneste (2013) put forth a theory based upon predictability of experience and the brain mechanisms underlying learning and free will. While this may seem superficially unrelated to the topic of repetitive stimuli, it is worth noting that a repetitive stimulus embodies a high degree of predictability. In a continuous stream of stimuli, each successive event arrives with absolute certainty as to its timing and characteristics. The authors note that stimuli that are completely predictable do not merit conscious processing. When faced with such an input, the system will respond in such a way that its mechanism for creating and evaluating multiple predictions will, correspondingly, be of lesser value. We may look, therefore, for brain responses that embody changes in dynamic allocation of resources, with respect to attentional and predictive mechanisms.

Various results are reported in the literature assessing EEG changes in association with rhythmic stimulation. In a prior study of photic stimulation,

Timmermann, Lubar, Rasey, and Frederick (1999) observed changes in beta that persisted after the stimulus was presented. Amirifalah, Mohammad, Firoozabad, Shafiei Darabi, and Assadi (2011), Amirifalah, Mohammad, Firoozabadi, and Ali Shafiei (2013), Ali Shafiei, Mohammad Firoozabadi, Rasoulzadeh Tabatabaie, and Ghabaee (2012), and Stahl and Collura (2012) demonstrated localized changes in EEG in response to repetitive low-intensity pulsed electromagnetic fields (pEMF). Generally, observed effects consisted of reductions in amplitude of particular bands. Stahl and Collura (2012), for example, reported localized reductions in alpha activity in response to pEMF, using standardized low-resolution brain electromagnetic tomography (sLORETA) to estimate current source density.

Further insight into the effects of small electromagnetic fields recently has been reported using microelectrode studies on rat hippocampus. Using this model, Zhang et al. (2014) and Qiu, Shivacharan, Zhang, and Durand (2015) recently have demonstrated that small electrical fields can in fact take part in neuronal communication. That is, even in the absence of synaptic, gap junction, or diffusion effects, neurons can respond to small extracellular fields. When applied as a form of rhythmic stimulation, even minute induced shifts in cellular activity (possibly via modulation of post synaptic potentials) can be expected to alter population dynamics. Depending on the dynamic stability of the networks, as well as how close transmembrane potentials are to threshold, a shift of a few millivolts could be expected to have effects at the network level, as small shifts in population dynamics shape the aggregate behavior of connected populations.

Toward a Model of Specificity in Rhythmic Stimulation

Given the richness and complexity embodied in a seemingly simple form of sensory information, the question arises as to how the brain responds to repetitive information, as a system. To this end, we are interested in how specificity can arise with regard to both the location of responses, as well as frequency-related effects. With the foregoing considerations in mind, the second author and associates undertook a brief pilot study to determine whether frequency-specific and location-specific changes in brain activation could be identified in response to repetitive stimulation.

We chose to investigate photic, auditory, and electromagnetic stimulation (EMS), each separately, in a sequence of experiments, some of which will be described later in greater detail. It is noted that the medium by which these stimuli are introduced to the brain are quite different. The visual system incorporates visual sensory and perceptual pathways that are activated by flash-elicited action potentials, primarily involving the relevant locations (optic nerve, lateral geniculate nucleus, primary visual cortex, etc.). Auditory stimulation enters the brain through lower-level pathways, including the inferior colliculus and olivary nuclei, thereafter projecting to temporal lobe locations. EMS produces small electrical fields in the brain tissue, which will have nonspecific, but local effects on transmembrane potentials and related local fields. Relevant to this, a link between EMS and thalamo-cortical dysrhythmia and neuropsychiatric disorders has been proposed to support clinical evidence of its efficacy (Fuggetta & Noh, 2012).

The purpose of this pilot study was to identify frequency-specific and location-specific changes in brain activation in short-term samples of EEG, using a statistical method. In order to maximize the likelihood of finding functionally meaningful location data, sLORETA was used to process whole-head EEG data. While surface EEG maps are subject to a variety of confounding effects, the use of sLORETA as a localizing procedure can be invaluable. Effects may be seen using sLORETA that are not visible on the surface due to blurring, paradoxical localization, effects of dipole orientation and multiple dipoles, and other distorting effects. The availability of the modality makes it possible to image the activation changes in response to situations or stimuli, with localization that is relevant to functional organization of the cortex. This can make it possible to differentiate the effects of different stimuli, based on their physiological properties.

In this study, we combined sLORETA with a new Z-score based procedure that uses the subjects' at-rest qEEG as a reference and identifies changes over the short-term based on statistical analysis. In order to take advantage of individualized EEG signatures and not rely on a "normative" database, a self-referenced model was employed. The procedure consisted of taking 2 minutes of at-rest EEG and computing over 16,000 variables and their variability (standard deviation) and using this as a statistical database for reference. In the same sitting as the acquisition of this baseline, stimuli were provided over 30-second intervals, separated by 10-second pauses, at frequencies of 3.5, 7.0, and 14.0 Hz. In this study, three forms of stimulation were used: photic, electromagnetic, and auditory. The procedure as shown in Table 3.3 was followed for each modality used.

Table 3.3 Stimulation Sequencing for Audio, Visual, and Electromagnetic Stimulation

2 min	Resting EEG
30 s	3.5 Hz stimulation
10 s	Resting EEG
30 s	7 Hz stimulation
10 s	Resting EEG
30 s	14 Hz stimulation
2 min	Resting EEG

All stimuli were presented with eyes closed. Stimulation consisted of (1) photic stimulation using DAVID PAL white LED glasses, (2) auditory stimulation using conventional ear buds, and (3) EMF stimulation using two MicroTesla (electromagnetic) paddles applied at C3 and C4. The Atlantis "photic" output channel was used for all modalities, providing flashes, pulsed EMF, or clicks, depending on the output devices.

In order to better measure and image changes in response to stimuli, a Z-score based imaging method was used. A reference data set was first constructed by taking 1 minute from a 2 minute at-rest baseline and processed using BrainAvatar Z-Builder signal processing (Collura, 2012, 2013, 2014a, 2014b) to produce amplitude means and standard deviations for all frequency bands, for all scalp locations, sLORETA voxels, and sLORETA Regions of Interest (ROI). ROIs were computed for 97 different homologous regions including the Brodmann areas, the named lobes and regions, and for the hubs described by Hagmann et al. (2008). Once this data set was computed, it was possible to compute metrics for any other selected samples, and by comparison, convert all measurements into Z-scores.

A 10-second sample was taken from the stimulus interval for each modality and analyzed using the subject's individual Z-score database, producing Z-score results. These Z-scores show which qEEG components and locations have changed. These Z-scores are not based on a normative reference database, but are instead based on the subject's own initial EEG. Thus, Z-scores reflect change from the initial state and do not reflect "normality" or "abnormality" in any way. As the use of Z-scores in this manner involves multiple comparisons of many ROIs, Bonferroni correction (Sankoh, Huque & Dubey, 1997) was applied to the results shown here.

The list of ROIs was sorted and examined to determine which ROIs have shown significant changes in activation, as evidenced by Z-score changes. The list of ROIs is taken from the sLORETA definitions, supplemented by custom ROIs designed based on published networks or hubs. Of the 97 ROIs on the list, there are 47 Brodmann areas, 40 named lobes, gyri, and structures, such as the insula, uncus, etc., the six Hagmann hubs, the salience network, central executive network, and two default-mode networks. The list therefore provides a broad range of possible detected structures, both anatomical and functional. The system produces data for all areas in both left, right, and L + R format. For the initial study, only the combined (L + R) data were examined.

For this pilot study report, the immediate responses to the 3.5 Hz stimulation were analyzed and only the delta and gamma bands were inspected. The intent here was to determine whether significant frequency-specific and location-specific changes would emerge in response to a particular stimulation. On inspection of the data, it was apparent that differences appeared in left versus right ROIs. That is, there was some lateral specificity as indicated in the figures. The results of this analysis are summarized later. It was seen that the different stimulation modalities had different effects on the subjects' EEG during stimulation and that the observed EEG changes did not persist into the period following the stimulation.

The effects of each type of stimulation in each band are illustrated in Figs. 3.24–3.31, which show the sLORETA images resulting from the individualized Z-score analysis described. It is evident that the immediate effects of the stimulation in this study demonstrate location-specific and frequency-specific changes and that these changes involve diverse brain regions and functional areas. It is interesting to note that one of the primary observed effects of repetitive stimulation was to inhibit unrelated frequency activity and in areas of the brain that are not directly related to sensory processing.

The following findings are of particular note:

1. The effect of all stimulation modalities was a decrease in the observed frequency bands.
2. The effect of photic stimulation was localized to the visual sensory cortex, whereas pEMF and auditory stimulation affected areas involved in higher-level functions.
3. The effect of EMF stimulation was primarily localized in the frontal areas.
4. Auditory stimulation produced specific network patterns in delta and gamma
5. During the at-rest condition after all stimulation, there were no significant changes.

The emergence of Hagmann's hubs (Hagmann et al., 2008; van den Heuvel & Sporns, 2013) in certain instances is notable. For example, photic stimulation, in addition to having specific location effects in delta in the occipital lobes, emerges with Hagmann Hub 1 as being affected. In the case of EMF, stimulation is seen to affect gamma in Hagmann Hub 3, as the most significant effect.

The fact that repetitive stimulation apparently can affect network, rather than simply locations, such as sensory areas, suggests that its effects go beyond simple changes in activation. This lends support to the idea that repetitive stimulation (RS) produces a response that is *functionally selective*, thus reflecting *specificity in both location and in time*. The effect of RS takes place in time and space, not simply in time. Based on the observation that different regions respond to different frequencies, the effect of the *RS* can be conceptualized as *a multidimensional trajectory in a multidimensional space*.

The effects observed, were highly location and network specific and were transient in nature. They are evidence of self-organization, in which the locations and patterns of activation at particular frequencies exhibit specificity and selectivity. Thus, the effects of stimulation are not well encapsulated in concepts, such as "speeds things up" or "slows things down" or "relaxes the system." Rather, the effects of stimulation are better expressed as dynamic, adaptive shifts in network component activation and interaction, as a response to an input that is minimal with respect to energy, but is more fully characterized with respect to information.

Fig. 3.24 shows that during Binocular Photic Stimulation at 3.5 Hz, we see a reduction in delta activity in Brodmann areas 17 and 18 and the cuneus. These are comprised of the primary and secondary visual sensory areas.

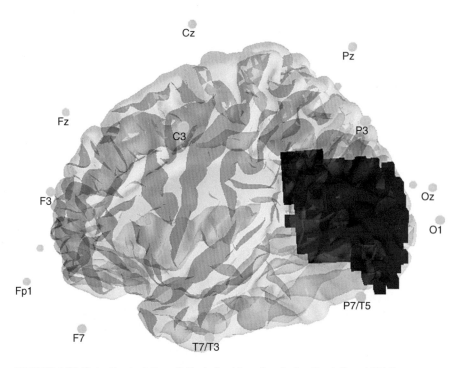

FIGURE 3.24 Reduction in delta activity during binocular photic stimulation at 3.5 Hz (left-side view).

Pulsed electromagnetic stimulation at 3.5 Hz produces a reduction in delta activity in Brodmann areas 45, 46, 10, and 44, as shown in Fig. 3.25. These areas are comprised of the frontal lobe areas associated with executive function, attention, emotion, and decision making.

Fig. 3.26 shows a reduction in gamma activity during pulsed electromagnetic stimulation at 3.5 Hz. This affects Brodmann areas 43, 20, 42, 21, and the inferior temporal gyrus. These areas are involved in hearing, memory, and sequential processing.

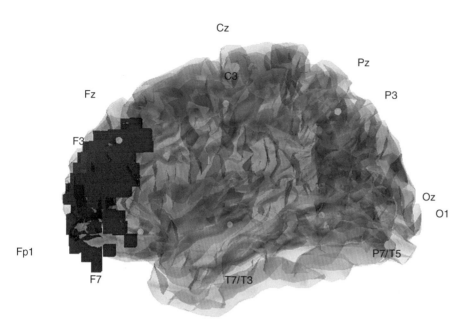

FIGURE 3.25 Reduction in delta activity during pulsed electromagnetic stimulation at 3.5 Hz (left-side view).

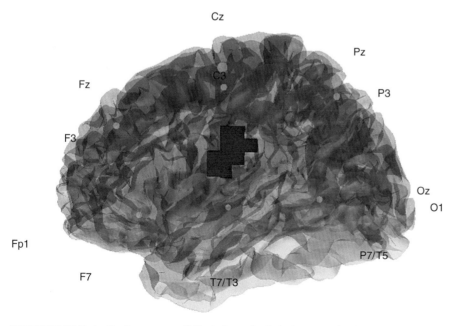

FIGURE 3.26 Reduction in gamma activity during pulsed electromagnetic stimulation at 3.5 Hz (left-side view).

Surprisingly, auditory stimulation (clicks) at 3.5 Hz produces a reduction in delta activity during stimulation. Fig. 3.27 depicts a left-side view showing the areas affected, which include Brodmann areas 13, 33, 24, 20, the insula, and sub lobar (lentiform nucleus, caudate, insula, and thalamus) areas. These areas are involved in auditory sensation, perception, awareness, and attentional control.

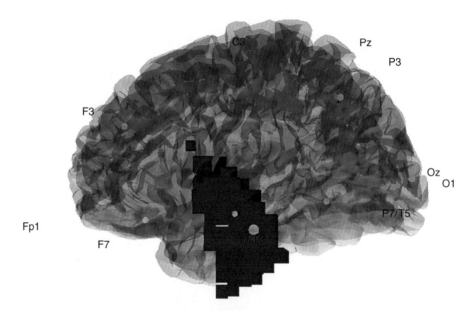

FIGURE 3.27 Reduction in delta activity during auditory stimulation (clicks) at 3.5 Hz (left-side view).

Fig. 3.28 depicts a superior view showing a reduction in delta activity during auditory stimulation (clicks) at 3.5 Hz. Affected areas include sub lobar areas, in particular, the insula, and Brodmann area 13. These areas are involved in perception and awareness, and attentional control.

Also, it's interesting that auditory clicks at 3.5 Hz produce a reduction in gamma activity. Affected areas include Brodmann areas 44 (R), 13 (L), and 34 (L), as shown in Fig. 3.29. These areas are involved in awareness, emotional sensation, and decision making.

Fig. 3.30 depicts a left-side view of the reduction in gamma activity during auditory stimulation (clicks) at 3.5 Hz. Left regions only from the left-side view shows Brodmann areas 13 (L) and 34 (L) are affected. These areas are involved in awareness, emotion, and olfaction.

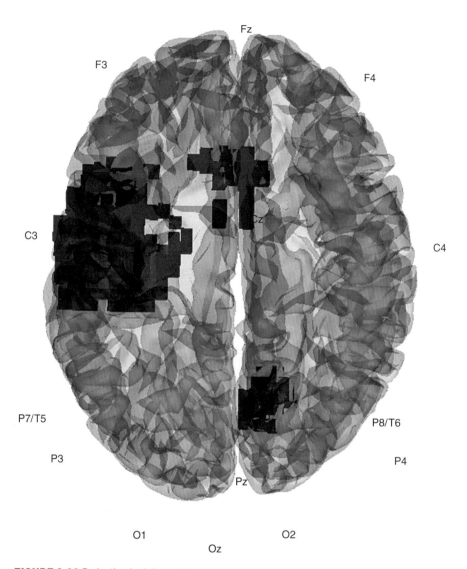

FIGURE 3.28 Reduction in delta activity during auditory stimulation (clicks) at 3.5 Hz (superior view).

Fig. 3.31 shows a reduction in gamma activity during auditory stimulation (clicks) at 3.5 Hz. This view depicts right regions only and as viewed from the right side. Affected areas include Brodmann 44, 34 and 28. These areas are involved in decision making, emotion, and facial recognition.

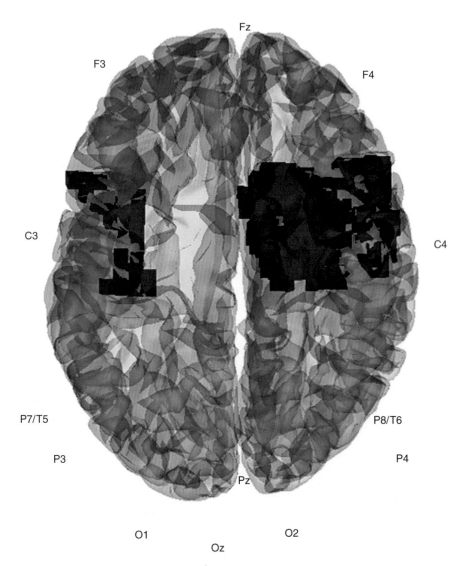

FIGURE 3.29 Reduction in gamma activity during auditory stimulation (clicks) at 3.5 Hz (superior view).

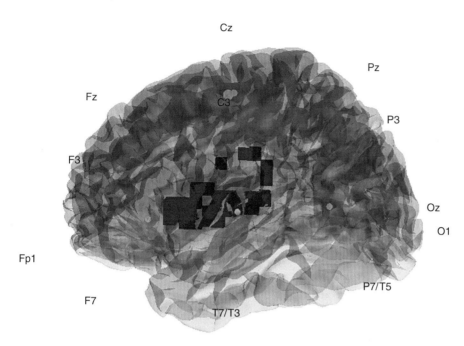

FIGURE 3.30 Reduction in gamma activity during auditory stimulation (clicks) at 3.5 Hz, left regions only (left-side view).

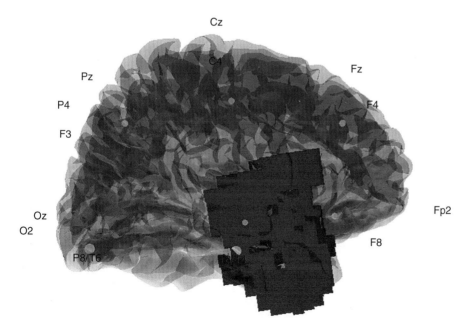

FIGURE 3.31 Reduction in gamma activity during auditory stimulation (clicks) at 3.5 Hz, right regions only (right-side view).

CONCLUSIONS

In conclusion, while there is a fair understanding of the effects of entrainment, there is yet more to explore in appreciating the underlying mechanisms by which repetitive stimuli affect the brain, as many events occur simultaneously. Passive light alone has far reaching effects within the brain. Entrained light possesses aspects of both passive light and frequency dependent activations within the brain. In regard to deep brain structures, many recent findings are both new and exciting. Various Veterans Affairs' centers in the USA have noted that war veterans using AVE, sometimes not only recover quickly from PTSD, but also improve cognitively from diffuse axonal injuries relating to shock waves from improvised explosives. However, this work requires more psychometric and imaging research [EEG, Functional Magnetic Resonance Imaging (fMRI) and Single Photon Emission Computerized Tomography (SPECT), etc.] as proof of concept. The ability of AVE to jump-start the thalamo-cortical loop has been compared with using a set of electrified paddles to jump-start a heart in fibrillation.

Other aspects of entrainment span well beyond traditional beliefs about its effects on the brain and clinical outcomes. A recent study on mice with Alzheimer's found that 1 hour of 40 Hz (gamma) photic stimulation reduced amyloid plaques by roughly 60 % within a few hours of stimulation while microglia diameters increased by 160% from consuming the plaques (Iaccarino et al., 2016). It is clear that the effects go well beyond simply "vibrating" the system and involves complex information relating to support structure processes. The fact that repetitive stimuli can have effects on mental and emotional states speaks to the fact that the involvement of functional networks is of paramount importance. Results shown in this chapter lend credence to the idea that we should be able to create a bottom-up analysis that puts repetitive stimuli in an objective and physiologically-based framework which will illuminate why such stimulation has the observed effects on how we think, feel, and perform in a myriad of situations.

References

Adrian, E., & Matthews, B. (1934). The Berger rhythm: potential changes from the occipital lobes in man. *Brain, 57*, 355–384.

Ali Shafiei, S., Mohammad Firoozabadi, S., Rasoulzadeh Tabatabaie, K., & Ghabaee, M. (2012). Evaluating the changes in alpha-1 band due to exposure to magnetic field. *Iranian Journal of Medical Physics, 9*(2), 141–152.

Amen, D. (1998). *Change your brain, change your life.* New York: Three Rivers Press.

Amirifalah, Z., Mohammad, S., Firoozabad, P., Shafiei Darabi, A., & Assadi, A. (2011). Scrutiny of brain signals variations in regions Cz, C3 and C4 under local exposure of extremely low frequency and weak pulsed magnetic field to promote neurofeedback systems. *Physiology and Pharmacology, 15*(1), 144–163 Spring 2011 [Article in Persian].

Amirifalah, Z., Mohammad, S., Firoozabadi, P., & Ali Shafiei, S. (2013). Local exposure of brain central areas to a pulsed ELF magnetic field for a purposeful change in EEG. *Clinical EEG and Neuroscience, 44*(1), 44–52.

Arco, A., & Mora, F. (2009). Neurotransmitters and prefrontal cortex-limbic system interactions: implications for plasticity and psychiatric disorders. *Journal of Neural Transmission, 116*, 941–952.

Balser, M., & Wagner, C. (1960). Observations of Earth–ionosphere cavity resonances. *Nature, 188*(4751), 638–641.

Basar, E. (2013). Brain Oscillations in neuropsychiatric disease. *Dialogues in Clinical Neuroscience, 15*(3), 291–300.

Berg, K., & Siever, D. (2004). The effect of audio-visual entrainment in depressed community-dwelling senior citizens who fall. *Unpublished manuscript.* Edmonton, Alberta, Canada, Mind Alive Inc.

Berg, K., & Siever, D. (2009). A controlled comparison of audio-visual entrainment for treating seasonal affective disorder (SAD). *Journal of Neurotherapy, 13*(3), 166–175.

Berg, K., Mueller, H., Seibel, D., & Siever, D. (1999). Outcome of medical methods, audio-visual entrainment, and nutritional supplementation in the treatment of fibromyalgia syndrome. *Unpublished manuscript.* Edmonton, Alberta, Canada, Mind Alive Inc.

Bertolero, M., Yeo, B., & D'Esposito, M. (2015). The modular and integrative functional architecture of the human brain. *Proceedings of the National Academy of Science, 112*(49), E6798–E6807.

Budzynski, T., & Tang, J. (1998). Bio-light effects on the electroencephalogram (EEG). *SynchroMed Report.* Seattle, WA.

Budzynski, T., Jordy, J., Budzynski, H., Tang, H., & Claypoole, K. (1999). Academic performance enhancement with photic stimulation and EDR feedback. *Journal of Neurotherapy, 3*, 11–21.

Budzynski, T., Budzynski, H., & Sherlin, L. (2002). Audio visual stimulation (AVS) in an Alzheimer's patient as documented by quantitative electroencephalography (QEEG) and low resolution electromagnetic brain tomography (LORETA). *Journal of Neurotherapy, 6*(1), 54.

Budzynski, T., Budzynski, H., & Tang, H. (2007). Brain brightening. In J. R. Evans (Ed.), *Handbook of neurofeedback: dynamics and clinical applications* (pp. 231–265). New York, NY: Haworth Press.

Carter, J., & Russell, H. (1993). A pilot investigation of auditory and visual entrainment of brain wave activity in learning disabled boys. *Texas Researcher, 4*, 65–72.

Chatrian, G., Petersen, M., & Lazarte, J. (1959). Response to clicks from the human brain: some depth electrographic observations. *Electroencephalography and Clinical Neurophysiology, 12*, 479–489.

Collura, T. (2012). Individualized assessment and treatment using advanced EEG and dynamic localization techniques with live sLORETA. *International Journal of Psychophysiology, 85*(3), 288–290.

Collura, T. (2013). BrainAvatar. US Patent Publication 20130303934, Filed November 5, 2012.

Collura, T. (2014a). *Technical Foundations of Neurofeedback.* New York: Routledge/Taylor & Francis.

Collura, T. (2014b). Specifying and developing references for live z-score neurofeedback. *Neuro-Connections, 26–39*, Spring 2014.

Collura, T., & Siever, D. (2009). Audio-visual entrainment in relation to mental health and EEG. In T. Budzynski, H. Budzynski, J. Evans, & A. Abarbanel (Eds.), *Introduction to quantitative EEG and neurofeedback: advanced theory and applications* (2nd ed., pp. 193–223). Boston: Elsevier.

De Ridder, D., Verplaetse, J., & Vanneste, S. (2013). The predictive brain and the "free will" illusion. *Frontiers in Psychology, 4*, 131.

Doidge, N. (2015). *The Brain's Way of Healing.* NY: Viking Books.

Donker, D., Njio, L., Storm Van Leewan, W., & Wieneke, G. (1978). Interhemispheric relationships of responses to sine wave modulated light in normal subjects and patients. *Encephalography and Clinical Neurophysiology, 44*, 479–489.

Emmons, H. (2010). The Chemistry of Calm. Simon & Schuster. Fort Lauderdale, FL.

Erba, G. (2001). Preventing seizures from "Pocket Monsters" a way to control reflex epilepsy. *Neurology, 57*(10), 1747–1748.

Fox, P., & Raichle, M. (1985). Stimulus rate determines regional blood flow in striate cortex. *Annals of Neurology, 17*(3), 303–305.

Fox, P., Raichle, M., Mintun, M., & Dence, C. (1988). Nonoxidative glucose consumption during focal physiologic neural activity. *Science, 241*, 462–464.

Frederick, J., Lubar, J., Rasey, H., Brim, S., & Blackburn, J. (1999). Effects of 18.5 Hz audiovisual stimulation on EEG amplitude at the vertex. *Journal of Neurotherapy, 3*(3), 23–27.

Frewin, C. L., Locke, C., Saddow, S. E., & Weeber, E. J. (2012). Biocompatibility of SiC for neurological applications. In S. E. Saddow (Ed.), *Silicon Carbide Biotechnology*. Waltham, MA: Elsevier Inc. (Fig. 6.1).

Fuggetta, G., & Noh, N. (2012). A neurophysiological insight into the potential link between transcranial magnetic stimulation, thalamocortical dysrhythmia and neuropsychiatric disorders. *Experimental Neurology, 245*, 87–95.

Gagnon, C., & Boersma, F. (1992). The use of repetitive audio-visual entrainment in the management of chronic pain. *Medical Hypnoanalysis Journal, 7*, 462–468.

Gervitz, R. (2000). Resonant frequency training to restore homeostasis for treatment of psychophysiological disorders. *Biofeedback, 27*, 7–9.

Glicksohn, J. (1986). Photic driving and altered states of consciousness: an exploratory study. *Imagination, Cognition, and Personality, 6*(2), .

Grossberg, S. (2013). Adaptive resonance theory: how a brain learns to consciously attend, learn, and recognize a changing world. *Neural Networks, 37*, 1–47.

Hagmann, P., Cammoun, L., Gigandet, X., Meuli, R., Honey, C., Wedeen, V., & Sporns, O. (2008). Mapping the structural core of human cerebral cortex. *PLoS Biology, 6*(7), 1479–1493.

Hawes, T. (2000). Chapter 14: Using light and sound technology to access "The Zone" in sports and beyond. In: D. Siever, *The Rediscovery of Audio-visual Entrainment Technology*. Edmonton, Alberta, Canada, Mind Alive Inc.

Hear, J. (1971). Field dependency in relation to altered states of consciousness produced by sensory-overload. *Perception and Motor Skills, 33*, 192–194.

Hebb, D. (1961). Distinctive features of learning in the higher animal. In J. F. Delafresnaye (Ed.), *Brain Mechanisms and Learning*. London: Oxford University Press.

Huang, T., & Charyton, C. (2008). A comprehensive review of the psychological effects of brainwave entrainment. *Alternative Therapies, 14*(5), 38–49.

Iaccarino, H., Singer, A., Martorel, A., Rudenko, A., Gao, F., Gillingham, T., Mathys, H., Seo, J., Kritskiy, O., Abdurrob, F., Adaikkan, C., Canter, R., Rueda, R., Brown, E., Boyden, E., & Tsai, L. (2016). Gamma frequency entrainment attenuates amyloid load and modifies microglia. *Nature, 540*, 230–251.

Joyce, M., & Siever, D. (2000). Audio-visual entrainment program as a treatment for behavior disorders in a school setting. *Journal of Neurotherapy, 4*(2), 9–25.

Kinney, J., McKay, C., Mensch, A., & Luria, S. (1973). Visual evoked responses elicited by rapid stimulation. *Encephalography and Clinical Neurophysiology, 34*, 7–13.

Kroger, W., & Schneider, S. (1959). An electronic aid for hypnotic induction: a preliminary report. *International Journal of Clinical and Experimental Hypnosis, 7*, 93–98.

Leonard, K., Telch, M., & Harrington, P. (1999). Dissociation in the laboratory: a comparison of strategies. *Behaviour Research and Therapy, 37*, 49–61.

Leonard, K., Telch, M., & Harrington, P. (2000). Fear response to dissociation challenge. *Anxiety, Stress, and Coping, 13*, 355–369.

Lewerenz, C. (1963). A factual report on the brain wave synchronizer. *Hypnosis Quarterly, 6*(4), 23.

Lim, L., (2013). The Potential of Intranasal Light Therapy for Brain Stimulation. Presented at the North American Association for Photobiomodulation Therapy (NAALT) Conference, Palm Beach Gardens, Florida, February 2, 2013.

Lipowsky, Z. (1975). Sensory and information inputs over-load: behavioral effects. *Comprehensive Psychiatry, 16*, 199–221.

Manns, A., Miralles, R., & Adrian, H. (1981). The application of audiostimulation and electromyographic biofeedback to bruxism and myofascial pain-dysfunction syndrome. *Oral Surgery, 52*(3), 247–252.

Margolis, B. (1966). A technique for rapidly inducing hypnosis. *Certified Akers Laboratories, 28*(12), 21–24.

Meier, T., Bellgowan, P., Singh, R., Kuplicki, R., Polanski, D., & Mayer, A. (2015). Recovery of cerebral blood flow following sports-related concussion. *JAMA Neurology, 72*(5), 530–538.

Meyer, J., Takashima, S., Terayama, Y., Obara, K., Muramatsu, K., & Weathers, S. (1994). CT changes associated with normal aging of the human brain. *Journal of the Neurological Sciences, 123*(1–2), 200–208.

Micheletti, L., (1999). The Use of Auditory and Visual Stimulation for the Treatment of Attention Deficit Hyperactivity Disorder in Children. University of Houston. Ph.D. dissertation, unpublished. Edmonton, Alberta, Canada, Mind Alive Inc.

Morse, D., & Chow, E. (1993). The effect of the Relaxodont™ brain wave synchronizer on endodontic anxiety: evaluation by galvanic skin resistance, pulse rate, physical reactions, and questionnaire responses. *International Journal of Psychosomatics, 40*(1–4), 68–76.

Murphy, D., Murphy, D., Abbas, M., Palazidou, B., Binnie, C., Arendt, J., Campos Costa, D., & Checkley, S. (1993). Seasonal affective disorder: response to light as measured by electroencephalogram, melatonin suppression and cerebral blood flow. *British Journal of Psychiatry, 163*, 327–331.

Naeser, M., Zafonte, R., Krengel, M., Martin, P., Frazier, J., Hamblin, M., Knight, J., Meehan, W., 3rd, & Baker, E. (2014). Significant improvements in cognitive performance post-transcranial, red/near-infrared light-emitting diode treatments in chronic, mild traumatic brain injury: open-protocol study. *Journal of Neurotrauma, 31*(11), 1008–1017.

Palmquist, C. (2014). Ph.D. Unpublished dissertation for Saybrook University.

Qiu, C., Shivacharan, R., Zhang, M., & Durand, D. (2015). Can neural activity propagate by endogenous electrical field? *Journal of Neuroscience, 35*(48), 15800–15811.

Regan, D. (1966). Some characteristics of average steady-state and transient responses evoked by modulated light. *Electroencephalogy and Clinical Neurophysiology, 20*, 238–248.

Ross-Swain, D. (2007). The effects of auditory stimulation on auditory processing disorder: a summary of the findings. *The International Journal of Listening, 27*(2), 140–155.

Russell, H. (1996). Entrainment combined with multimodal rehabilitation of a 43-year-old severely impaired postaneurysm patient. *Biofeedback and Self Regulation, 21*, 4.

Ruuskanen-Uoti, H., & Salmi, T. (1994). Epileptic seizure induced by a product marketed as a "Brainwave Synchronizer". *Neurology, 44*, 180.

Sadove, M. (1963). Hypnosis in anaesthesiology. *Illinois Medical Journal*, 39–42.

Sankoh, A. J., Huque, M. F., & Dubey, S. D. (1997). Some comments on frequently used multiple endpoint adjustment methods in clinical trials. *Statistics In Medicine, 16*(22), 2529–2542.

Sappey-Marinier, D., Calabrese, G., Fein, G., Hugg, J., Biggins, C., & Weiner, M. (1992). Effect of photic stimulation on human visual cortex lactate and phosphates using 1H and 31P magnetic resonance spectroscopy. *Journal of Cerebral Blood Flow and Metabolism, 12*(4), 584–592.

Schreckenberger, M., Lange-Asschenfeld, C., Lochmanna, M., Mann, K., Siessmeiera, T., Buchholza, H., Bartensteina, P., & Gründer, G. (2004). The thalamus as the generator and modulator of EEG alpha rhythm: a combined PET/EEG study with lorazepam challenge in humans. *NeuroImage, 22*(2), 637–644.

Shealy, N., Cady, R., Cox, R., Liss, S., Clossen, W., & Veehoff, D. (1989). A comparison of depths of relaxation produced by various techniques and neurotransmitters produced by brainwave entrainment. *Shealy and Forest Institute of Professional Psychology*, Unpublished manuscript.

Siever, D. (2003a). Audio-visual entrainment: I. History and physiological mechanisms. *Biofeedback*, *31*(2), 21–27.

Siever, D. (2003b). Audio visual entrainment: II. Dental studies. *Biofeedback*, *31*(3), 29–32.

Siever, D. (2003c). Applying audio-visual entrainment technology for attention and learning-part III. *Biofeedback*, *31*(4), 24–29.

Siever, D. (2013). Transcranial DC Stimulation. *NeuroConnections*, Spring Issue, 33–40. http://www.mindalive.com/1_0/article%2010.pdf

Stahl, C., & Collura, T. (2012). Assessing the effects of subthreshold magnetic stimulation on brain activation using sLORETA. *NeuroConnections*, 34–36.

Steriade, M., Nuñez, A., & Amzica, F. (1993). Intracellular analysis of relations between the slow (<1 Hz) neocortical oscillations and other sleep rhythms of electroencephalogram. *Journal of Neuroscience*, *13*, 3266–3283.

Sterman, M., & Kaiser, D. (1999). Topographic analysis of spectral density co-variation: normative database & clinical assessment. *Clinical Neurophysiology*, *110*(S1), S80.

Tang, H., Riegel, B., McCurry, S., & Vitiello, M. (2016). Open-loop audio-visual stimulation (AVS): a useful tool for management of insomnia? *Applied Psychophysiology and Biofeedback*, *41*(1), 39–46.

Teicher, M., Anderson, C., Polcari, A., Glod, C., Maas, L., & Renshaw, P. (2000). Functional deficits in basal ganglia of children with attention-deficit/hyperactivity disorder shown with functional magnetic resonance imaging relaxometry. *Nature Medicine*, *6*(4), 470–473.

Thatcher, R., Walker, R., Gerson, I., & Geisler, F. (1989). EEG discriminant analyses of mild head trauma. *Electroencephalography and Clinical Neurophysiology*, *73*(2), 94–106.

Thomas, N., & Siever, D. (1989). The effect of repetitive audio/visual stimulation on skeletomotor and vasomotor activity. In D. Waxman, D. Pederson, I. Wilkie, & P. Meller (Eds.), *Hypnosis: 4th European Congress at Oxford* (pp. 238–245). London, UK: Whurr Publishers.

Timmermann, D., Lubar, J., Rasey, H., & Frederick, J. (1999). Effects of 20-min audio-visual stimulation (AVS) at dominant alpha frequency and twice dominant alpha frequency on the cortical EEG. *International Journal of Psychophysiology*, *32*(1), 55–61.

Toman, J. (1941). Flicker potentials and the alpha rhythm in man. *Journal of Neurophysiology*, *4*, 51–61.

Townsend, R. (1973). A device for generation and presentation of modulated light stimuli. *Electroencephalography and Clinical Neurophysiology*, *34*, 97–99.

Trans Cranial Technologies (2012). 10–20 system positioning manual. http://www.trans-cranial.com/local/manuals/10_20_pos_man_v1_0_pdf.pdf

Trenité, D., Guerrini, R., Binnie, C., & Genton, P. (2001). Visual sensitivity and epilepsy: a proposed terminology and classification for clinical and EEG phenomenology. *Epilepsia*, *42*(5), 692–701.

Trudeau, D. (1999). A Trial of 18 Hz Audio-Visual Stimulation (AVS) on Attention and Concentration In Chronic Fatigue Syndrome (CFS). Proceedings of the Annual Conference for the International Society for Neuronal Regulation. Haworth Press, Binghamton, NY.

Van den Heuvel, M., & Sporns, O. (2013). Network hubs in the human brain. *Trends in Cognitive Sciences*, *17*(12), 683–696.

Van Der Tweel, L., & Lunel, H. (1965). Human visual responses to sinusoidally modulated light. *Encephalography and Clinical Neurophysiology*, *18*, 587–598.

Walter, W. (1956). Color illusions and aberrations during stimulation by flickering light. *Nature, 177,* 710.

Williams, J., Ramaswamy, D., & Oulhaj, A. (2006). 10 Hz flicker improves recognition memory in older people. *BMC Neuroscience, 7*(21), 1–7.

Wolitzky-Taylor, K., & Telch, M. (2010). Efficacy of self-administered treatments for pathological academic worry: a randomized controlled trial. *Behaviour Research and Therapy, 48,* 840–850.

Wuchrer, V. (2009). Study on memory and concentration. *Conducted at the Psychological Institute of the Friedrich-Alexander University Erlangen-Nürnberg.* Unpublished manuscript. Edmonton, Alberta, Canada, Mind Alive Inc.

Zhang, M., Ladas, T., Qiu, C., Shivacharan, R., Gonzalez-Reyes, L., & Durand, D. (2014). Propagation of epileptiform activity can be independent of synaptic transmission, gap junctions, or diffusion and is consistent with electrical field transmission. *Neuroscience, 34*(4), 1409–1419.

CHAPTER 4

Photobiomodulation and Other Light Stimulation Procedures

Marvin H. Berman*, Michael R. Hamblin**, Paul Chazot†

*Quietmind Foundation, Elkins Park, PA, United States; **Harvard Medical School, Wellman Center for Photomedicine, Massachusetts General Hospital, Boston, MA, United States; †Durham University, Durham, United Kingdom

INTRODUCTION

This chapter considers neural rhythms and oscillations as observed, and how they might be affected by the light spectrum of electromagnetic energy. We will consider both electrophysiological and biochemical impact of visible and near infrared (NIR) light on brain functioning as it relates to health, and specifically neuromodulation. Particular attention is given to brain network activation as measured by changes in brain electrical activity influenced by directed electromagnetic energy in the visible and NIR wavelength bands. Current methods for intervening noninvasively on neuroelectrical and metabolic activity are reviewed, as well as how light can influence these processes. We also will consider the systemic effects light engenders on neural connectivity throughout the brain, and subsequent effects on cognitive processing and physical/emotional behavior.

PHOTOBIOMODULATION AND PHOTOTHERAPY PRINCIPLES

Photobiomodulation as described by Huang, Mroz, and Hamblin (2009) "involves the transformation of light energy to chemical, kinetic or heat energy to achieve a desired physiological result. As stated by the First Law of Photobiology, light energy must be absorbed by a bond or a molecule in order to initiate a physical or chemical process. Therefore, light that is used for therapeutic applications must be absorbed by a specific chromophore in the biological tissue. Therapeutic applications of photobiomodulation are defined by Hamblin as, the use of low (nonthermal levels of visible or NIR light to stimulate or inhibit

Rhythmic Stimulation Procedures in Neuromodulation. http://dx.doi.org/10.1016/B978-0-12-803726-3.00004-3

biological cells and tissues via a photochemical mechanism (without the addition of an external photosensitizer) (Hamblin, 2016a).

Several transcription factors are regulated by changes in cellular redox state. Among them are redox factor-1 (Ref-1) dependent activator protein-1 (AP-1) (Fos and Jun), nuclear factor κB (NF-κB), activating transcription factor/cAMP-response element-binding protein (ATF/CREB), hypoxia-inducible factor (HIF)-1α, and HIF-like factor. However, it also has been shown that low levels of oxidants appear to stimulate proliferation and differentiation of some type(s) of cells (Alaluf, Muir-Howie, Hu, Evans, & Green, 2000; Kirlin et al., 1999; Yang et al., 1996). These effects, in turn, lead to increased cell proliferation and migration (particularly by fibroblasts), modulation in the levels of cytokines, growth factors and inflammatory mediators, and increased tissue oxygenation (Pastore, Greco, Petragallo, & Passarella, 1994). The results of these biochemical and cellular changes in animals and patients include such benefits as increased healing in chronic wounds, improvements in sports injuries and carpal tunnel syndrome, pain reduction in arthritis and neuropathies, and amelioration of damage after heart attacks, stroke, nerve injury, and retinal toxicity (Hamblin & Demidova, 2006). Hamblin described the mechanism of action as centering around cytochrome c oxidase (CCO), which is unit four of the mitochondrial respiratory chain, responsible for the final reduction of oxygen to water using the electrons generated from glucose metabolism (De Frietas & Hamblin, 2016). The theory is that CCO enzyme activity may be inhibited by nitric oxide (NO) (especially in hypoxic or damaged cells). This inhibitory NO can be dissociated by photons of light that are absorbed by CCO (which contains two heme and two copper centers with different absorption spectra) (Lane, 2006). These absorption peaks are mainly in the red (600–700 nm) and NIR (760–940 nm) spectral regions. When NO is dissociated, the mitochondrial membrane potential is increased, more oxygen is consumed, more glucose is metabolized and more ATP is produced by the mitochondria.

CLINICAL APPLICATIONS

Clinical applications have been evolving since 1981 with attention focused on optimizing the wavelength (color of the light), the rate at which light energy is delivered and the overall quantity being absorbed. Colored laser light therapy at relatively high powers are often seen in dermatology where, for example, green light is used to treat vascular lesions, including port wine stains and varicose veins, while NIR lasers, for example, Q-switched lasers like the Nd:YAG 1064 nm, will remove discolored pigmentations. NIR LEDs within the 1060–1080 nm range have now been developed to help reduce acne scarring lesions and discoloration (Barolet & Boucher, 2010).

LOW LEVEL LASER THERAPY

Low-level laser (LLLT) or light emitting diode therapy is an emerging medical and veterinary technique in which exposure to LLLT light or light emitting diodes (at levels that do not heat the tissue) might stimulate or inhibit cellular function, possibly leading to beneficial clinical effects. The use of low levels of visible or NIR light for reducing pain, inflammation, and edema, promoting healing of wounds, deeper tissues and nerves, and preventing tissue damage has been known for almost 40 years since the invention of lasers (Karu, 2008). A summary of history of phototherapy, including LLLT, is provided in Chapter 1 of the text Laser Function Medicine and its Applications (Cheng-YiLiu & Ping Zhu, 2008). Despite many reports of positive findings from experiments conducted in vitro, in animal models, and in randomized controlled clinical trials, LLLT remains controversial. Karu has proposed (see Photobiological Sciences Online unit entitled "Action Spectra: Their Importance for Low Level Light Therapy") that mitochondria are a likely site for the initial effects of light, specifically that the enzyme CCO (unit four in the mitochondrial respiratory chain) absorbs photons and increases its activity leading to increased ATP production, modulation of reactive oxygen species and induction of transcription factors (Turrens, 2003) (Fig. 4.1).

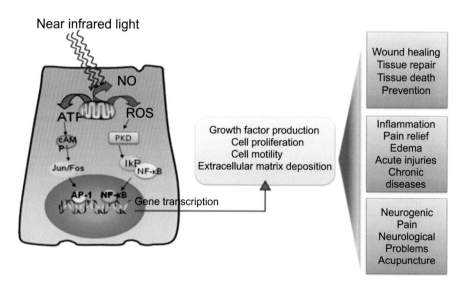

FIGURE 4.1 LLLT mechanism and application.
Incoming red and NIR photons are absorbed in cell mitochondria, producing reactive oxygen species (ROS) and releasing nitric oxide (NO), which leads to gene transcription via activation of transcription factors (NF-κB and AP1) (Huang et al., 2009).

Low-level laser/light therapy (LLLT) for neurological disorders in the central nervous system (CNS) currently is an experimental concept. The broad goals for clinical utilization are the prevention and/or repair of damage, relief of symptoms, slowing of disease progression, and correction of genetic abnormalities. Experimental studies have tested and continue to test these goals by investigating LLLT in animal models of diseases and injuries that affect the brain and spinal cord. Successful clinical trials have been carried out for transcranial laser therapy for stroke (Henderson & Morries, 2015). Discoveries concerning the molecular basis of various neurological diseases, combined with advances that have been made in understanding the molecular and cellular mechanisms in LLLT, both in vitro and in vivo, have allowed rational light-based therapeutic approaches for a wide variety of CNS disorders to be investigated. There are an increasingly wide range of treatment applications (Fig. 4.2) involving NIR stimulation including Parkinson's and Alzheimer's disease (Saltmarche et al., 2016).

Neurodegenerative diseases are caused by the deterioration of certain populations of nerve cells (neurons), as seen for example in Alzheimer's disease, Parkinson's disease (Trimmer et al., 2009), and Amyotrophic Lateral Sclerosis (Moges et al., 2009). All are due to neuronal degeneration in the CNS

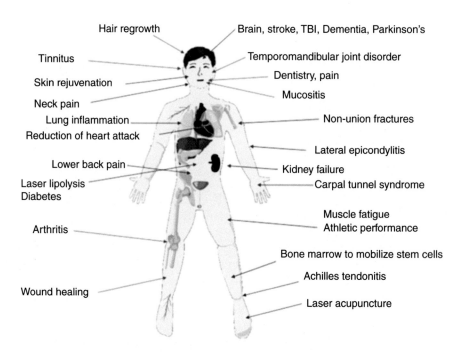

FIGURE 4.2 Therapeutic targets employing photobiomodulation (Hamblin, 2016b). *(With permission from Michael Hamblin, PhD.)*

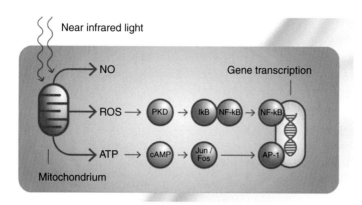

FIGURE 4.3 LLLT for central nervous system (CNS) neurological disorders.
NIR light can penetrate through the skull into the brain, reducing neuronal cell death, reducing inflammation and increasing the likelihood of neurogenesis. The retinal nerves and the spinal cord are classified as part of the CNS, and light can be delivered into the eye or to the neck or back, at the site of a spinal cord lesion (Huang et al., 2009).

(Friedlander et al., 2003). The chronic, unrelenting, progressive nature of these devastating degenerative diseases has motivated the search for therapies that could slow down or arrest the downward course experienced by most patients; and even more desirable would be a therapy that could reverse neuronal damage. Transcranial light therapy is considered to have the potential to accomplish these goals (Fig. 4.3). Limitations in knowledge are still apparent, however, regarding variables such as the optimal wavelength, light source, doses, pulsed compared to continuous stimulation, polarization state, treatment timing, and repetition frequency. Collaborative efforts between clinicians and basic researchers likely will increase the usage and understanding of effective laser-based therapies involving the CNS (Lampl et al., 2007). Additionally, light-emitting diodes (LED) are revolutionizing the whole lighting industry. Their availability in almost any wavelength and with steadily increasing total output power means that light delivery applications, previously thought to require an expensive laser, can now be performed at a fraction of the cost (less than 1%) by LEDs compared with the equivalent laser source. Not surprisingly, LEDs are becoming much more widely used in medical applications (Barolet, 2008).

NEAR INFRARED STIMULATION IN THE TREATMENT OF NEURODEGENERATIVE DISORDERS

NIR therapy gained considerable credence with the insight into its biochemical mechanism of action related to the vasodilating effects of NO. Red blood cells carry NO and it was found through studies at Duke University that blue and red light stimulated the release of NO at the site of wounds and increased the rate of healing. The basic research on NO as a cardiovascular signaling molecule earned Furchgott, Ignarro, and Murad the Nobel Prize in Medicine and Physiology (Furchgott, Ignarro, & Murad, 2016). NIR stimulation was initially employed as a diagnostic technique and only later as a therapy, and is now being used by physicians to address a variety of conditions primarily in the areas of pain management and circulation.

Neurodegeneration and dementia are speculated to be related to pathologies of cerebral blood flow (Chaudhary et al., 2013; Spilt et al., 2005). The accumulation of beta-amyloid (Aβ) and hyper-phosphorylated tau peptides contribute to a signaling cascade leading to synaptic impairments, intraneuronal Aβ42 aggregates, and correlated cognitive deficits (Teipel et al., 2016). An argument for a possible role of impaired blood flow in the etiology of dementia is that Alzheimer's patients with brain damage [regions of magnetic resonance imaging (MRI) signal hyperintensity] have increased oxygen extraction per mL/min. The central problem then is blood supply rather than demand. Oxygen extraction would be expected to be unaltered if reduced blood flow were secondary to tissue damage (Spilt et al., 2005; Yamaji et al., 1997). Single photon emission computed tomography (SPECT) studies have shown amyloid accumulation to be similar to autopsy-based data (Thal, Attems, & Ewers, 2014) and regional cerebral blood flow (RCBF) to be significantly reduced in the frontal and temporal regions in fronto-temporal dementia (FTD) patients (Miller et al., 1997; Read et al., 1995). The anatomical distribution of reduced RCBF corresponds to observed patterns of neuropsychological deficits (McMurtray et al., 2006). Unsurprisingly, MRI and computed tomography (CT) in FTD patients shows atrophy in the frontal and temporal regions (Mendez et al., 1996; Neary & Snowden, 1996; Neary, Snowden & Mann, 2005). RCBF in Alzheimer's disease patients is 20% lower than normal (Roher et al., 2012). Positron emission tomography (PET) imaging in FTD patients reveals reduced glucose metabolism in the frontal and anterior temporal lobes, and in the cingulate gyrus, insula, uncus, and subcortical structures (Garraux, Salmon, Degueldre, Lemaire, & Franck, 1999; Ishii, Sakamoto, & Sasaki, 1998; Jeong, Cho, & Park, 2005). Grimmer, Diehl, Drzezga, Forstl, and Kurz (2003) performed a longitudinal study on ten patients diagnosed with FTD. At the initial assessment, FTD patients had reduced metabolic activity compared to controls in frontal cortical areas, the caudate nuclei, and the thalami. On a 1- to 2-year follow-up, significant progression of the original deficits was observed in the orbitofrontal cortex and the subcortical structures. Based on the above-mentioned research, a treatment mode that increases cerebral blood flow should prove beneficial for patients with neurodegenerative disorders.

It has been shown that NIR light stimulation increases NO synthase (iNOS) expression and cerebral blood flow (Bradford, Barlow, & Chazot, 2005; Uozumi et al., 2010) and increases cell viability against oxidative stressors, for example,

UV exposure (Bradford et al., 2005), H_2O_2 (Duggett & Chazot, 2014). NIR exposure then may be therapeutic in two modes: it can increase cerebral blood flow, thus preventing degeneration, and also can increase cell viability.

RESEARCH AT QUIETMIND FOUNDATION

Studies conducted at Quietmind Foundation (QMF) a not-for-profit clinical research and consultation group founded in 2000, has sought to integrate noninvasive, nondrug treatment modalities in public healthcare and educational programs. QMF had been conducting research on the efficacy of EEG biofeedback or neurofeedback training in remediation of ADHD, learning and behavioral challenges in school aged children (Monastra, 2005), and in 2007 the research agenda shifted to include studies on the effects of NIR stimulation as a component of treating neurodegeneration.

An initial pilot (N = 11), placebo controlled, double blind study began in 2007 in QMF's suburban Philadelphia clinic. It involved subjects diagnosed with early to mid-stage dementia being stimulated daily with NIR light for six minutes for 28 consecutive days (Berman et al., 2017). Control subjects were randomly assigned to treatment or control conditions as they were referred into the program. The placebo treatment was identical to the active using a unit that did not produce NIR light during the activation periods but had the same sounds and external indicator lights as the active treatment unit.

The purpose of the study was to investigate feasibility of NIR stimulation for improving cognitive, behavioral, and memory dysfunctions resulting from a dementing illness. The primary interest was to evaluate the potential of NIR phototherapy to stimulate neuroplasticity in the brain, not to remediate the specific pathology which, even with the best diagnosis, very often is unknown. Given the substantial evidence linking dementia to reduced RCBF and the cytoprotective and neuronal rehabilitative features of NIR light stimulation (Bradford et al., 2005; Duggett & Chazot, 2014; Grillo, Duggett, Ennaceur, & Chazot, 2013; Uozumi et al., 2010), we hypothesized that repeated exposure to NIR stimulation might increase cerebral blood oxygenation, increasing cell viability and improving cognitive and behavioral functioning. It also was noted that NIR stimulation had been shown to improve the healing rate of infection. This was an early application of 1072 nm infrared light and was shown to be of greater efficacy in lesion reduction than topical antimicrobial agents (Dougal & Kelly, 2001; Hargate, 2006). Furthermore, subsequent studies have shown that NIR exposure improved rates of healing in rats exposed to Methicillin-resistant *Staphylococcus aureus* (MRSA) (Lee et al., 2011). Such findings, and Itzhaki and Wozniak's (2008) findings regarding the relationship between herpes I virus and Alzheimer's disease have substantively influenced the direction of our investigations of intensive NIR exposure and its potential influence on neurodegeneration.

The stimulation device used was originally designed by Gordon Dougal, MD, BSEE, with subsequent technical design input from Marvin Berman, PhD and James Halper, MD of Quietmind Foundation. The original unit design had two rigid circumferences to fit smaller and larger subjects' heads and fans to move heat away from the scalp while the current design is flexible and adjustable allowing for a more consistent and comfortable fit (Fig. 4.4).

Each of the device's plastic housings contained printed circuit boards with LEDs matched to produce 1060–1080 nm, pulsed at 10 hz, stimulation running on a 5-min duty cycle. Informed consent was obtained upon study approval by QMF's institutional review board. Subjects underwent 19-channel Quantitative EEG (QEEG) recordings with eyes open and closed (Lubar, 2004; Thatcher, 2005), and were administered the Alzheimer's Disease Assessment Scale-Cognitive subscale (ADAS-cog) (Verma, Beretvas, Pascual, Masdeu, & Markey, 2015), the Boston Naming Test and the Trail Making Test. Sessions were conducted daily for 28 days and included 2-min baseline frontal cortical blood flow measurement using a red and infrared photospectroscopy headband developed at the Biocomp Research Institute (Toomim, Mize, & Kwong, 2005). The red and NIR light that penetrated 1.5 cm into the cortical tissue and reflected back to the scalp surface was digitally transformed and recorded before and after each stimulation session. The ratio of received red

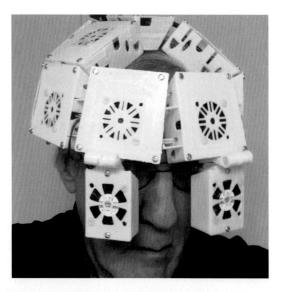

FIGURE 4.4 Cognitolite infrared (1068 nm) phototherapy devices.
(From Quietmind Foundation, 2016.)

(660 nm) to NIR (850 nm) light served as an indicator of cortical perfusion under the sensors. A 2-min baseline recordings were made before and after each 6-min NIR stimulation period and the data averaged, yielding measures of net change in perfusion for each session.

These aggregated perfusion changes then were averaged over the study period, and differences between treatment and control subjects recorded. Results indicated significant changes in cortical blood flow in the treatment group compared to the control group. Control subjects averaged 0.5% increase on the pre/postmeasure of cortical perfusion compared to 4.07% for the active treatment. The lack of increased blood flow in the control group, along with the absence of improved mood, energy, or memory functions offers indirect support for the hypothesis that decreased RCBF is a central element in dementia's etiology and progression (Spilt et al., 2005).

Interestingly, one subject in the control group, a housewife in her mid-80s, living with her husband in a small row home and speaking very little English, responded at a comparable level to the treatment group. She had spent most of her time watching television with her husband who did not interact with her much, leaving her with minimal social contact or stimulation. She then came to the clinic everyday, was greeted with a friendly smile and was shown genuine concern and support. This can be considered a very good example of the impact of human contact on neurophysiological functioning, and also could be considered a significant placebo response. The other placebo group members were in socially and emotionally stimulating environments, (engaged and supportive caregivers) similar to those of the active treatment cohort.

T-tests of group means showed significant pre-posttreatment differences on components of the ADAS-cog between active treatment and placebo groups as well as on measures of ideational praxis (P = .03), Boston Naming Test (P = .035) and Trail Making Test: Part A, improved (fewer) omission errors (P = .044). The latter suggests more active and efficient engagement with the task. Increased commission errors (P = .034) were seen in the treatment group on Trail Making Test, parts A&B. While objectively this infers that more errors were made in the completion of the task, this may have been because treatment group subjects completed more of the test within the 5-min time limit.

Quantitative EEG changes in amplitude and coherence believed to result from the NIRS were noteworthy as indicated by circled numbers in Fig. 4.5. Abnormal (deficient) delta in absolute and relative power (#s 1–2) normalized over the study period, as did alpha relative power (#3). Decreased z-scores were observed in cortical connectivity measures, for example, delta amplitude asymmetry (#4), delta, theta and alpha coherence (#5), and phase lag (#6).

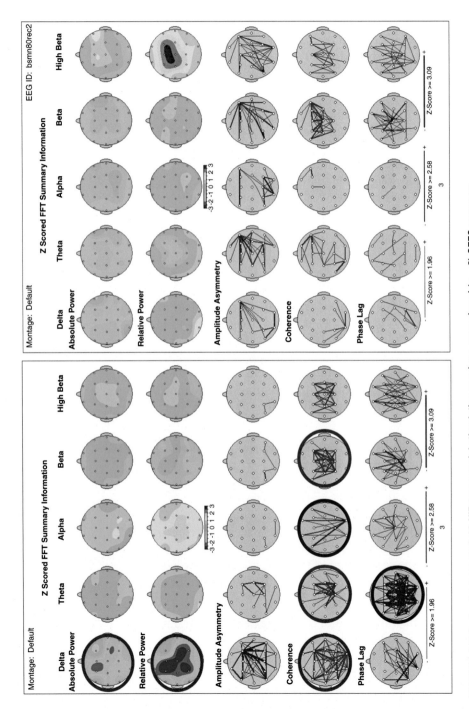

FIGURE 4.5 Pre/post aggregated, 28 NIR transcranial stimulation sessions, eyes closed (*n* = 9) QEEG summary.

Possible implications of the circled areas in Fig. 4.5 include: #1&2: Increased slow wave activity may contribute to improved sleep quality and daytime energy level. #3: Normalized excessive alpha amplitude generally correlates with decreased anxiety and "sundowning" symptoms. #4: Less hypocoherence may support improved memory, expressive, and receptive language processing. #5: Decreased phase lag correlates with improved cognitive processing speed. (Berman & Frederick, 2009)

NEAR-INFRARED LIGHT STIMULATION TECHNOLOGY

An example of current commercially available NIR stimulation technology is the "Vielight 810 Infrared" from Vielight Inc. in Toronto, Canada. Their intranasal NIR delivery system has been commercially available for 10 years, and is one of the more established devices for delivering red and infrared light energy to the brain and body. This is the first commercially available device to combine transcranial and intranasal delivery of NIR stimulation specifically targeting decreased connectivity in the default mode network (DMN), which has been implicated in Alzheimer's disease and other neurodegenerative disorders (Blautzik et al., 2016).

The Vielight "810 Infrared" (intranasal only) and the "Neuro" (intranasal–transcranial combination) are for use in photobiomodulation (PBM) of the brain (Fig. 4.6).

Both devices transmit NIR light at 810 nm, pulsing at 10 Hz. The neuro is designed to deliver more power throughout the brain, emphasizing prefrontal and parietal regions, that is, default mode network. The wavelength of 810 nm was selected because it has been found to provide optimal penetration and response out of living nerve tissues (Byrnes et al., 2005). Although the devices pulse at 10 Hz (alpha oscillation) often sought for entrainment in the neurofeedback field, the basis of the invention is primarily for neuromodulation. The choice of 10 Hz is based on studies that demonstrated the superior restorative quality of this frequency for brain injuries when compared with other frequencies (Ando et al., 2011; Wu et al., 2012). Intranasal administration was chosen based on the accessible rich blood capillary network in the nasal region for blood irradiation (Chien, Su, & Chang, 1989), and as a path for efficient light penetration into the brain (Pitzschke et al., 2015).

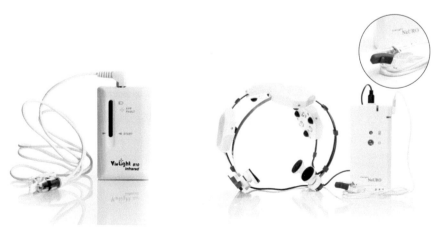

FIGURE 4.6 (Left) Vielight 810 intranasal near infrared (NIR) stimulation and (right) neuro transcranial/intranasal system (www.vielight.com).

There has been considerable anecdotal and small pilot research supporting the therapeutic efficacy of these devices. Larger studies are being undertaken and will provide more validation related to NIR stimulation's impact on cognitive and behavioral functioning.

Improved Mini Mental State Examination (MMSE) scores were obtained after one year of regular use of the Vielight 810 in a case report of "Rudy," a patient with a clinical diagnosis of Alzheimer's disease (Fig. 4.7) (Lim, 2014).

Recent findings from a pilot 16-week, single blinded, randomized, placebo-controlled study (*N* = 19) using combined intranasal and transcranial NIR stimulation (810 nm), produced significant positive changes in cognitive functioning, and behavioral regulation, with no adverse reactions (Saltmarche et al., 2016). Transcranial NIR stimulation was provided using the Vielight Neuro device (Fig. 4.6) twice weekly for 2 weeks and then once weekly for 10 weeks. Subjects also would self-administer treatment at home

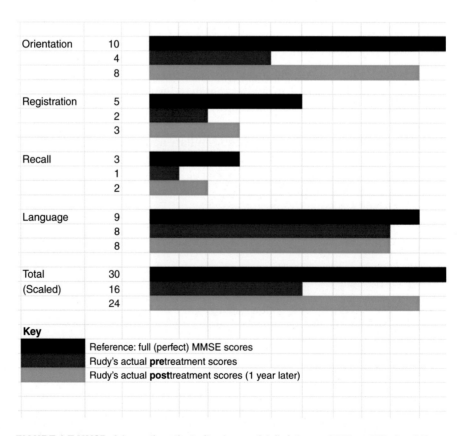

FIGURE 4.7 MMSE of dementia patient after 1 year of daily intranasal 810 nm NIR stimulation (Lim, 2014).

using the Vielight 810 nm intranasal device (Fig. 4.6) which delivered 13 mW of energy into the nasal cavity. Cognitive assessments were conducted at baseline, end of week 6, end of week 12, and after a 4-week follow-up period of no treatment.

MMSE and ADAS-Cog mean baseline scores were: active treatment group ($N = 13$) 18.4 (SD 9.37) and 32.1 (SD 21.41), and placebo group ($N = 6$); 25.8 (SD 4.36) and 14.8 (SD 7.91). The MMSE scores increased two points for the eight subjects in the moderate-severely impaired active treatment group (baseline MMSE 524) after 12-week treatment cycle ($P = .03$, 2-tailed), and their ADAS-cog scores decreased 5.00 ($P = .03$). A placebo participant in the more impaired group dropped out before completing the trial. Overall decline in both groups' functioning was noted at the 4-week follow-up assessment. No significant differences were noted in the higher functioning group (MMSE = 25–30) between active and placebo treatment. However, qualitative feedback from caregivers in both groups noted improved sleep quality, decreased aggression, agitation, anxiety and wandering events (Saltmarche et al., 2016). Alternate forms of the MMSE were not used, thereby introducing possible history confounds (McLeod, 2008) to the data. Home-based treatments were monitored with a daily journal that was reviewed weekly for compliance and changes in quality of life factors.

Rhythmic visible light stimulation results in EEG alpha wave entrainment, and has been the basis for numerous approaches to analyzing and modifying brain activity (Notbohm, Kurths, & Herrmann, 2016). Transcranial photobiomodulation (PBM) with NIR light (which is nonvisible) may not stimulate the photoreceptors in the eyes, but there is evidence from EEG that the brain responds to transcranial PBM (Hesse, Werner, & Byhahn, 2015), thus lending additional support for its potential clinical utility. Our efforts sought to take advantage of the Quantitative EEG (QEEG), which has been recognized as a highly effective neuroimaging technique to identify neuropsychiatric disorders, (Hughes & John, 1999), and to discern effects of these treatments in neurodegenerative conditions, for example, Alzheimer's and Parkinson's disease (Fonseca, Tedrus, Carvas, & Machado, 2013). A case review of 250 subjects in one clinic's treatment of dementia using LORETA z-score neurofeedback has been published, showing positive improvements in both cognitive and behavioral symptoms (Koberda, 2013).

A recently conducted pilot study conducted at Quietmind Foundation, measured the response of several student volunteers to a single 25-min exposure using a Vielight 810 nm LED driven intranasal device. Subjects completed a 6–8 min 19-channel EEG recording before and after the 25 min Vielight 810 exposure. In each instance, the device was placed in the right nostril. Notably, the observed changes are consistent with the region receiving the most direct stimulation.

The following Quantitative EEG (QEEG) images (Figs. 4.8–4.12) indicate positive modification in both absolute power and coherence measures. Green areas

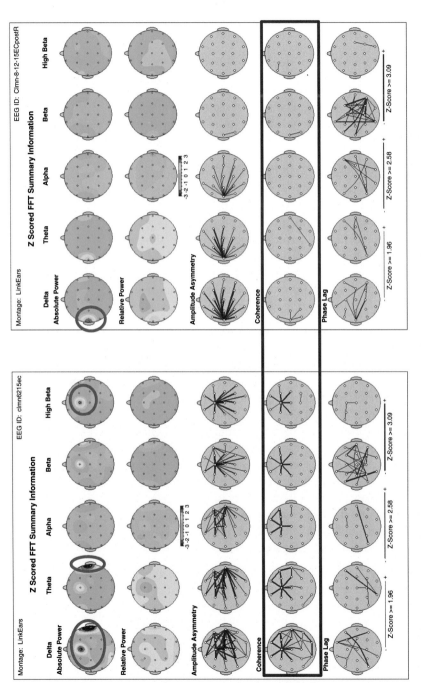

FIGURE 4.8 Pre-/postintranasal Vielight 810 nm IR stimulation (20 min/day for 2 mo) eyes closed QEEG summary of 19-year-old male with right temporal traumatic brain injury (TBI).

Significant hypocoherence changes from the treatment, which positioned the transcranial 810 nm IR-LED arrays across the parietal region (P3, PZ, P4) and at the prefrontal midline (FZ). The intranasal applicator was placed in the right nostril which appears to contribute to the elimination of excessive delta, theta and high beta activity in this subject's right prefrontal and temporal region. Posttreatment, subject reported decreased social anxiety and improved concentration and ability to complete schoolwork.

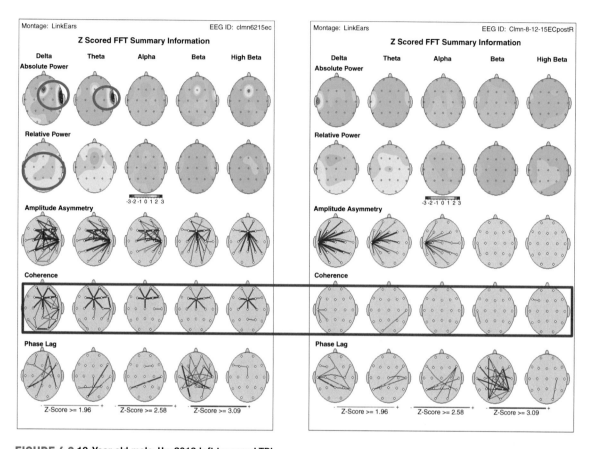

FIGURE 4.9 19-Year old male, Hx: 2013 left temporal TBI.

These images are post 18 months LORETA neurofeedback training. Fully recovered 1st year premed student.

on the head maps indicate normative comparisons of below 1 SD from average, with red and blue outlined areas indicating 2–3 SD above or below mean normative values, respectively. Circled regions indicate locations of significant pre- to postchange following 25 min of intranasal NIR stimulation.

CURRENT THERAPEUTIC APPLICATIONS OF NEAR INFRARED LIGHT

Quietmind Foundation is focusing on the application of NIR stimulation in combination with EEG biofeedback (neurofeedback) training in the treatment of neurodegeneration stemming from age-related decreases in RCBF, TBI, Parkinson's and Lyme's disease and other tick-borne infectious diseases. We have seen in our treatment of clinic patients, improvement in both physical and

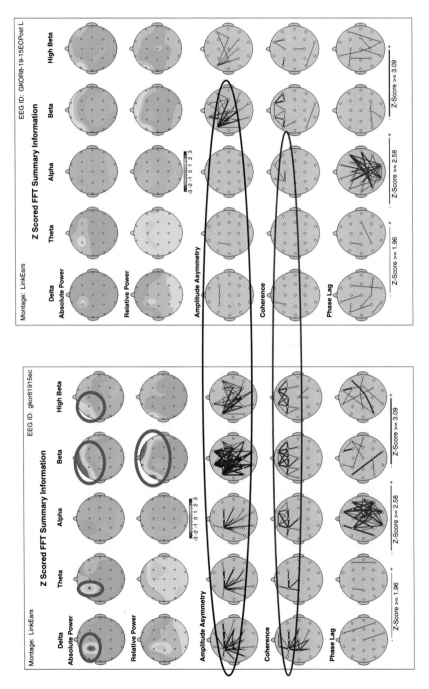

FIGURE 4.10 19-Year old female with Dx: mixed state bipolar disorder, stable with Rx: Seroquel 100 mg 3 years' duration. Notable decrease of left prefrontal slow and frontal beta activity and significantly reduced asymmetry, frontal hypocoherence and beta phase lag.

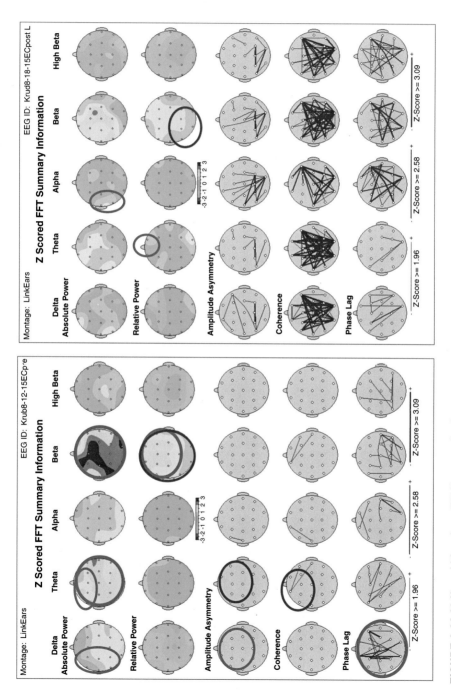

FIGURE 4.11 22-Year old female, no TBI Hx, no medication, graduate student.
Notable decrease in left side beta amplitude and appearance of increased coherence and phase lag across all bands (except delta).

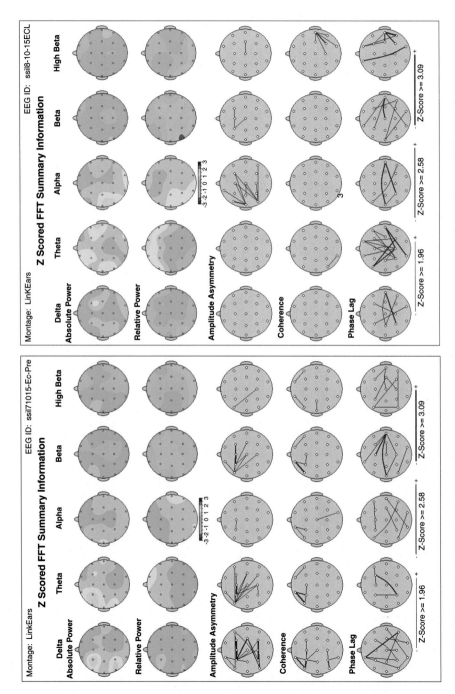

FIGURE 4.12 22-Year-old male, with social anxiety and Hx congenital orthopedic (foot) abnormality requiring multiple surgeries (four with general anesthesia).

Note decreased left frontal and parietal delta and theta, decreased left occipital beta and left prefrontal hypocoherence. Subject within 2 months of this treatment was able to obtain his first full-time professionally relevant and socially challenging employment and manage two jobs to save up for school in the coming year.

cognitive functioning with either or both treatments, used in combination with individual body-centered, psychotherapy and peripheral biofeedback, particularly respiration training (Berman & Frederick, 2009). These findings have been reinforced by Naeser et al. (2014) pilot study data using red and near-infrared LED transcranial stimulation in the treatment of traumatic brain injury.

Quietmind Foundation's research efforts regarding NIR stimulation and neuro-modulation evolved from cell line and animal work showing the effects of varying wavelengths of red and NIR stimulation. One of the few comparative studies showed that 665 and 810 nm, but not 730 and 980 nm, improved postconcussion symptoms in mice, especially oxidative stress, ionic imbalance, changes in vascular permeability and mitochondrial dysfunction (Wu et al., 2012). It now has been shown that the effectiveness of certain shorter NIR frequencies is related to the absorption spectrum of cytochrome oxidase that stimulates mitochondrial chromophore activity (Hayworth et al., 2010). In addition, neuronal mitochondrial Complex II appears to be positively modulated by using 1072 and 1068 nm transcranial NIR-LED stimulation (Grillo, Duggett & Chazot, unpublished). Interestingly, Complex II recently has been shown to be a key determinant in skin aging (Bowman & Birch-Machin, 2016) (Fig. 4.13).

Quietmind Foundation's interest in noninvasive treatment of neurodegeneration began in 2008 during its clinical trial of home- and office-based 2-channel EEG biofeedback training for early to mid-stage dementia (Berman & Frederick, 2009). This investigation was initiated after the researchers regularly noted differences in slow wave amplitudes in patients seen in our clinic at different stages of dementia, and with traumatic brain injury, PTSD, memory loss, (e.g., probable Alzheimer's), frontotemporal dementia, vascular disorder, chronic Lyme's disease, or other tick-borne infections (Hodges, Garrard, & Patterson, 2009; Rojas & Gonzalex-Lima, 2013; Lee et al., 2011).

Searching the PubMed database using search terms "EEG," "electroencephalography and dementia" resulted in 3130 citations, and "dementia severity and EEG" resulted in an extensive literature of 280 citations related to the progression of dementia and discrimination from other neurodegenerative conditions, for example, Parkinson's disease. Some typical EEG changes common to dementia include the following:

- Resting alpha power declines in early stages.
- Predictably increasing slow and decreasing faster wave amplitudes.
- Impairment of callosal (splenium), thalamic, and anterior–posterior white matter bundles.
- Reduced correlation of resting state blood oxygen level-dependent activity across several intrinsic brain circuits including default mode and attention-related networks. Abnormal power and functional coupling of resting state cortical EEG rhythms (Teipel et al., 2016).

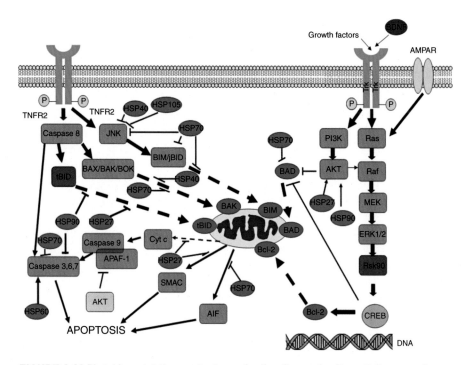

FIGURE 4.13 Photobiomodulation mechanisms of action (Burroughs, Duggett, Ennaceur, & Chazot, unpublished and published material).

A range of selective HSPs have been shown to be upregulated following in vivo treatment of Alzheimer mice (HSP27, 60, 70, 90, and 105) (Duggett & Chazot, unpublished; Schirmera et al., 2016); the role of these proteins in mitochondrial function, apoptosis and chaperone-mediated protein folding are highlighted. *(With permission from Prof. Paul Chazot, PhD.)*

While these findings were recognized and replicated in hundreds of studies in neurology and neurophysiology over the last 5 decades, it is only recently that clinical applications of these kinds of electrophysiological data are being proposed. For example, Teipel suggested, "From a clinical point of view, the study of network connectivity may provide a diagnostic marker of early disease, as well as a prognostic marker at an individual level where the integrity of key functional networks will influence the likelihood of cognitive decline at a given level of molecular pathology (Teipel et al., 2016). It is this type of background which motivated and continues to motivate, our current three-pronged research emphasis on using neurofeedback and NIR-based photobiomodulation to optimize neuronal repair, regrowth, and renormalization (Rochkind, El-Ani, Nevo, & Shahar, 2009). Studies are now underway using QEEG assessment methods to evaluate therapeutic efficacy of directed energy stimulation combined with real time control using both LORETA z-score neurofeedback and heart rate variability (HRV).

The incorporation of NIR stimulation came after QMF was informed by John Haas, a well-known corporate executive and philanthropist with an interest in improving human services in the greater Philadelphia region and early advocate of integrative health research. He shared a news story concerning United Kingdom–based researchers having reversed dementia in mice using brief exposures to NIR light (Derbyshire, 2008). The article reported on experiments conducted at the Sunderland School of Pharmacy and Durham University that demonstrated improved learning performance and diminished behavioral dysregulation in mice that innately displayed premature ageing memory deficits. The mice receiving the 1072 nm infrared light stimulation showed significant improvement in both behavior and navigational memory in radial arm memory trials (Michalikova, Ennaceur, van Rensburg, & Chazot, 2008). These results helped coalesce our group's understanding that the NIR stimulation was serving a "reparative" function for the neural tissue and biological systems that were being compromised by the neurodegenerative processes, and that the neurofeedback was providing a retraining function by supporting the renormalization of neural connectivity. Further studies with the 1060–1080 nm stimulation have now added support to this model with evidence of amyloid load suppression in an Alzheimer's disease mouse and some neuroprotective capabilities (Duggett & Chazot, 2014). Vielight technology was discovered in 2014 and quickly became a treatment option for clients seeking to continue using transcranial NIR stimulation. The Vielight Neuro was introduced, at that time and Quietmind Foundation became a research affiliate to provide QEEG outcome data and technical application development support.

OTHER CLINICAL APPLICATIONS OF LIGHT STIMULATION

Evidence from recent studies of blue-enriched white light stimulation (469 nm) demonstrated the potential for improved cognitive functioning (improved working memory) for 40 min after a single 30-min exposure. Subjects in this comparative study ($N = 35$) showed enhanced activity on fMRI scans in both the dorsolateral prefrontal cortex and ventromedial prefrontal cortex after a single exposure using goggles emitting pulsed blue light compared to subjects similarly exposed to amber light. Changes were noted in improved reaction time ($P = .04$) and enhanced response efficiency in a standard N-back working memory task (Alkozei, 2016).

Circadian or diurnal rhythmic changes are deeply ingrained in our physiology, organizing in large measure all ongoing physiological and psychosocial responses. Recent investigations have concluded that the internal pacemakers in the superchiasmic nuclei (SCN) are entrained to the 24-h circadian light–dark rhythm in the external environment (Kohaska, Waki, Cui, Gouraud, &

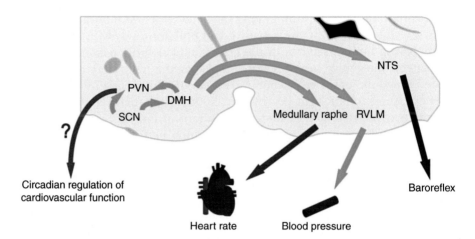

FIGURE 4.14 Hypothetical neural network that governs the cardiovascular diurnal rhythm, including the SCN and the DMH, which regulates several sections of the medulla oblongata including the raphe pallidus nucleus. *(From Kohaska, A., Waki, H., Cui, H., Gouraud, S. S., & Maeda, M. (2012) Integration of metabolic cardiovascular diurnal rhythms by circadian clock. Endocrine Journal, 59(6), 447–456.)*

Maeda, 2012). Recognizing the influence of circadian rhythms and sleep–wake cycles on perceived emotional states has been a focus of considerable study in psychiatry, particularly in depression research (Nechita, Pirlog, & Chirita, 2015; Boivin, 2000; Boivin & Boudreau, 2014). Disturbances in REM and slow-wave sleep have been recognized as core elements in the pathophysiological features of depression and other major psychiatric illnesses. The connection between the SCN and the dorsomedial hypothalamic nucleus (DMH) is of particular interest as it may relate to cardiovascular regulation, that is, heart rate variability. Since the SCN is contributing significant efferent inputs to the DMH there is evidence that this network influences heart rate modulation (Samuels, Zaretsky, & DiMicco, 2002) (Fig. 4.14).

In addition to the indirect projection from the SCN to the paraventricular nucleus (PVN) via the DMH, the PVN also receives a direct projection from the SCN; however, this direct pathway is thought to regulate the melatonin secretion rhythm by activating sympathetic preganglionic neurons in the upper thoracic spinal cord, and it is unknown whether this projection is involved in the circadian regulation of cardiovascular function. Taken together, the current findings suggest that the DMH serves as a nodal area mediating the transmission of the SCN signal to the cardiovascular tissues, but the precise neural and/or neuroendocrine pathways critical for the 24-h blood-pressure rhythm and other cardiovascular function has yet to be determined (Kohaska et al., 2012).

LIGHT'S RELATIONSHIP TO PSYCHIATRIC DISORDERS

Seasonal affective disorder (SAD) is widely recognized as responsive to bright light therapy, especially full sunlight, although very intense artificial light stimulation has also helped remediate SAD symptoms. It has been noted that lower intensity blue wavelengths (460–485 nm) may provide comparable results to high intensity white light, though this distinction diminishes with age (Hodges, Garrard, & Patterson, 1998). This biological response has been identified as being mediated by melanopsin, a molecule that absorbs blue light (450–480 nm) and functions as a light-gated ion channel. Melanopsin is a 7-transmembrane G-protein-coupled receptor (Kalsbeek, Merrow, Roenneberg, & Foster, 2012) that can be used as an optogenetic tool to control neural and cellular activities in various tissues. Compared with other optogenetic pigments, melanopsin is more sensitive to light, shows long-lasting activation, and can also control intracellular Ca^{2+} dynamics. Melanopsin is a naturally occurring mammalian protein (also present in humans) in contrast to channel rhodopsin which was originally isolated from bacteria. It has recently been discovered that melanopsin is expressed in tissue outside the eyes and the brain, in endothelial cells of blood vessels and in the skin confirming that blue (and bright white) light exposure can indeed have a multitude of biological effects (Iyengar, 2013; Sikka et al., 2014). How the brain responds to light is wavelength dependent, adding to the importance of the melanopsin-expressing retinal ganglion cells (Pandi-Perumal, Kramer, & Hobson, 2010).

Several other psychiatric disorders have been studied as to their response to light therapy, including premenstrual dysphoric disorder, antepartum, and postpartum major depressive disorder, bulimia nervosa and adult attention-deficit disorder. Combining light therapy with antidepressant medication has been shown to help accelerate the rate at which patients notice therapeutic benefit. It should be noted that mood stabilizing agents are suggested, as well as shifting light therapy exposure to afternoon from early morning when treating bipolar depression patients with bright light therapy (Geoffroy et al., 2015). Elderly dementia patients' sleep architecture improved when exposed to full daylight sun for several hours in the morning and afternoon (Fetveit & Bjorvatn, 2006); and light therapy (10,000 lux) thirty minutes daily is suggested along with daily 2,500 lux indirect light therapy to increase melatonin levels and help reduce "sundowning" agitation, and excessive sleeping (Pandi-Perumal et al., 2010).

Delayed sleep onset syndrome or circadian rhythm sleep disorder is responsive to light therapy; however, timing the delivery is important, that is, immediately upon awakening (Figueiro, 2016). Exposure to bright light through the ear canal during the appropriate time periods before, during and after air travel has been reported to reduce the symptoms of jet lag and accelerate the recalibration of the body clock (Jurvelin, Jokelainen, & Takala, 2015).

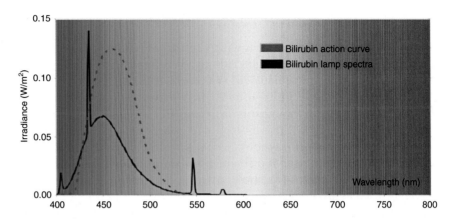

FIGURE 4.15 Blue light enhances the excretion of bilirubin, because it photoisomerizes the molecule.

The photoisomer is more readily excreted than the unisomerized form. *(From http://www.light-sources. com/solutions/specialty-fluorescent/applications/medical-phototherapy/#Neonatal_Jaundice.)*

LIGHT THERAPY FOR JAUNDICE

Neonatal jaundice is a condition that can occur in newborn infants, particularly when they are born prematurely. If the ratio of bilirubin to albumin becomes too high, then unbound bilirubin in the bloodstream can enter tissues of the brain, and this can cause damage. Neonatal jaundice can be treated with phototherapy. Blue light (450–500 nm) is applied to the skin of babies in order to keep the bilirubin concentration at a safe level and allow excretion through urine and stool rather than overburdening the liver (Fig. 4.15).

THERAPEUTIC APPLICATION OF COLOR: SYNTONIC OPTOMETRY

Evolving from the work of Edwin Babbit (Babbitt & Birren, 2000) on color and light in nature and health, Spitler completed a doctoral dissertation which he developed into *The Syntonic Principle*, outlining the effects of light related to health and behavior (Spitler, 2011). His thesis centered on the use of colored light stimulation to the eyes to stimulate biochemical changes in the body that would promote healing and optimize healthy functioning. Syntonics treatment sought to improve visual processing deficits by influencing sympathetic and parasympathetic CNS activation, with higher (red–yellow) frequency spectrum stimulating sympathetic arousal, and (blue–violet) frequencies stimulating parasympathetic activation. Green light was expected to equally stimulate both sympathetic and parasympathetic pathways. A basic principle as described by Gottlieb and Wallace (2001) was that chronic systemic, mental, emotional, and visual ailments were caused primarily by autonomic nervous system and endocrine imbalances, and that shining specific light frequencies into the eyes could stimulate or suppress autonomic and endocrine functioning. This would restore balance via direct retinal input to thalamic and hypothalamic regulatory centers, thereby also correcting visual dysfunctions at their source. This work gained a following among optometrists and led to the creation of the College of Syntonic Optometry. It is still being practiced and promoted through the College of Syntonic Optometry (see http://www.collegeofsyntonicoptometry.com).

According to Spitler, Syntonic Optometry (as taken from *The Gale Encyclopedia of Alternative Medicine*) considers illness largely to be caused by imbalances in the body's endocrine and nervous systems (Helwig, 2014). Balance is believed to be restorable, and healing achieved by exposing the eyes to visible frequencies of light. Syntonic treatment is conducted such that, "the patient is exposed to one or more colors of light for a fixed period of time. This is done in a darkened room, with colors generated by a machine known as a syntonizer. In a typical session, a patient might absorb one color for 10 min, then another for an additional 10 min. Alternatively, just one color might be absorbed for 20 min. Treatment typically could involve three and five sessions a week, for a period of 4–8 weeks." (Helwig, 2014) (Fig. 4.16).

Syntonics offers a light-based, color specific treatment for a range of visual and CNS-mediated disorders including convergence insufficiency, and visual stress/fatigue that could be related to neurotrauma, and emotional and learning

FIGURE 4.16 Syntonic syntonizer.

disorders rooted in visual processing dysfunction. Considerable controversy has existed regarding the scientific validation of Syntonic Optometry, and, as with all such approaches involving colored light, more research is needed to better understand the clinical successes that have been reported anecdotally, as well as in the small number of peer-reviewed publications in the literature.

OTHER LIGHT STIMULATION RESEARCH

In one study involving light stimulation, PET imaging documented RCBF and cerebral blood volume modification in response to flickering white light stimulation (Ito, Takahashi, Hatazawa, Kim, & Kanno, 2001). Clinical therapeutic application of colored light was employed by Macdonald in the Woman's Hospital of Philadelphia to treat women with rheumatoid arthritis pain. Results were significant when subjects' bodies were exposed to blue light (no light was directed into the eyes), and the analgesic response was dose-dependent in terms of length of exposure (Macdonald, 1982). As noted earlier, there is new evidence that brief passive exposure to blue enriched light is now showing positive effects on memory task performance (Alkozei, 2016). The growing acceptance of sunlight's effects on seasonal affective disorder has provided additional credence to this approach. However, there is considerably more evidence now for LLLT and LED-based light therapies with the latter being recognized recently as "a neuro-restorative and or neuroprotective therapy for the treatment of injury and diseases of the CNS (Anders, 2009). Next generation treatments for memory disorders will include the application of NIR phototherapeutic techniques now being developed that can identify precise neuronal locations of concentrated calcium that drives neurotransmitter production using "carrier molecules" that respond to NIR stimulation (Jakkampudi et al., 2016). This will lead investigators to develop more finely targeted intervention strategies as well as grow our understanding of memory formation, stability, and adaptive capacity.

LIGHT-BASED TREATMENT'S FUTURE

The search for the integration of mind and body was prophesied by Goethe near the beginning of the 19th century, saying that in 100 years the world would become one great hospital with the goal of achieving immortality (Haldane, 2014). The tools we now employ offer unparalleled opportunities. As Harris and Gordon (2015) recently described, "The rich tapestry formed by the trillions of connections between far-flung brain regions demonstrates the complexity of the brain...." Yet the connectivity of the brain does not derive statically from these anatomical pathways; it fluctuates dynamically with mood and cognitive states, influenced by stimuli and influencing behavior. Detecting and quantifying connectivity provide a key to understanding this dynamism. We are coming to appreciate neural network activity and the broader system-as-a-whole dynamics in the brain, and can expect to see ever more sophisticated and creatively adaptive clinical applications to treating physical and neuropsychiatric disorders.

It is likely that the next wave of technologies will more fully integrate the use of noninvasive light-based and directed energy therapies guided by QEEG, Heart Rate Variability, sLORETA, real-time functional (rtFMRI) and transcranial ultrasound imaging, (Roher et al., 2011; van de Haar et al., 2016), and TUS therapeutic applications (Hameroff et al., 2013). Integrated QEEG-based z-score and heart rate variability (HRV) neurotherapy and neuromodulation systems, employing pulsed electromagnetic fields, and transcranial direct and alternating current already are available to clinicians under 510 K exemption (Dogris & Wiitala, 2016). Based on the rate of technological advancement, data aggregation and integration, the field can expect more intelligently automated, remotely deployable systems that will provide real-time, cloud-based assessment and dynamically responsive treatment for patients in the clinic and at home. The convergence of these new applications with the growing international demand for more personalized care utilizing telemedicine service delivery systems will further support decentralization and expand service delivery network capability to achieve universal accessibility to the highest levels of evidence-based healthcare.

Acknowledgments

Trent Nichols, MD and James Halper, MD for review and commentary on the manuscript. Steven Silverman, BS, and Christopher Lehman for coordinating clinical data collection and analysis.

References

Alaluf, S., Muir-Howie, H., Hu, H. L., Evans, A., & Green, M. R. (2000). Atmospheric oxygen accelerates the induction of a post-mitotic phenotype in human dermal fibroblasts: the key protective role of glutathione. *Differentiation, 66*(2–3), 147–155.

Alkozei, A. (2016). Exposure to blue wavelength light is associated with increased dorsolateral prefrontal cortex responses and increases in response times during a working memory task. University of Arizona poster presentation at Sleep 2016, Denver, CO., June 15, 2016.

Anders, J. J. (2009). The potential of light therapy for central nervous system injury and disease. *Photomedicine and Laser Surgery, 27*(3), 379–380.

Ando, T., Xuan, W., Xu, T., Dai, T., Sharma, S. K., Kharkwal, G. B., Huang, Y. Y., Wu, Q., Whalen, M. J., Sato, S., Obara, M., & Hamblin, M. R. (2011). Comparison of therapeutic effects between pulsed and continuous wave 810 nm wavelength laser irradiation for traumatic brain injury in mice. *PLoS One, 6*(10), e26212.

Babbitt, E. S., & Birren, F. (2000). *The principles of light and color: the healing power of color*. New York, NY: Citadel Press.

Barolet, D. (2008). Light-emitting diodes (LEDs) in dermatology. *Seminars in Cutaneous Medicine and Surgery, 27*(4), 227–238.

Barolet, D., & Boucher, A. (2010). Radiant near infrared light emitting diode exposure as skin preparation to enhance photodynamic therapy inflammatory type acne treatment outcome. *Lasers in Surgery and Medicine, 42*(2), 171–178.

Berman, M. H., Halper, J. P., Nichols, T. W., Henderson, J., Lundy, A., & Huang, J. H. (2017). Photobiomodulation with Near Infrared Light Helmet in a Pilot, Placebo Controlled Clinical Trial in Dementia Patients Testing Memory and Cognition. *Journal of Neurology and Neuroscience, 8*(1), 176.

Berman, M., & Frederick, J. (2009). Efficacy of neurofeedback for executive and memory functioning dementia. Poster presented at *Alzheimer's Association International Conference (Hot Topics)*. Vienna, Austria.

Blautzik, J., Keeser, D., Paolini, M., Kirsch, V., Berman, A., Coates, U., Reiser, M., Teipel, S. J., & Meindl, T. (2016). Functional connectivity increase in the default-mode network of patients with Alzheimer's disease after long-term treatment with galantamine. *European Neuropsychopharmacology, 26*(3), 602–613.

Boivin, D. B. (2000). Influence of sleep wake and circadian rhythm disturbances in psychiatric disorders. *Journal of Psychiatry Neuroscience, 25*(5), 446–458.

Boivin, D. B., & Boudreau, P. (2014). Impacts of shift work on sleep and circadian rhythms. *Pathologie Biologie (Paris), 62*(5), 292–301.

Bowman, A., & Birch-Machin, M. A. (2016). Age-dependent decrease of mitochondrial complex II activity in human skin fibroblasts. *Journal of Investigative Dermatology, 136*(5), 912–919.

Bradford, A., Barlow, A., & Chazot, P. L. (2005). Links probing the differential effects of infrared light sources IR1072 and IR880 on human lymphocytes: evidence of selective cytoprotection by IR1072. *The Journal of Photochemistry and Photobiology B, 81*(1), 9–14.

Byrnes, K. R., Waynant, R. W., Ilev, I. K., Wu, X., Barna, L., Smith, K., Heckett, R., Gerst, H., & Anders, J. J. (2005). Light promotes regeneration and functional recovery and alters the immune system after spinal cord injury. *Lasers in Surgery and Medicine, 36*(3), 171–185.

Chaudhary, S., Scouten, A., Schwindt, G., Janik, R., Lee, W., Sled, J. G., Black, S. E., & Stefanovic, B. (2013). Hemodynamic effects of cholinesterase inhibition in mild Alzheimer's disease. *The Journal of Magnetic Resonance Imaging, 38*(1), 26–35.

Cheng-Yi, Liu, & Ping, Zhu (2008). *Laser function medicine and its applications*. Guangzhou, China: South China Normal University.

Chien, Y. W., Su, K. S. E., & Chang, S. F. (1989). Chapter 1: anatomy and physiology of the nose. *Nasal systemic drug deliver*. New York, NY: Dekker, pp.1–26..

De Freitas, L. F., & Hamblin, M. R. (2016). Proposed mechanisms of photobiomodulation or low-level light therapy. *IEEE Journal of Selected Topics in Quantum Electronics, 22*, 7000417.

Derbyshire, D. (2008). Dementia patient makes 'amazing' progress after using infra-red helmet. Available from: http://www.dailymail.co.uk/health/article-510172/The-helmet-turn-symptoms-Alzheimers.html

Dogris, N., & Wiitala, B. (2016). *Neurofield Training Manual Bishop*. CA: Neurofield, Inc. Available from: www.neurofield.com

Dougal, G., & Kelly, P. (2001). *A pilot study of treatment of herpes labialis with 1072 nm narrow waveband light*. Stockton-on-Tees, UK: Occupational Health Department, North Tees Hospital.

Duggett, N. A., & Chazot, P. (2014). Low-intensity light therapy (1068 nm) protects CAD neuroblastoma cells from B-amyloid-mediated cell death. *Biology and Medicine, 1*, S1.

Fetveit, A., & Bjorvatn, B. (2006). Sleep duration during 24-hour day is associated with the severity of dementia in nursing home patients. *Journal of Geriatric Psychiatry, 21*(10), 945–950.

Figueiro, M. G. (2016). Delayed sleep phase disorder: clinical perspective with a focus on light therapy. *Nature and Science of Sleep, 8*, 91–106.

Fonseca, L. C., Tedrus, G. M., Carvas, P. N., & Machado, E. C. (2013). Comparison of quantitative EEG between patients with Alzheimer's disease and those with Parkinson's disease dementia. *Clinical Neurophysiology, 124*(10), 1970–1974.

Friedlander, R. M., Zhang, Y., Li, M., Drozda, M., Chen, M., Ren, S., Mejia Sanchez, R. O., Leavitt, B. R., Cattaneo, E., Ferrante, R. J., & Hayden, M. R. (2003). Depletion of wild-type huntingtin in mouse models of neurologic diseases. *Journal of Neurochemistry, 87*(1), 101–106.

Furchgott, R. F., Ignarro, L. J., & Murad, F. (2016). The Nobel Prize in Physiology or Medicine 1998. Nobel Media AB 2014. See acceptance speech available from: http://www.nobelprize.org/nobel_prizes/medicine/laureates/1998/

Garraux, G., Salmon, E., Degueldre, C., Lemaire, C., & Franck, G. (1999). Medial temporal lobe metabolic impairment in dementia associated with motor neuron disease. *Journal of the Neurological Sciences, 168*(2), 145–150.

Geoffroy, P. A., Fovet, T., Micoulaud-Franchi, J. A., Boudebesse, C., Thomas, P., Etain, B., & Amad, A. (2015). Bright light therapy in seasonal bipolar depressions. *Encephale, 41*(6), 527–533.

Gottlieb, R., & Wallace, L. (2001). Syntonic phototherapy. *Journal of Behavioral Optometry, 12*(2), 31–38.

Grillo, S. L., Duggett, N. A., Ennaceur, A., & Chazot, P. L. (2013). Non-invasive infra-red therapy (1072 nm) reduces β-amyloid protein levels. *Journal of Behavioral Optometry, 123*, 13–22.

Grimmer, T., Diehl, J., Drzezga, A., Forstl, H., & Kurz, A. (2003). Region-specific decline of cerebral glucose metabolism in patients with frontotemporal dementia: a prospective 18F-FDG-PET study. *Dementia and Geriatric Cognitive Disorders, 18*, 32–36.

Haldane, S. (2014). *Pulsation*. Cork, Ireland: Parmenides Books, Run Press.

Hamblin, M. R. (2016a). Photobiomodulation and the brain—has the light dawned? *Biochemical Society, 38*(6), 24–33.

Hamblin, M. R. (2016b). Shining light on the head: photobiomodulation for brain disorders. *BBA Clinical, 6*, 113–124.

Hamblin, M. R., & Demidova, T. N. (2006). Mechanisms of Low Level Light Therapy—an Introduction. In M. R. Hamblin, J. J. Anders, & R. W. Waynant (Eds.), *Mechanisms for Low-Light Therapy*. Bellingham, WA: The International Society for Optical Engineering, Proc SPIE 6140. art. no. 61001 1-12.

Hameroff, S., Trakas, M., Duffield, Annabi, E., Gerace, M. B., Boyle, P., Lucas, A., Amos, Q., Buadu, A., & Badal, J. J. (2013). Transcranial ultrasound (TUS) effects on mental states: a pilot study. *Brain Stimulation, 6*(3), 409–415.

Hargate, G. (2006). A randomized double-blind study comparing the effect of 1072-nm light against placebo for the treatment of herpes labialis. *Clinical and Experimental Dermatology, 31*(5), 638–641.

Harris, A. Z., & Gordon, J. A. (2015). Long range neural synchrony in behavior. *The Annual Review of Neuroscience, 38*, 171–194.

Hayworth, C. R., Rojas, J. C., Padilla, E., Holmes, G. M., Sheridan, E. C., & Gonzalez-Lima, F. (2010). In vivo low-level light therapy increases cytochrome oxidase in skeletal muscle. *Photochemistry Photobiology, 86*(3), 673–680.

Helwig, D. (2014). Syntonic optometry. In L. J. Fundukian (Ed.), *The Gale encyclopedia of alternative medicine* (4th ed., pp. 1962–1963). Independence, KY: Gale.

Henderson, T. A., & Morries, L. D. (2015). SPECT perfusion imaging demonstrates improvement of traumatic brain injury with transcranial near-infrared laser phototherapy. *Advances in Mind-Body Medicine, 29*(4), 27–33.

Hesse, S., Werner, C., & Byhahn, M. (2015). Transcranial low-level laser therapy may improve alertness and awareness in traumatic brain injured subjects with severe disorders of consciousness: a case series. *Journal of Neurology and Neuroscience, 6*(2), 10.

Hodges, J. R., Garrard, P., & Patterson, K. (1998). Semantic dementia. In A. Kertesz, & D. G. Munoz (Eds.), *Pick's Disease and Pick Complex* (pp. 83–104). New York, N.Y.: Wiley-Liss Inc.

Hodges, J. R., Garrard, P., & Patterson, K. (2009). Semantic dementia. In A. Kertesz, & D. G. Munoz (Eds.), *Pick's disease and Pick complex* (pp. 83–104). New York, NY: Wiley-Liss.

Huang, Y. Y., Mroz, P., & Hamblin, M. R. (2009). *Basic photomedicine.* Available from: http://photobiology.info/Photomed.html

Hughes, J. R., & John, E. R. (1999). Conventional and quantitative electroencephalography in psychiatry. *Journal of Neuropsychiatry and Clinical Neuroscience, 11*(2), Spring.

Ishii, K., Sakamoto, S., Sasaki, M., et al. (1998). Cerebral glucose metabolism in patients with frontotemporal dementia. *Journal of Nuclear Medicine, 39*(11), 1875–1878.

Ito, H., Takahashi, K., Hatazawa, J., Kim, S. G., & Kanno, I. (2001). Changes in human regional cerebral blood flow and cerebral blood volume during visual stimulation measured by positron emission tomography. *The Journal of Cerebral Blood Flow & Metabolism, 21*(5), 608–612.

Itzhaki, R. F., & Wozniak, M. A. (2008). Herpes simplex virus type 1 in Alzheimer's disease: the enemy within. *Journal of Alzheimer's Disease, 13*(4), 393–405.

Iyengar, B. (2013). The melanocyte photosensory system in the human skin. *Springerplus, 2*(1), 158.

Jakkampudi, S., Abe, M., Komori, N., Takagi, R., Furukawa, K., Katan, C., Sawada, W., Takahashi, N., & Kasai, H. (2016). Design and synthesis of a 4-nitrobromobenzene derivative bearing an ethylene glycol tetraacetic acid unit for a new generation of caged calcium compounds with two-photon absorption properties in the near-ir region and their application in vivo. *ACS Omega, 1*(2), 193.

Jeong, Y., Cho, S. S., Park, J. M., et al. (2005). 18F-FDG PET findings in frontotemporal dementia: an SPM analysis of 29 patients. *Journal of Nuclear Medicine, 46*(2), 233–239.

Jurvelin, H., Jokelainen, J., & Takala, T. (2015). Transcranial bright light and symptoms of jet lag: a randomized, placebo-controlled trial. *Aerospace Medicine and Human Performance, 86*(4), 344–350.

Kalsbeek, A., Merrow, M., Roenneberg, T., & Foster, R. G. (2012). *The neurobiology of circadian timing.* Oxford, England: Elsevier.

Karu, T. (2008) ACTION SPECTRA: Their importance for low level light therapy. Laboratory of Laser Biomedicine Institute of Laser and Information Technologies, Russian Academy of Sciences, Troitsk 142190, Moscow Region, Russian Federation (See tkaru@isan.troitsk.ru www.isan.troitsk.ru/dls/karu.html). Also see T. I. Karu, (1999). Primary and secondary mechanisms of action of visible-to-near IR radiation on cells. *The Journal of Photochemistry and Photobiology B, 49*(1), 1–17.

Kirlin, W. G., Cai, J., Thompson, S. A., Diaz, D., Kavanagh, T. J., & Jones, D. P. (1999). Glutathione redox potential in response to differentiation and enzyme inducers. *Free Radical Biology and Medicine, 27*(11–12), 1208–1218.

Koberda, L. (2013). Alzheimer's dementia as a potential target of Z-score LORETA 19-electrode neurofeedback. *Neuroconnections, Winter*, 30–31.

Kohaska, A., Waki, H., Cui, H., Gouraud, S. S., & Maeda, M. (2012). Integration of metabolic cardiovascular diurnal rhythms by circadian clock. *Endocrine Journal, 59*(6), 447–456.

Lampl, Y., Zivin, J. A., Fisher, M., Lew, R., Welin, L., Dahlof, B., Borenstein, P., Andersson, B., Perez, J., Caparo, C., Ilic, S., & Oron, U. (2007). Infrared laser therapy for ischemic stroke: a new treatment strategy: results of the thera effectiveness and safety trial-1 (NEST-1). *Stroke, 38*(6), 1843–1849.

Lane, N. (2006). Cell biology: power games. *Nature, 443*, 901–903.

Lee, S. Y., Seong, I. W., Kim, J. S., Cheon, K. -A., Gu, S. H., Kim, H. H., & Park, K. H. (2011). Enhancement of cutaneous immune response to bacterial infection after low-level light therawith 1072 nm infrared light: a preliminary study. *Journal of Photochemistry and Photobiology B, 105*(3), 175–182.

Lim, L. (2014). Intranasal photobiomodulation improves Alzheimer's conditions in case studies. NAALT/WALT Arlington, Virginia, September 9–12, 2014, *Conference poster presentation.*

Lubar, J. F. (Ed.). (2004). *Quantitative electroencephalographic analysis (QEEG) databases for neurotherapy.* Binghamton, NY: Haworth Press.

Macdonald, S. (1982). Color therapy for rheumatoid arthritis. *International Journal Biosocial Research, 3*(2), 49–54.

McLeod, S. A. (2008). Serial position effect. Available from: www.simplypsychology.org/primacy-recency.html

McMurtray, A. M., Chen, A. K., Shapira, J. S., Chow, T. W., Mishkin, F., Miller, B. L., & Mendez, M. F. (2006). Variations in regional SPECT hypoperfusion and clinical features in frontotemporal dementia. *Neurology, 66*(4), 517–522.

Mendez, M. F., Cherrier, M., Perryman, K. M., Pachana, N., Miller, B. L., & Cummings, J. L. (1996). Frontotemporal dementia versus Alzheimer's disease. Differential cognitive features. *Neurology, 47*(5), 1189–1194.

Michalikova, S., Ennaceur, A., van Rensburg, R., & Chazot, P. L. (2008). Emotional responses and memory performance of middle-aged CD1 mice in a 3D maze: effects of low infrared light. *Neurobiology of Learning and Memory, 89*(4), 480–488.

Miller, B. L., Ikonte, C., Ponton, M., Levy, M., Boone, K., Darby, A., Berman, N., Mena, I., & Cummings, J. L. (1997). A study of the Lund-Manchester research criteria for frontotemporal dementia: clinical and single-photon emission CT correlations. *Neurology, 48*(4), 937–942.

Moges, H., Vasconcelos, O. M., Campbell, W. W., Borke, R. C., McCoy, J. A., Kaczmarczyk, Feng, J., & Anders, J. J. (2009). Light therapy and supplementary Riboflavin in the SOD1 transgenic mouse model of familial amyotrophic lateral sclerosis (FALS). *Lasers in Surgery and Medicine, 41*(1), 52–59.

Monastra, V. J. (2005). Electroencephalographic biofeedback (neurotherapy) as a treatment for attention deficit hyperactivity disorder: Rationale and empirical foundation. *Child Adolescent Psychiatrics Clinics of North America, 14*(1), 55–82vi.

Naeser, M., Zafonte, R., Krengel, M. H., Martin, P. I., Frazier, J., Hamblin, M. R., Knight, J. A., Meehan, W. P., & Baker, E. H. (2014). Significant Improvements in cognitive performance post-transcranial, red/near-infrared light-emitting diode treatments in chronic, mild traumatic brain injury: open-protocol study. *Journal of Neurotrauma, 31*, 1008–1017.

Neary, D., & Snowden, J. (1996). Frontotemporal dementia: nosology, neuropsychology, and neuropathology. *Brain and Cognition, 31*, 176–187.

Neary, D., Snowden, J., & Mann, D. (2005). Frontotemporal dementia. *Lancet Neurology, 4*(11), 771–780.

Nechita, F., Pirlog, M. C., & Chirita, A. L. (2015). Circadian malfunctions in depression-neurobiological and psychosocial approaches. *Romanian Journal of Morphology and Embryology, 56*(3), 949–955.

Notbohm, A., Kurths, J., & Herrmann, C. S. (2016). Modification of brain oscillations via rhythmic light stimulation provides evidence for entrainment but not for superposition of event-related responses. *Frontiers in Human Neuroscience, 10,* 10.

Pandi-Perumal, S. R., Kramer, M., & Hobson, J. A. (2010). *Sleep and mental illness.* Cambridge, England: Cambridge University Press.

Pastore, D., Greco, M., Petragallo, V. A., & Passarella, S. (1994). Increase in H⁺/e⁻ ratio of the cytochrome c oxidase reaction in mitochondria irradiated with helium-neon laser. *Biochemistry and Molecular Biology International, 34*(4), 817–826.

Pitzschke, A., Lovisa, B., Seydoux, O., Zellweger, M., Pfleiderer, M., Tardy, Y., & Wagnières, G. (2015). Red and NIR light dosimetry in the human deep brain. *Physics in Medicine & Biology, 60,* 2921–2937.

Read, S. L., Miller, B. L., Mena, L., Kim, R., Itabashi, H., & Darby, A. (1995). SPECT in dementia: clinical and pathological correlation. *Journal of the American Geriatrics Society, 43*(11), 1243–1247.

Rochkind, S., El-Ani, D., Nevo, Z., & Shahar, A. (2009). Increase of neuronal sprouting and migration using 780 nm laser phototherapy as procedure for cell therapy. *Lasers in Surgery and Medicine, 41*(4), 277–281.

Roher, A. E., Garami, Z., Tyas, S. L., Maarouf, C. L., Kokjohn, T. A., Belohlavek, M., Vedders, L., Connor, D., Sabbagh, M. N., Beach, T. G., & Emmerling, M. R. (2011). Transcranial doppler ultrasound blood flow velocity and pulsatility index as systemic indicators for Alzheimer's disease. *Alzheimer's & Dementia, 7*(4), 445–455.

Roher, A. E., Debbins, J. P., Malek-Ahmadi, M., Chen, K., Pipe, J. G., Maze, S., Belden, C., Maarouf, C. L., Thiyyagura, P., Mo, H., Hunter, J. M., Kokjohn, T. A., Walker, D. G., Kruchowsky, J. C., Belohlavek, M., Sabbagh, M. N., & Beach, T. G. (2012). Cerebral blood flow in Alzheimer's disease. *Vascular Health and Risk Management, 8,* 599–611.

Rojas, J. C., & Gonzalez-Lima, F. (2013). Neurological and psychological applications of transcranial lasers and LEDs. *Biochemical Pharmacology, 86*(4), 447–457.

Saltmarche, A. E., Naeser, M. A., Ho, K. F., Michael, R., Hamblin, M. R., & Lim, L. (2016). Significant improvement in memory and quality of life after transcranial and intranasal photobiomodulation: a randomized, controlled, single-blind pilot study with dementia. *Alzheimer's Association International Conference,* Toronto, July 24, 2016, poster presentation.

Samuels, B. C., Zaretsky, D. V., & DiMicco, J. A. (2002). Tachycardia evoked by disinhibition of the dorsomedial hypothalamus in rats is mediated through medullary raphe. *Journal of Physiology, 538*(Pt 3), 941–946.

Schirmera, C., Eléonore Lepvriera, E., Duchesneb, L., Decauxa, O., Thomasa, D., Delamarchea, C., & Garniera, C., (2016) Hsp90 directly interacts, in vitro, with amyloid structures and modulates their assembly and disassembly. Biochimica et Biophysica Acta (BBA), 1860, 2598–2609.

Sikka, G., Hussmann, G. P., Pandey, D., Cao, S., Hori, D., Park, J. T., Steppan, J., Kim, J. H., Barodka, V., Myers, A. C., Santhanam, L., Nyhan, D., Halushka, M. K., Koehler, R. C., Snyder, S. H., Shimoda, L. A., & Berkowitz, D. E. (2014). Melanopsin mediates light-dependent relaxation in blood vessels. *Proceedings of the National Academy of Sciences USA, 111*(50), 17977–17982.

Spilt, A., Weverling-Rijnsburger, A. W. E., Middelkoop, H. A. M., van Der Flier, W. M., Gussekloo, J., de Craen, A. J. M., Bollen, E. L. E. M., Blauw, G. J., van Buchem, M. A., & Westendorp, R. G. J. (2005). Late-onset dementia: structural brain damage and total cerebral blood flow. *Radiology, 236*(3), 990–995.

Spitler, H. R. (2011). The syntonic principle: Its relation to health and ocular problems. Resource Publications. Wipf and Stock Publishers: Eugene, OR.

Teipel, S., Grothe, M. J., Zhou, J., Sepulcre, J., Dyrba, M., Sorg, C., & Babiloni, C. (2016). Measuring cortical connectivity in Alzheimer's disease as a brain neural network pathology: toward clinical applications. *Journal of the International Neuropsychological Society, 22*(2), 138–163. Cambridge, England: Cambridge University Press.

Thal, D. R., Attems, J., & Ewers, M. (2014). Spreading of amyloid, tau, and microvascular pathology in Alzheimer's disease: findings from neuropathological and neuroimaging studies. *Journal of Alzheimer's Disease, 42*(Suppl. 4), S421–S429.

Thatcher, R. (2005). Evaluation and validity of a LORETA normative EEG database. *Clinical EEG Neuroscience, 35*(2), 116–122.

Toomim, H., Mize, W., & Kwong, P. C. (2005). Intentional increase of cerebral blood oxygenation using hemoencephalography (HEG): an efficient brain exercise therapy. *Journal of Neurotherapy, 8*(3), 5–21.

Trimmer, P. A., Schwartz, K. M., Borland, M. K., De Taboada, L., Streeter, J., & Oron, U. (2009). Reduced axonal transport in Parkinson's disease cybrid neurites is restored by light therapy. *Molecular Neurodegeneration, 4*, 26.

Turrens, J. F. (2003). Mitochondrial formation of reactive oxygen species. *The Journal of Physiology, 552*(Pt 2), 335–344.

Uozumi, Y. I., Nawashiro, H., Sato, S., Kawauchi, S., Shima, K., & Kikuchi, M. (2010). Targeted increase in cerebral blood flow by transcranial near-infrared laser irradiation. *Lasers in Surgery and Medicine, 42*(6), 566–576.

Van de Haar, H. J., Burgmans, S., Jansen, J. F., van Osch, M. J., van Buchem, M. A., Muller, M., Hofman, P. A., Verhey, F. R., & Backes, W. H. (2016). Blood–brain barrier leakage in patients with early Alzheimer disease. *Radiology, 281*(2), 527–535.

Verma, N., Beretvas, S. N., Pascual, B., Masdeu, J. C., & Markey, M. K. (2015). New scoring methodology improves the sensitivity of the Alzheimer's disease assessment scale-cognitive subscale (ADAS-Cog) in clinical trials. *Alzheimer's Research & Therapy, 7*(1), 64.

Wu, Q., Xuan, W., Ando, T., Xu, T., Huang, L., Huang, Y. Y., Dai, T., Dhital, S., Sharma, S. K., Whalen, M. J., & Hamblin, M. R. (2012). Low-level laser therapy for closed-head traumatic brain injury in mice: Effect of different wavelengths. *Lasers in Surgery and Medicine, 44*(3), 218–226.

Yamaji, S., Ishii, K., Sasaki, M., Imamura, T., Kitagaki, H., Sakamoto, S., & Mori, E. (1997). Changes in cerebral blood flow and oxygen metabolism related to magnetic resonance imaging white matter hyperintensities in Alzheimer's disease. *Journal of Nuclear Medicine, 38*(9), 1471–1474.

Yang, M., Nazhat, N. B., Jiang, X., Kelsey, S. M., Blake, D. R., Newland, A. C., & Morris, C. J. (1996). Adriamycin stimulates proliferation of human lymphoblastic leukaemic cells via a mechanism of hydrogen peroxide (H_2O_2) production. *British Journal of Haematology, 95*(2), 339–344.

Further Reading

Helwig (2009). *Gale Encyclopedia of Alternative Medicine* (3rd ed). Independence, KY: Gale, pp. 2178–2179.

Lee, H., Brekelmans, G., & Roks, G. (2015). The EEG as a diagnostic tool in distinguishing between dementia with Lewy bodies and Alzheimer's disease. *Clinical Neurophysiology, 126*(9), 1735–1739.

CHAPTER 5

Nexalin and Related Forms of Subcortical Electrical Stimulation

Nancy E. White, Leonard M. Richards

Unique MindCare, Houston, Texas, United States

INTRODUCTION

Transcranial electrical stimulation (tCES) has a long history and for our purposes this history may best be told by looking at the evolution of its waveforms. In this paper we will be looking at the evolution of certain advanced forms of cortical/subcortical stimulation that use specific delivery frequencies and waveforms based in alternating current (AC), and discuss what is currently known about AC's mechanisms of action and how its mode of stimulation differs from that of direct current (DC).

We say *advanced forms* not just because of the ways in which they utilize AC, but because these forms represent a truly transcranial modality delivering a fixed-frequency subcortical/cortical stimulation that is proving to be highly effective and seemingly more lasting than existing forms of cranial stimulation. In that respect we examine means by which these advanced AC waveforms might increase neural plasticity, alter neural activity and have resulting affective and behavioral correlates. These advanced waveforms are currently delivered by way of several devices designed for differing symptoms, but they all work on the same general principles arising from the same seminal research. Consequently, for the purposes of this investigation they may be considered variations on a single modality.

Currently, only one form of advanced AC stimulation is cleared for use in the US and, more to the point, its treatment focus can provide unique potential for the practitioner, which is why we have the temerity to mention Nexalin by name.

Once we have provided an overview of the modality in general, its waveforms and what current research tells us may be its mechanisms of action, we look

131

Rhythmic Stimulation Procedures in Neuromodulation. http://dx.doi.org/10.1016/B978-0-12-803726-3.00005-5

at Nexalin Advanced tCES, both as a proxy for the transcranial modality and as a valid treatment form in itself. We discuss salient aspects of its history, its waveform and certain means by which its mechanisms of action, still under investigation, may work. Then, using case histories, we look into both expected outcomes and outcomes not directly related to the diagnosed disorder treated, and offer some possible explanations for such events based on what neuroscience currently understands about brain function (particularly that of the hypothalamus and related limbic structures).

ORIGINS OF THE NEXALIN tCES WAVEFORM

Coming to Terms

From Joannis Aldini's 18th century experiments with dead frogs' legs using procedures based on his uncle Luigi Galvani's work (Giovanni Aldini, 1794), through Leduc's early experiments with rhythm and periodicity (Leduc & Rouxeau, 1903) and, later, Limoge's analgesic stimulation studies (Limoge & Dixmerias-Iskandar, 2004), most all of the published brain stimulation research has been conducted using either transcranial magnetic stimulation (TMS) current or direct current stimulation (DCS), either cranial or transcranial. (The term *transcranial* indicates that electrodes are applied in ways that deliver current directly to specific underlying brain structures). Consequently, the literature contains a large number of studies and reviews of TMS' and DCS' neurophysiologic effects and clinical outcomes (e.g., Nitsche et al., 2008).

The 1960's saw an increasing interest in neurostimulation in the US—along with newly-minted waveforms and protocols-encouraged by earlier reports of electrosleep's (ES) effectiveness with insomnia and electroanalgesia's (EA) success. Advances in technology and learning stoked interest as well.

During this time the fledgling US field of neurostimulation was working to make sense of these various waveforms and dosages through a series of symposia and conferences focused mainly on ES and EA. ES employed mostly AC current (with a DC bias added in 1957), and EA used mainly pulsed DC current in differing doses accompanied by various unique names and initials to describe them. In the late 1960's tCES was proposed as an umbrella term to describe the field's overall practice. The name was changed again in the late 1970s to *cranial electrotherapy stimulation* (CES), and this is the term most used today, although a proliferation of subsidiary terms persist along with their sometimes confusing initials.

After the US Food and Drug Administration (FDA) assumed regulation of the medical device market in the mid-1970s, early CES devices, ancestors of today's earlobe-based cranial stimulation systems, were cleared for Anxiety, Depression and Insomnia. Although newer devices employed forms of AC current, published research through the late 1990s still seemed to focus mainly on waveforms associated with transcranial magnetic stimulation (TMS) and direct current stimulation (DCS) (Guleyupoglu, Schestatsky, Edwards, Fregni, & Bikson, 2013).

From DC to AC and Beyond

Most of the studies on the effects of AC stimulation have been published since the turn of this century, riding on a continuing resurgence of interest in CES and further advances in neurostimulation technology. According to reviewers, however, most CES studies have tended to be vague about the specific techniques and systems used for stimulation and this also includes those AC studies which met reviewers' eyes. For instance, Zaghi and others (2009) reported that in the AC studies that qualified for inclusion in their review, many researchers failed to specify the type of equipment and montage(s) they employed. Where researchers did specify the equipment used in their studies, it consisted of hand-held devices delivering low intensity (100 μA to 4 mA) current to the brain, mainly indirectly, with electrodes attached to earlobes or mastoid. This means the term *transcranial* frequently did not apply. However, this detracts nothing from the treatment outcomes of AC studies, which have been shown to at least equal those of DC stimulation (Zaghi, Acar, Hultgren, Boggio, & Fregni, 2009; Hamid, Gall, Speck, Antal, & Sabel, 2015). Rather, it speaks to the need for more rigor in alternating current stimulation (ACS) studies, as called for early on by Klawansky and others (1995), and later by Kavirajan, Lueck, and Chuang (2013), and perhaps including the simultaneous use of neuroimaging methods. This has been done in several of the recent CES studies (e.g., Datta, Dmochowski, Guleyupoglu, Bikson, & Fregni, 2013), to give researchers more precise information about how and why AC current gets the quality of responses it does (Hamid et al., 2015).

Recognize as well that the advanced, more robust systems of AC transcranial stimulation discussed in this paper already were established in Europe, but were not available for use in this country when domestic reviewers originally assembled their studies (Zaghi et al., 2009). More recent reviews (e.g., Guleyupoglu et al., 2013) offer more definitive information on AC stimulation and its possible mechanisms of action.

The literature generally shows that both DC and AC stimulation support effective neurostimulation outcomes (e.g., Fuscà, Ruhnau, Demarchi, Weisz, & Neuling, 2015), but they achieve results by different mechanisms. However, the rapidly expanding volume of AC research tends to indicate that, because of unique aspects of its stimulation properties, AC may command the wider range of clinical applications going forward (Antal & Paulus 2013; Herrmann, Rach, Neuling, & Strüber, 2013).

Briefly, DC stimulation tends to modulate neuronal activity in a polarity-dependent way, either increasing local neural excitation (anodal stimulation) or decreasing it (cathodal stimulation), depending on which of two electrodes is placed over a target area. That is, DC stimulation mainly affects the firing rate of neurons in the target area. Stimulation may be continuous or intermittent (pulsed). EEG findings support the notion that DC stimulation's site-specific effects also can provoke sustained and widespread changes in other parts of the brain via the brain's multicircuit neural networks. (Zaghi et al., 2009; Tanaka & Watanabe 2009; Jaberzadeh & Zoghi, 2013).

Instead of polarizing neurons, AC rhythmic stimulation is seen to work by synchronizing and entraining neuronal activity (Zaghi et al., 2009; Herrmann et al., 2013). AC rhythmic stimulation may consist of a continuous sine wave or pulsed current (Limoge, Robert, & Stanley, 1999; Zaghi et al., 2009).

Studies providing transcranial AC stimulation (tACS) at or near the brain's endogenous frequencies (theta or alpha bands, for instance), suggest that rhythmic AC stimulation entrains neurons with similar intrinsic frequencies (Helfrich et al., 2014), modulating rather than overriding natural network dynamics (Fröhlich & McCormick 2010). This tends to induce shifts in brain function by increasing neural plasticity, as indicated by shifts in neuronal firing sequences (Gerstner, Kempter, Hemmen, & Wagner, 1996; Caporale & Dan 2008; Herrmann et al., 2013).

Further, newer studies suggest that (tACS) can modulate cross-frequency interactions (Herrmann, Strüber, Helfrich, & Engel, 2016; Vosskuhl, Huster, & Herrmann, 2015) and may help enhance weakened synchrony underlying cognitive deficits such as those from Traumatic Brain Injury (TBI) (Fröhlich, Sellers, & Cordle, 2015). This ability of tACS to broadly modulate neuronal oscillations also can offer important information on causal links to cognitive processes as well as significant insights into human brain function in general. In addition, there are indications that tACS can provide insight into subcortical/cortical interactions (Mindes, Dubin, & Altemus, 2015).

From AC to the Advanced tCES Waveform

Guleyupoglu and others (2015) trace the descendency of neurostimulation waveforms over time from ES to cranial electrical stimulation (CES), during which the field's practitioners experimented with increasingly complex, sometimes idiosyncratic, waveforms and montages, motivated by a perceived need for increased efficacy, safety or patient tolerability. Out of this perception two important waveforms have been developed, giving rise to what is now known as *tCES*. It has been a game-changer for the practice of neurostimulation, as we will see.

tCES is the process of using a low-frequency DC, AC, or combined AC–DC "envelope" to deliver high-frequency balanced current pulses through the scalp directly into the brain. Two main waveforms, as well as the placement montage associated with them, have emerged to become fundamental to today's advanced forms of tCES: *Limoge Current* and *Lebedev Current*.

In an attempt to address directly the problem of patient discomfort and safety, Dr. Aimé Limoge, professor of physiology at René Descartes University in Paris, introduced a completely AC waveform in 1963 (Limoge et al., 1999), which later was deliverable by a system he patented in 1974, the *Anesthelec* (Limoge, 1974). Limoge's Current, originally employed as a method of transcranial analgesia, is used today in a range of neurostimulation applications.

Limoge's Current is a composite waveform consisting of two AC signals in sequence, the first of which is a bipolar pulse delivered at 77 Hz and the second of which consists of frequencies in a neutral, relatively white noise spectrum, each signal with a duration of 3–4 milliseconds, that is, a pulsed "square" wave (Limoge & Dixmerias-Iskandar, 2004; Kovalyov, 1998). Limoge's montage consisted of a negative electrode placed on the forehead between the eyebrows, and two positive electrodes placed behind the ears on each mastoid. Not only did this waveform and montage turn out to be more comfortable for the patient, but the frequency pattern and the placement turned out to be a foundation for the development of today's advanced tCES waveform and montage.

Dr. Valery Lebedev of the Pavlov Physiology Institute at the Russian Academy of Sciences, another pioneer in Electroanalgesia, developed a waveform after Limoge's, now known as *Lebedev Current* that proved quite effective for pain. Using an electrode placement similar to Limoge, he introduced a combination of AC and DC current in a 2:1 ratio, that is, two pulses of AC stimulation followed by a DC pulse of similar duration (Lebedev, 1990).

Later, Lebedev and others (2002) conducted studies designed to come up with an optimal protocol for stimulating the brain's pain suppression system. With a newly-developed device using Lebedev Current, they experimented with the range of frequencies and impulse durations previously used in the field. They found, first, that 77.5 Hz elicited the highest level of β-endorphins, and that β-endorphin levels fell sharply at frequencies above or below 77–78 Hz. It has been established that the level of β-endorphins secreted at a given delivery frequency can give an indication of that frequency's neuromodulation potential.

Second, they showed the optimal impulse duration to be 3.5–4.0 milliseconds. Third, the studies showed a reduction in well-validated stress-related markers (re: Dragunow & Faull, 1989) in certain nuclei of the Thalamus and Hypothalamus, strengthening the premise that tCES could have application for "stress and other accompanied psychophysiological disturbances" (Lebedev et al., 2002, p. 251). Not-so-much-by-chance, these outcomes and the device used to get them became forerunners of an advanced device and protocol design that could effectively treat disorders such as Depression, Anxiety and Insomnia, as well as pain.

Limoge Current and Lebedev Current remain primary tCES neurostimulation waveforms used in Europe today. So, in that respect, to know the waveform is to understand the delivery system: electrode placement on the forehead and behind the ears on left and right mastoid; delivering a 77–78 Hz current "envelope" either by way of a bipolar AC pulsed "square" wave (Limoge) with a duration of 3–4 ms, or by way of pulsed AC followed by DC 2:1 (Lebedev) also with a pulse duration of 3–4 ms.

Dr. Aimé Limoge's *Anesthelec* device, produced in France, was the main tCES system used in Europe for years to deliver Limoge Current, i.e., bipolar AC pulses without a DC component (Leosko, Schlemis, & Baranovsky, 1998). A number of subsequent systems claim heritage from the Anesthelec. For instance, one newly patented tCES device, the *FutraMed*, developed in collaboration with Dr. Limoge and his associates, claims to have replicated the Limoge Current. (Dr. Limoge passed away in 2009). This device is in the final stages of European regulatory approval for use in the treatment of pain and opioid dependence (FutraMed, 2015).

Devices using Lebedev Current were among several developed in Russia during the last decades of the 20th century. Among them were a series of tCES devices developed by Dr. Lebedev and his associates from their research at the Pavlov Institute of Physiology in St. Petersburg, and later produced commercially by the TES Center, Ltd. in several configurations as the TRANSAIR (Lebedev et al., 2002). Updated versions of these devices are in use in Europe today for a variety of tCES applications.

Providing earlier support for Lebedev's observation that tCES could be helpful with affective disorders, Krupitsky et al. (1991), Lebedev and others conducted a double blind placebo-controlled study that documented a high degree of improvement in the depression and anxiety symptoms experienced by alcoholic patients. The study's authors noted significant similarities between outcomes of the experimental group and outcomes reported in psychotherapeutic literature. In addition, they saw statistically significant improvements in certain biochemical and physiological indices. Based on this evidence they recognized a potential for the application of tCES to affective disorders in a broader (i.e., nonaddicted) population (Krupitsky et al., 1991).

Enter Dr. Yakov Katsnelson. Katsnelson, a protégé of Lebedov at the Pavlov Institute, had come to the US and was developing a tCES device, called *TESA*, a device that delivered stimulation with an advanced waveform consisting solely of pulsed AC current. He presented the results of feasibility studies conducted with a prototype of this device in the fall of 2003 (Katsnelson, Lapshin, Claude, & Bartoo, n.d.). Further, at a conference in Prague that year his associates and he provided data showing that the TESA device "brings about not only an analgesic effect but also the normalization of psychological status" (Lapshina et al., 2003, p. 3).

From Electroanalgesia to Emotional Well-Being

Katsnelsons work is part of a sea change in the application of tCES. A range of application hinted at more than two decades ago by Lebedev and his coworkers, now is taking a promising role in the remediation of affective disorders.

As mentioned earlier, several *trans*cranial devices have been cleared in Europe for treatment of stress-related disorders, depression and hypertension, as well as pain, using Limoge or Lebedev waveforms (Guleyupoglu et al., 2013). A number of effective *cranial* devices (e.g., pulses delivered via ear clips) also are in use internationally (Gilula & Kirsch, 2005). However, no system that pulses AC current directly across the brain at an optimal fixed frequency was available in the US prior to FDA clearance of the Nexalin system. This last sentence is important because it emphasizes that there was a strong need here for a system, which was based largely on the work of latter-day tCES researchers such as Limoge and Lebedev, who recognized that a delivery frequency of 77–78 Hz was optimal for eliciting highest levels of β-endorphins (Lebedev et al., 2002). The appearance of high levels of β-endorphins indicated that this frequency was stimulating the subcortical structures that produce them (Sprouse-Blum, Smith, Sugai, & Parsa, 2010). In the central nervous system β-endorphins inhibit the production of GABA, resulting in excess production of Dopamine. Dopamine is helpful not only in pain suppression, but in mood elevation as well (Sprouse-Blum et al., 2010). We follow this train of thought further in a later section on tCES mechanisms of action.

Out of Anesthelec, and from Transair through TESA has come today's Nexalin Advanced tCES, cleared by the FDA in this country for the treatment of depression, anxiety, and insomnia. Nexalin's waveform is an AC pulsed "square" wave delivering current in a 77.5 Hz envelope, shown to maximize production of β-endorphins during treatment (Lebedev, 2000). The patient sits quietly with electrode pads on the forehead and each mastoid, without pain or discomfort, while the hypothalamic region is gently stimulated to rebalance and normalize neurotransmitter activity.

POSSIBLE WAYS NEXALIN tCES MAY INDUCE CHANGE

Mechanisms by which tCES induces changes in mood and behavior are still under investigation, but we can look at some of the *means* through which these barely sketched-out mechanisms could operate. Here we are offering up two well-researched prospects capable of acting as avenues for mechanisms of change: the massive connective pathways of the hypothalamus, and the effects of AC oscillations on neuronal entrainment, phase-reset and synchronization.

The Well-Connected Hypothalamus

Katsnelson et al. (n.d.) demonstrated that, in addition to inducing release of β-endorphins from the Periacqueductal Gray and Raphe Nuclei, the waveform he developed (and now incorporated in Nexalin tCES) stimulated areas of the hypothalamus, activating certain neurotransmitter mechanisms.

The hypothalamus seems to be an appropriate target for Nexalin tCES in the treatment of mood disorders because of its wide role in maintaining homeostasis in the brain–body system, including the use of various neurotransmitters distributed among its nuclei, or through its control of the pituitary (Brodal, 2010). The hypothalamic-pituitary-adrenal (HPA) axis assumes particular importance in depression and anxiety, which have been characterized by overactivity in the HPA Axis as part of an adaptive response to stress (Tsigos & Chrousos, 2002; Pariante, 2003).

The hypothalamus is well-connected locally. The Russian researcher L'vovich (1980) mapped direct pathways linking other structures within the limbic region with the hypothalamus.

The hypothalamus is also well interconnected thalamocortically. Almost 80 years ago James Papez (1937), an American neuroanatomist, postulated a distinct neural circuit for emotion, one originating in the amygdala, which signaled, via the thalamus (the brain's switchboard), the orbitofrontal cortex for directions which then were relayed back to the hypothalamus for action. Subsequent research (e.g., Barbas, 2000, 2007) filled out Papez' proposals; specifically, Helen Barbas (2000) identified pathways in the primate brain similar to Papez' postulation that ran from the medial prefrontal and orbitofrontal cortices to the hypothalamus for emotional expression. Clifford Saper (2000) indicated that autonomic control by the cerebral cortex is directed from the medial prefrontal cortex, the insular cortex and the amygdale to the hypothalamus, which has projections back to the cerebral cortex.

Even more broadly, the hypothalamus is shown not only to maintain direct projections to and from the prefrontal cortex (Rempel-Clower & Barbas, 1998; Ongur, An, & Price, 1998), including specific axonal projections from the orbitofrontal and medial prefrontal regions to the hypothalamus, but then maintains projections reaching to brainstem and spinal autonomic centers (Barbas, Saha, Rempel-Clower, & Ghashghaei, 2003).

Clearly then, the hypothalamus is identified as an important player not only in factors relating to emotions generally present in anxiety and depression, but plays a significant role in regulation of the autonomic nervous system. Based on these findings we can postulate that the hypothalamus has ample internetwork circuitry to accommodate tCES mechanisms modulating depression, anxiety and insomnia, and their neural and behavioral correlates—and potentially much more.

With respect to "more," our clinical work with the Nexalin tCES has yielded outcomes not well explained simply by rebalancing neurochemistry or quieting the HPA axis. Even though the treatment had been prescribed for diagnoses of depression, anxiety and/or insomnia, a number of patients have come away with attenuation of seemingly unrelated symptoms, such as reduction in Parkinson's symptoms, remission of bipolar symptoms and mediation of anger issues such as those we comment about in case studies later on.

Alternating Current as Neural Instigator

Earlier we mentioned research suggesting that rhythmic AC stimulation entrains neurons with similar natural frequencies (Helfrich et al., 2014) and modulates, rather than overrides natural network dynamics (Fröhlich & McCormick 2010), providing possible causal links from stimulation to brain function. In support of these findings, extensive research conducted by Dr. Rodolfo Llinas (2015) gives insight into just that kind of action at the neuronal level. Studying the behavior of individual neurons, he describes how stimulated neurons entrain others to a frequency like theirs in order to accomplish specific tasks in a given direction. At the same time other researchers have been exploring intrinsic, or natural, frequencies by which neurons in corticothalamic circuits communicate to manage functional activity (von Stein & Sarnthein, 2000; Rosanova et al., 2009). Hillebrand et al. (2016) demonstrated that the direction of information flow in large scale neural networks is frequency dependent. If we integrate Llinas' findings and those of Hillebrand's group with those of previously cited researchers who describe the ability of tACS to entrain neurons of similar intrinsic frequency, we can imply that tCES may act on neurons in a way similar to the way neurons act on each other.

This implication allows us to take things a step further. Some time back, von Stein and Sarnthein (2000) proposed that large scale integration of neural activity requires synchronization among neurons and neuron assemblies that operate in different frequency ranges. The mechanisms here remain unclear, but the idea of large scale synchronization implies that (1) neurons and neuronal circuits must be recruited and coordinated both within and among neural networks, and that, (2) this synchronization has to occur in several domains: phase must be aligned, often among several frequencies (Varela, Lachaux, Rodriguez, & Martinerie, 2001; Clayton, Yeung, & Kadosh, 2015; Voloh & Womelsdorf 2016), burst spikes must be choreographed (Caporale & Dan,

2008; Womelsdorf, Ardid, Everling, & Valiante, 2014); and functional cross-frequency coupling has to occur in temporal and spatial dimensions (Canolty & Knight, 2010; Jirsa & Müller 2013). Summing up this complex process, Song, Meng, Chen, Zhou, and Luo (2014) found accumulated evidence, including their own, suggesting that the dynamics of brain oscillations act as "an internal temporal context, based on which neural ensembles are dynamically formed and dissolved to mediate sensory processing, perception, memory, consciousness and attention" (p. 4842). Similarly, Saarimaki et al. (2015) found that specific emotional states were represented by distinct neural signatures, regardless of how the emotions were induced.

What one may see from all this—our words and our postulations here—is that the brain exchanges information within and among its networks by way of a sophisticated electrical language, complete with grammar and syntax, transmitted with pulsed words built into intricately ordered and precisely synchronized multifrequency phrases, the many combinations of which, like our spoken language, have specific meanings and evoke specific responses. It follows, then, that the disordering of these intricate neural conversations by whatever means—trauma, injury, disease, chronic stress—can give rise to problems of mood, behavior, and bodily function.

This gives us a putative means by which the Nexalin pulsed AC waveform may affect neural function. And, at a level far beyond what it's originators may have imagined, make use of Nexalin's access to the Hypothalamus' massive neural highway system to assist in this process. Posit this: if neural oscillations carry electrical conversations, one can speculate that AC pulsed waveforms used by Nexalin are able to affect brain function—and, by extension, mood and behavior—by "hooking in" to and influencing short-and long-range frequency-dependent neural communication patterns. That is, by joining a neural conversation in the brain's neural language, Nexalin may serve to alter both the neural conversation and its functional correlates.

TREATMENT CONSIDERATIONS

While Nexalin has shown its value as a stand-alone therapy, experience shows that it is equally effective as the first treatment modality in a broader program that may include psychotherapy, neurotherapy, or biofeedback or a 12-step program. Experience also shows that, unless contra-indicated, Nexalin can help subsequent therapies achieve faster results.

Every clinic has its own process of intake, testing and diagnosis. The case studies that follow used pretreatment and posttreatment quantitative EEG and/or 3-D body scan (Esteck Systems, 2011) to measure change and its effect on the overall treatment plan. We have found that it has helped to do appropriate pre- and posttesting.

Nexalin tCES is an intensive therapy, as are the related forms of AC transcranial stimulation. The usual protocol is to administer at least ten sessions consisting of two groups of 5 weekday sessions each. Some patients require more—up to 20 sessions in some cases. At the conclusion of the ten sessions a qEEG, Scan or related form of testing can help determine if the patient continues with Nexalin or moves on to another form of therapy. If additional sessions seem indicated, we sometimes intersperse those sessions with the treatments that follow, usually administered twice or three times weekly at that point. In these cases Nexalin seems best administered prior to any psychotherapy, counseling, neurofeedback, or biofeedback session. Overall, the Nexalin form of tCES has integrated well with a practice that offers an array of brain–body treatment modalities (see, White & Richards, 2016).

During our first year with the Nexalin tCES, and before we fully integrated the modality into the practice, we decided to see if we could come close to replicating some original efficacy research in clinical practice (Nexalin Technology/Clinical Trials, 2011). We asked each patient, before undergoing Nexalin tCES each day, to rate the intensity of their symptoms and mood as elements on a Likert Psychometric Scale (Mogey, 1999). At the end of that time we compared the responses offered prior to the first session with those offered before the final session. We found that:

1. 90% of the patients treated had a greater than 50% improvement in their diagnosed condition, and
2. the average percentage of improvement for the group ($N = 24$) was 74% for depression, 77% for anxiety and 84% for insomnia.

These simply are our outcomes during the period in question; other practitioners may experience different results based on any number of factors unique to their patient population and its array of presenting problems.

While these outcomes gave us the confidence we needed to include Nexalin tCES as an important component of our brain-body model, we also found that a small number of patients undergoing Nexalin tCES experienced significant improvement in conditions unrelated to their diagnosis. We describe two in particular in the case studies below, because they may provide further insight into the workings of AC stimulation and the Nexalin advanced tCES waveform.

CASE STUDIES AND COMMENTARY

The following case studies are presented to provide useful information on the neural and behavioral effects of tCES using a pulsed AC waveform. Most persons we work with present with more moderate symptoms, and thus probably are more likely to have less dramatic outcomes.

Case Study #1

This case involves a high functioning 50-year-old medical doctor who wanted to experience the tCES therapy before she began referring patients. Prior to therapy she was described as distractible. After 10 sessions of tCES therapy with Nexalin her dominant brain wave frequencies, that is, the frequencies in which her brain was spending the most time, increased from 4–5 Hz to 11–12 Hz, which is generally considered desirable for an awake, alert, and high functioning adult (Fig. 5.1). Moreover, her brain wave amplitudes (power) moved toward a more optimal balance; slow wave power decreased significantly (36%–51%; [p< 0.000]) while the power of those frequency bands supporting concentration and on-task behavior ("alpha," particularly high alpha, and "beta," particularly low beta) increased significantly (78%–87%; [$P < 0.000$]), (see Fig. 5.2).

She is now even more functional and more highly productive than before, evidenced not only by her writing four training manuals in a short period without effort, but by significant strides she has made—and is making—to enhance her medical practice. She reports being more organized and functional. She feels great and thinks the planet could benefit from having everyone do this treatment.

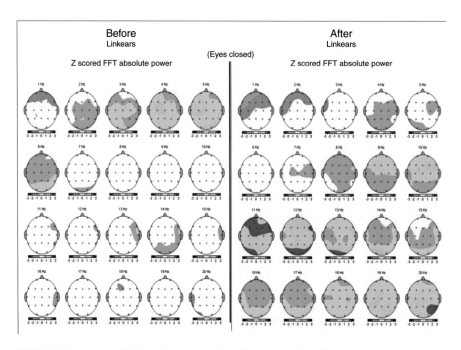

FIGURE 5.1 Case 1. qEEG absolute power before (L) and after (R) tCES treatment.

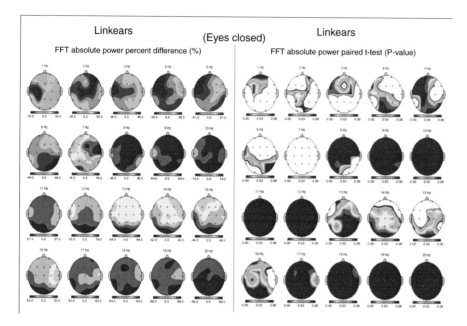

Linkears
(Eyes closed)
Linkears

FFT absolute power percent difference (%)

FFT absolute power paired t-test (P-value)

FIGURE 5.2 Case 1. qEEG pre- and posttreatment percent difference (L) and paired t-test results (R).

Case Study #2

This 17 year-old male had been a typical healthy and functional teenager, well-liked by most who knew him. He had a recreational marijuana habit and had been offered some "pot" which apparently had been laced with a powerful foreign narcotic. Within 36 h he was in complete psychosis. He spent 2 weeks locked up in the psychiatric department of a major hospital from which he emerged no better. He then spent 4 weeks in the psychiatric unit of another major hospital where he was diagnosed with bipolar disorder and given a list of medications that his father said made him a "zombie." He was sent to still a third hospital where doctors told his father and his grandparents that they could do no more for him and that he would have to be cared for the rest of his life. He was then sent home with four medications.

Of necessity, he had dropped out of school during his senior year when his psychosis began and was subsequently unable to return to classes. Once home, this previously popular teenager became a recluse with extreme social anxiety and psychotic episodes (hearing voices, etc.). At one point he was found walking down the street in the middle of the night. He did not know his name or where he was.

At the suggestion of his best friend's mother, who had begun investigating treatments that could be helpful to him, his father and grandparents called to

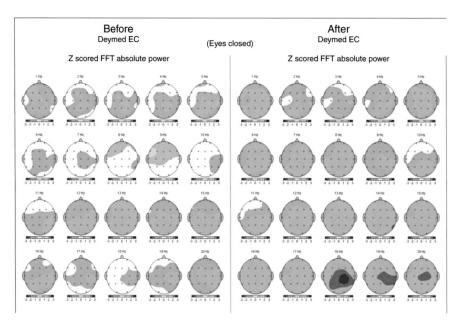

FIGURE 5.3 Case 2. qEEG Absolute Power before and after tCES treatment.

investigate the possibility that tCES could provide hope for recovery. His family drove him 80 miles each way for extensive testing and evaluation, then later on for the 3-week period of five consecutive treatments each week.

Initially he was severely depressed, highly anxious and heavily affected by prescription drugs. He threw up prior to the first and second days' treatments and barely spoke during the first week.

After the first week it was like working with a completely different person. Our medical director was reducing his meds as fast as was safe. He had become friendlier, had gained near-normal affect and carried on effortless conversations with staff, but still reported extreme social anxiety. For instance, he refused to go into public places, such as a restaurant; rather, he would sit in the car while his grandparents got food to take with them. By the last day of his treatment he was willing to go into a restaurant to eat. There was no further socially anxious behavior a month after the therapy ended and a 6-month follow-up confirmed that there had been no repeat episodes of psychotic behavior.

A remarkable feature of the young man's post-treatment qEEG compared with his pre-treatment readings was the reduction in left frontal alpha (8–9 Hz) (see Fig. 5.3). A pattern of left frontal hypoactivation, particularly in the alpha range, is a pattern found by Davidson's research to be indicative of depression (Henriques & Davidson, 1991). Measured improvements ranged from 54%–67% [$P < 0.000$] (see Fig. 5.4).

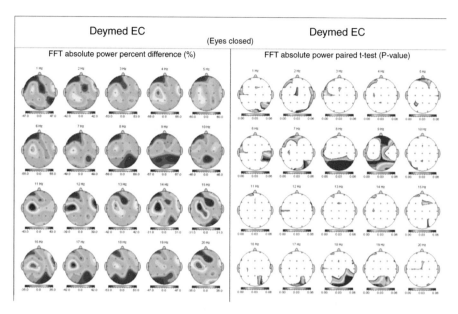

FIGURE 5.4 Case 2. qEEG, Percent Difference and Paired t-Test.

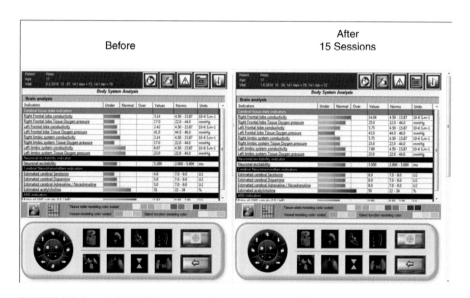

FIGURE 5.5 Case 2. Estek 3D body scan, before and after 15 TES treatments.

Figure 5.5 illustrates differences in the pre and post analysis of the 3D Body Scan. Right and left frontal lobe conductivity, right and left frontal lobe tissue oxygen pressure, neuronal excitability, estimated cerebral Serotonin, estimated cerebral Dopamine, estimated cerebral Adrenaline/Noradrenaline and estimated acetylcholine are measured. Initially there is a predominance of

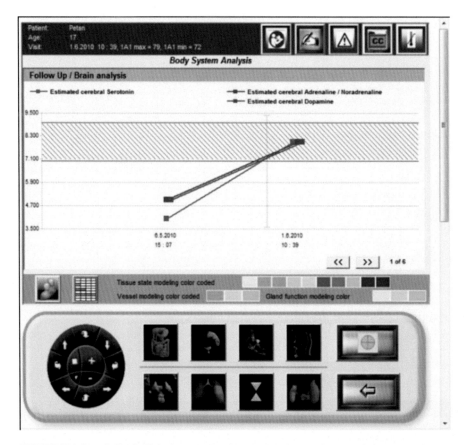

FIGURE 5.6 Case 2. Estek 3D body scan showing 3-weeks' neurochemistry improvement.

out-of-range readings. After 15 sessions of the tCES all the identical readings are in range except left frontal tissue oxygen pressure and estimated acetylcholine. Actually, left frontal tissue oxygen pressure is only very slightly low, with a measurement score of 43 where the norm is 44–46, leaving only estimated acetylcholine clearly out of range with a measurement of 50 versus a normal range of 22–34 (See Fig. 5.5).

Fig. 5.6 offers a graphic representation of estimated cerebral Serotonin, estimated cerebral Adrenaline/Noradrenalin and estimated cerebral Dopamine. The green bar represents normal range. The first measurements were well out of range, while all three of these neurochemicals normalized posttreatment, measuring in the center of the normal range.

This client's family is thrilled to have their son and grandson back to normal functioning. In the 5-month follow-up, he had no further psychotic episodes, remained quite normal without medication and has overcome his social anxiety. He took the GED evaluation, passed it, and now has successfully entered college.

Comment

This patient's pre- and postqEEG absolute power scores (Fig. 5.3) provided no indication of his significant shift in affect and functioning, or of the considerable shift in neurotransmitters. A paired t-Test (Fig. 5.4), however, indicates that a statistically significant change did occur pretreatment to posttreatment, which is more in keeping with the observed outcomes. This may reflect the potentially deep effects of the Nexalin tCES' AC waveform on subcortical regions.

Case Study #3

This case involves a 19-year-old college student diagnosed with depression, TBI, and bipolar disorder. He had been referred based on a prior evaluation using the 3D body scan. When he arrived he would not look at anyone, and exhibited an apparent inability to talk. Although he had been a jogger in the past, his body was so out of balance that he was unable to run at the time he came for therapy. He was hearing voices and exhibiting bipolar tendencies, evidenced by the first three data points on the charts in Figs. 5.10 and 5.11 and the "Before" chart in Fig. 5.7.

His pretreatment scan and those taken during the first week of treatment showed unstable neurochemistry and conductivity, along with excessive neuronal excitability, all in support of his diagnosis of bipolar disorder. (In Figs. 5.10 and 5.11 the green band represents the normal range). This instability continued through the first week of treatment even though the neurochemistry

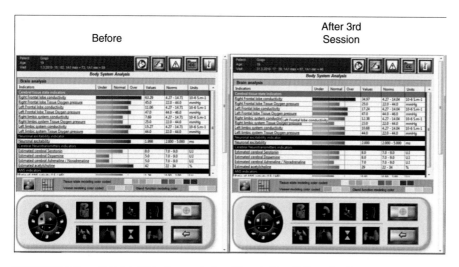

FIGURE 5.7 Case 3. Estek 3D body scan before (L) and after (R) three treatments.

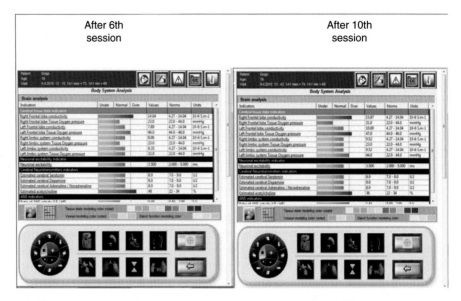

FIGURE 5.8 Case 3. Estek 3D body scan after the 6th (L) and 10th (R) tCES treatment.

was beginning to normalize as evidenced by charts shown in Figs. 5.7 and 5.8. Beginning with the second week of treatment his measurements of neuronal excitability moved to the center of the normal range as shown in Fig. 5.8. Estimated cerebral Serotonin, Dopamine, and Adrenaline/Noradrenalin, which had exhibited pretreatment instability similar to that of neuronal excitability, showed similar improvement at the beginning of the second week. The conductivity measure in Fig. 5.11 also showed initial instability, visible in the right frontal lobe, left frontal lobe and left limbic region, which also began to normalize during the second week of treatment. The right limbic system remained normal pre and post treatment. The normal levels seen in some measures at the beginning of the second week persisted through the remainder of treatment and at 3 months follow up as evidenced in Fig. 5.9. Although not included in the measures shown, normal measures also were maintained at a 5-month follow-up Figs. 5.10 and 5.11.

After the end of the first week the patient became more communicative and his activities were returning to normal. He returned to jogging in the mornings, as had been his preillness routine, even though he was in a strange city. No further bipolar type behavior was observed, and this has been consistent through the 5 months of follow-up.

In his posttreatment interview, the patient reported that when he first arrived he was unable to find words, which created his inability to talk. He reported that during the first few days of treatment he had been unable to think of words. He then progressed to finding the words, but was unable to form them

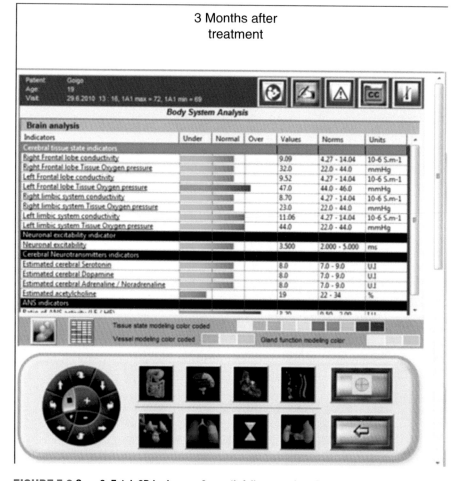

FIGURE 5.9 Case 3. Estek 3D body scan 3-month follow-up after tCES treatment.

or speak them. Toward the end of the first week he began speaking fluently and then told us that he could now find words, form words and then speak the words. His depression was also resolved, and upon returning home he began living a more normal life and was better able to interact with his friends.

Comment

TBI generally occurs during extremely rapid deceleration, such as striking the head, and frequently results in diffuse axonal injury within discrete brain regions, mainly those of the brainstem and the cortex. This disrupts brain signaling, including subcortical/cortical messaging, thereby affecting mood, mental function, and even physical functioning (Meythaler, Peduzzi, Eleftheriou, & Novack, 2001).

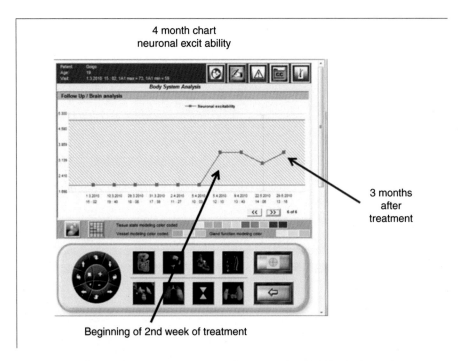

FIGURE 5.10 Case 3. Estek 3D body scan showing neuronal excitability for 3-months posttreatment.

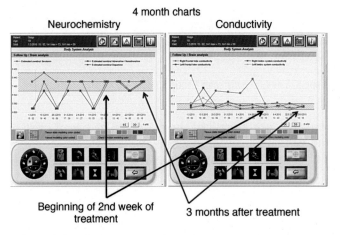

FIGURE 5.11 Case 3. 3D body scan 4 month chart of progress viewing changes in neurochemistry and brain conductivity.

Bipolar disorder, on the other hand, is mediated by a complex network of sub-cortical/cortical neurotransmitter pathways. And, for several decades treatment of bipolar disorder has rested on the foundation of belief that problems of neurotransmitter signaling and disruption of the HPA axis are central to both anxiety and depression (Manji et al., 2003). However, experimental evidence suggests that more primary causes may lie with abnormalities of signal trans-duction and neuroplasticity (Manji et al., 2003).

Consideration of primary causes of dysfunction in both bipolar disorder and TBI seems to lead back to an earlier discussion of the brain's language and what can happen when it becomes garbled (re: Song et al., 2014). So, it may be that treating depression by gently activating the hypothalamus with Nexalin's AC waveform helped the brain normalize symptoms common to both of this client's presenting diagnoses.

Case Study #4

This daughter of ranchers, and the wife of a rancher, developed Parkinson's and at age 81 was effectively confined to a wheelchair; she had trouble swallowing and it was difficult for her to speak. She worried over her future and had become severe-ly depressed. A friend of the family, who also had Parkinson's, lived in California and had been helped by Nexalin, told this lady about his experience. She talked to her husband and son and they agreed that she should explore the therapy.

This client, whom we will refer to as Ranch Lady, began a thirteen session program of Nexalin tCES, which was to take all the time she could devote to the program before her presence was required back at the ranch. Because of her presenting condition her workup included both a quantitative EEG and an evaluation by the clinic's Medical Director based on results of an Estek 3D body scan (Esteck Systems, 2011), plus tests for heavy metals, food sensitivi-ties and nutritional deficiencies, along with comprehensive blood, metabolic, hormone, and neurotransmitter tests, all aimed at maximizing physiological support for rapid improvement.

The day Ranch Lady presented for her blood draws she was able to take just three or four steps to complete a brief intake at the receptionist window before collapsing back into her wheel chair. By the second day of treatment she was able to assist herself some in her transfer to the special Nexalin chair, and af-ter the third session her husband mentioned that she seemed more alert and that she looked at him "more directly." However, she still was sleeping poorly. Intravenous nutrients, ordered midtreatment because she had trouble swallow-ing and ate very little, helped speed onset of better sleep from about five restless hours per night to almost seven restful hours.

As treatment progressed the staff noticed a completely unexpected side effect of this client's Nexalin protocol: her Parkinson symptoms were abating to the extent that she began using her wheel chair as a walker. By the last day of treatment she was smiling, pleasant and free of her symptoms of depression. Further, her "side effects" had intensified: she had given up her wheel chair and was walking with a cane. As she was checking out of her hotel her practitioner came in to offer her goodbyes. A desk clerk took the practitioner aside and asked: "What have you done to this woman? She's like a different person!" That afternoon she flew back to the ranch and to an enhanced quality of life.

Comment

In this case we see broad positive effects of tCES subcortical stimulation with a pulsed AC current and they raise two questions: how durable may be the outcomes achieved and by what plausible means could tCES attain such an outcome?

Taking the second question first, we return to the well-connected hypothalamus, which in this case was stimulated over thirteen sessions with Nexalin's 77.5 Hz pulsed AC current. Limousin et al. (1998) found that deep brain stimulation of the subthalamus reduced the severity of Parkinson's symptoms. This view was strongly supported in a more recent review by Benabid, Chabardes, Mitrofanis, and Pollak (2009), although both expressed concern over the high level of morbidity arising from that procedure (Limousin and Martinez-Torres, 2008; Benabid et al., 2009). This opens the question of whether the hypothalamus has a connection with the subthalamus, its neighbor in the Diencephalon, thus sharing that body's connections with areas heavily implicated in Parkinson's, including the substantia nigra. In fact, L'vovich (1980) found that such connections do exist, making it theoretically possible for the NEXALIN tCES waveform to modify Parkinson's symptoms, perhaps even bypassing the morbidity problems said to be encountered occasionally after deep brain stimulation procedures.

There is little published research on tCES treatment for Parkinson's symptoms. One randomized, double-blind study (Shill, Obradov, Katsnelson, & Pizinger, 2011) treated 23 participants with ten sessions of either active or placebo Nexalin tCES, and then followed their condition for 14 weeks, ultimately finding no significant difference between the groups, based on the Unified Parkinson's Disease Rating Scale. This squares with outcomes of one or two private studies we have heard of. So, until there is further well-constructed research in this area we continue to code Ranch Lady's outcome as an anomalous event that opens speculation about what might be possible in regard to positive and enduring symptom changes.

As for durability, Parkinson's is a progressive disease for which there is no presently known cure. Frequent, or at least periodic, treatment by the many available means may only slow the progress of the disease and improve quality of life. While it is too early to include tCES among those means, still, the potential seems to exist.

CONCLUSIONS

In these pages we have attempted to examine the process by which an advanced AC waveform, incorporated in the Nexalin System and cleared for treatment of Depression, Anxiety and Insomnia in the US, has come about, and explored some means by which it may achieve effectiveness. None of what we have discussed is meant to imply that any one type of waveform or system is better or worse than another. That determination is for future researchers and practitioners to make based on study results or anticipated needs.

While some of the speculations we have made may follow from existing research, much has yet to be investigated and confirmed. But the groundwork for what we write has been well-laid, and the outcomes of tCES to date indicate that these speculations have merit. Meantime, the trend of neuroscientific research continues to move in the direction of further defining and deciphering the brain's intricate operating language.

We would be remiss if we failed to acknowledge our own biases in this area. They are simple: from our experience we have reason to believe that AC stimulation has an edge over DC stimulation in most, but not all, applications; that *trans*cranial stimulation seems to have the stronger effect on subcortical regions, thus bolstering the potential effect of AC stimulation. For these reasons at least, tCES tends to provide the stronger, more persistent outcomes. This leads us to conclude that Nexalin tCES, using an advanced waveform and montage a century in the making, merits a well-earned position in the field of Neuromodulation to go along with similar advanced forms that may be developed in the future.

References

Aldini, G. (1794). *Joannis Aldini De animali electricitate dissertationes duae.* Bononiæ: Ex typographia Instituti Scientiarum.

Antal, A., & Paulus, W. (2013). Transcranial alternating current stimulation (tACS). *Frontiers in Human Neuroscience, 7,* 317.

Barbas, H. (2000). Connections underlying the synthesis of cognition, memory, and emotion in primate prefrontal cortices. *Brain Research Bulletin, 52*(5), 319–330.

Barbas, H. (2007). Flow of information for emotions through temporal and orbitofrontal pathways. *J Anatomy Journal of Anatomy, 211*(2), 237–249.

Barbas, H., Saha, S., Rempel-Clower, N., & Ghashghaei, T. (2003). Serial pathways from primate prefrontal cortex to autonomic areas may influence emotional expression. *BMC Neuroscience, 4*(1), 25.

Benabid, A. L., Chabardes, S., Mitrofanis, J., & Pollak, P. (2009). Deep brain stimulation of the subthalamic nucleus for the treatment of Parkinson's disease. *The Lancet Neurology*, *8*(1), 67–81.

Brodal, P. (2010). *The central nervous system: Structure and function* (4th ed.). New York, NY: Oxford University Press, pp. 440, 442–443.

Canolty, R. T., & Knight, R. T. (2010). The functional role of cross-frequency coupling. *Trends in Cognitive Sciences*, *14*(11), 506–515.

Caporale, N., & Dan, Y. (2008). Spike timing–dependent plasticity: a hebbian learning rule. *Annu. Rev. Neurosci. Annual Review of Neuroscience*, *31*(1), 25–46.

Clayton, M. S., Yeung, N., & Kadosh, R. C. (2015). The roles of cortical oscillations in sustained attention. *Trends in Cognitive Sciences*, *19*(4), 188–195.

Datta, A., Dmochowski, J. P., Guleyupoglu, B., Bikson, M., & Fregni, F. (2013). Cranial electrotherapy stimulation and transcranial pulsed current stimulation: a computer based high-resolution modeling study. *NeuroImage*, *65*, 280–287.

Dragunow, M., & Faull, R. (1989). The use of c-fos as a metabolic marker in neuronal pathway tracing. *Journal of Neuroscience Methods*, *29*(3), 261–265.

Esteck Systems. (2011). Available from: http://uniquemindcare.com/wp-content/uploads/2011/04/esteck-Compatibility-Mode.pdf

Fröhlich, F., & McCormick, D. A. (2010). Endogenous electric fields may guide neocortical network activity. *Neuron*, *67*(1), 129–143.

Fröhlich, F., Sellers, K. K., & Cordle, A. L. (2015). Targeting the neurophysiology of cognitive systems with transcranial alternating current stimulation. *Expert Review of Neurotherapeutics*, *15*(2), 145–167.

Fuscà, M., Ruhnau, P., Demarchi, G., Weisz, N., & Neuling, T. (2015). Brain modulation during transcranial alternating current stimulation recorded with magnetoencephalography. *Brain Stimulation*, *8*(2), 386.

FutraMed. (2015). Limoge tCES for the reduction of opioid consumption and dependence in the treatment of chronic pain. AvailableRetrieved January 22, 2016, from: http://www.futramed.com/

Gerstner, W., Kempter, R., Hemmen, J. L., & Wagner, H. (1996). A neuronal learning rule for sub-millisecond temporal coding. *Nature*, *383*(6595), 76–81.

Gilula, M. F., & Kirsch, D. L. (2005). Cranial electrotherapy stimulation review: a safer alternative to psychopharmaceuticals in the treatment of depression. *Journal of Neurotherapy*, *9*(2), 7–17.

Guleyupoglu, B., Schestatsky, P., Edwards, D., Fregni, F., & Bikson, M. (2013). Classification of methods in transcranial electrical stimulation (tES) and evolving strategy from historical approaches to contemporary innovations. *Journal of Neuroscience Methods*, *219*(2), 297–311.

Hamid, A. I., Gall, C., Speck, O., Antal, A., & Sabel, B. A. (2015). Effects of alternating current stimulation on the healthy and diseased brain. *Frontiers in Neuroscience*, *9*, 391.

Helfrich, R., Schneider, T., Rach, S., Trautmann-Lengsfeld, S., Engel, A., & Herrmann, C. (2014). Entrainment of brain oscillations by transcranial alternating current stimulation. *Current Biology*, *24*(3), 333–339.

Henriques, J. B., & Davidson, R. J. (1991). Left frontal hypoactivation in depression. *Journal of Abnormal Psychology*, *100*(4), 535–545.

Herrmann, C. S., Rach, S., Neuling, T., & Strüber, D. (2013). Transcranial alternating current stimulation: a review of the underlying mechanisms and modulation of cognitive processes. *Frontiers in Human Neuroscience*, *7*(June), .

Herrmann, C. S., Strüber, D., Helfrich, R. F., & Engel, A. K. (2016). EEG oscillations: From correlation to causality. *International Journal of Psychophysiology*, *103*, 12–21.

Hillebrand, A., Tewarie, P., Dellen, E. V., Yu, M., Carbo, E. W., Douw, L., & Stam, C. J. (2016). Direction of information flow in large-scale resting-state networks is frequency-dependent. *Proceedings of the National Academy of Sciences USA, 113*(14), 3867–3872.

Jaberzadeh, S., & Zoghi, M. (2013). Non-invasive brain stimulation for enhancement of corticospinal excitability and motor performance. *Basic and Clinical Neuroscience, 4*(3), 257–265.

Jirsa, V., & Müller, V. (2013). Cross-frequency coupling in real and virtual brain networks. *Frontiers in Computational Neuroscience, 7*, 78.

Katsnelson, Y., Lapshin, V., Claude, J., & Bartoo, G. (n.d.). Transcranial electrotherapy stimulation device for temporary reduction of pain. *Proceedings of the 25th Annual International Conference of the IEEE Engineering in Medicine and Biology Society (IEEE Cat. No.03CH37439)*. Available form: doi:10.1109/iembs.2003.1280371

Kavirajan, H. C., Lueck, K., & Chuang, K. (2014). Alternating current cranial electrotherapy stimulation (CES) for depression. *Protocols Cochrane Database of Systematic Reviews*, Issue 7. Article No.: CD010521. DOI: 10.1002/14651858.CD010521.pub2.

Klawansky, S., Yeung, A., Berkey, C., Shah, N., Phan, H., & Chalmers, T. C. (1995). Meta-analysis of randomized controlled trials of cranial electrostimulation. *The Journal of Nervous and Mental Disease, 183*(7), 478–484.

Kovalyov, M.G. (1998). A comparative experimental study of the analgesic effects produced by a new transcranial stimulation metnod and by Limoge's method. In V.P., Lebedev (Ed.), *A New Method of Transcranial Analgesia* (Part I, pp. 127–128). St. Petersburg, Russia.

Krupitsky, E., Burakov, A., Karan-Ova, G., Katsnelson, J., Lebedev, V., Grinenko, A., & Borodkin, J. (1991). The administration of transcranial electric treatment for affective disturbances therapy in alcoholic patients. *Drug and Alcohol Dependence, 27*(1), 1–6.

Lapshina, L., Katsnelson, Y., Lapshin, V., Strahov, I., Razumov, A., Katsnelson, R. (2003). Psychoemotional condition while treating pain syndromes using "TESA" device. *Pain in Europe, Prague, Czech Republic.*

Lebedev, V. P. (1990). Transcranial Electroanalgesia. In V. A. Mikhailovich, & Y. D. Ignatov (Eds.), *Pain Syndrome* (pp. 162–172). Leningrad: Meditsina.

Lebedev, V.P. (2000). Non-invasive transcranial electrostimulation of the brain antinociceptive system as method of TES: An overview. *Proceedings of the 5th Annual Conference of the International Functional Electrical Stimulation Society*, Aalborg University, Denmark, pp. 123–126.

Lebedev, V. P., Malygin, A., Kovalevski, A., Rychkova, S., Sisoev, V., Kropotov, S., & Kozlowski, G. (2002). Devices for noninvasive transcranial electrostimulation of the brain endorphinergic system: application for improvement of human psycho-physiological status. *Artificial Organs, 26*(3), 248–251.

Leduc, S., & Rouxeau, A. (1903). Influence de la rythme et de la periode sur la production de l'inhibition par les courants intermittents de basse tension. *L'Annee Biologique, 8*, 231–232.

Leosko, V. A., Schlemis, G. I., & Baranovsky, A. L. (1998). The main electrical parameters and instrumentation for transcranial electroanalgesia. *A New Method of Transcranial Analgesia, Part I*. Leningrad, Russia: A Review: All-Union Scientific Research Institute of Pulmonology, 123–124.

Limoge, A., & Dixmerias-Iskandar, F. (2004). A Personal Experience Using Limoge's Current During a Major Surgery. *Anesthesia & Analgesia, 99*(1), 309.

Limoge, A. (1974). Method for obtaining neurophysiological effects. *U.S. Patent Office Publication*, US 3,835,833 A. Available from http://www.google.com/patents/US3835833#backward-citations

Limoge, A., Robert, C., & Stanley, T. (1999). Transcutaneous cranial electrical stimulation (TCES): a review 1998. *Neuroscience & Biobehavioral Reviews, 23*(4), 529–538.

Limousin, P., Krack, P., Pollak, P., Benazzouz, A., Ardouin, C., Hoffmann, D., & Benabid, A. (1998). Electrical Stimulation of the Subthalamic Nucleus in Advanced Parkinson's Disease. *New England Journal of Medicine, 339*(16), 1105–1111.

Limousin, P., & Martinez-Torres, I. (2008). Deep brain stimulation for Parkinson's disease. *Neurotherapeutics*, 5(2), 309–319.

Llinas, R.R. (2015). Llinas Intrinsic Neuronal properties. Available from: http://researchgate.net/publication/271074758

L'vovich, A. I. (1980). Connections of the globus pallidus and putamen with the hypothalamus and subthalamus. *Neuroscience and Behavioral Physiology*, 10(4), 285–290.

Manji, H. K., Quiroz, J. A., Payne, J. L., Singh, J., Lopez, B. P., Viegas, J. S., & Zarate, C. A. (2003). The underlying neurobiology of bipolar disorder. *World Psychiatry*, 2(3), 136–146.

Meythaler, J. M., Peduzzi, J. D., Eleftheriou, E., & Novack, T. A. (2001). Current concepts: diffuse axonal injury—associated traumatic brain injury. *Archives of Physical Medicine and Rehabilitation*, 82(10), 1461–1471.

Mindes, J., Dubin, M. J., Altemus, M. (2015). Cranial Electrical stimulation, in textbook of neuromodulation: Principles, methods and clinical applications. Part II, Chapter 11, (pp.127-150), New York: Springer.

Mogey, N. (1999). So you want to use a Likert Scale. Available from: http://www.icbl.hw.ac.uk/lydi/cookbook/info_likert_scale/index.html

Nexalin Technology/Clinical Trials. (2011). Available from: http://www.nexalin.com/nexalin-technology-resources-for-doctor-and-health-professionals/clinical-trials-nexalin-technology/

Nitsche, M. A., Cohen, L. G., Wassermann, E. M., Priori, A., Lang, N., Antal, A., & Pascual-Leone, A. (2008). Transcranial direct current stimulation: State of the art 2008. *Brain Stimulation*, 1(3), 206–223.

Ongur, D., An, X., & Price, J. (1998). Prefrontal cortical projections to the hypothalamus in Macaque monkeys. *The Journal of Comparative Neurology*, 401(4), 480–505.

Papez, J. W. (1937). A Proposed Mechanism of Emotion. *Archives of Neurology and Psychiatry*, 38(4), 725.

Pariante, C. (2003). Depression, Stress and the Adrenal axis. *Journal of Neuroendocrinology*, 15(8), 811–812.

Priori, A. (2003). Brain polarization in humans: a reappraisal of an old tool for prolonged noninvasive modulation of brain excitability. *Clinical Neurophysiology*, 114(4), 589–595.

Rempel-Clower, N., & Barbas, H. (1998). Topographic organization of connections between the hypothalamus and prefrontal cortex in the rhesus monkey. *The Journal of Comparative Neurology J. Comp. Neurol.*, 398(3), 393–419.

Rosanova, M., Casali, A., Bellina, V., Resta, F., Mariotti, M., & Massimini, M. (2009). Natural frequencies of human corticothalamic circuits. *Journal of Neuroscience*, 29(24), 7679–7685.

Saper, C. B. (2000). Hypothalamic connections with the cerebral cortex. *Progress in Brain Research*, 126(1), 39–48.

Shill, H. A., Obradov, S., Katsnelson, Y., & Pizinger, R. (2011). A randomized, double-blind trial of transcranial electrostimulation in early Parkinson's disease. *Movement Disorders*, 26(8), 1477–1480.

Song, K., Meng, M., Chen, L., Zhou, K., & Luo, H. (2014). Behavioral oscillations in attention: rhythmic alpha pulses mediated through theta band. *Journal of Neuroscience*, 34(14), 4837–4844.

Sprouse-Blum, A. S., Smith, G., Sugai, D., & Parsa, F. D. (2010). Understanding endorphins and their importance in pain management. *Hawai'i Medical Journal*, 69(March), 70-S.

Stein, A. V., & Sarnthein, J. (2000). Different frequencies for different scales of cortical integration: from local gamma to long range alpha/theta synchronization. *International Journal of Psychophysiology*, 38(3), 301–313.

Tanaka, S., & Watanabe, K. (2009). Transcranial direct current stimulation—a new tool for human cognitive neuroscience. *Brain and Nerve*, 61(1), 53–64.

Tsigos, C., & Chrousos, G. P. (2002). Hypothalamic–pituitary–adrenal axis, neuroendocrine factors and stress. *Journal of Psychosomatic Research, 53*(4), 865–871.

Varela, F., Lachaux, J., Rodriguez, E., & Martinerie, J. (2001). The brainweb: Phase synchronization and large-scale integration. *Nature Reviews Neuroscience, 2*(4), 229–239.

Voloh, B., & Womelsdorf, T. (2016). A role of phase-resetting in coordinating large scale neural networks during attention and goal-directed behavior. *Frontiers in Systems Neuroscience, 18,* http://doi.org/10.3389/fnsys.2016.00018.

Vosskuhl, J., Huster, R. J., & Herrmann, C. S. (2015). Increase in short-term memory capacity induced by down-regulating individual theta frequency via transcranial alternating current stimulation. *Frontiers in Human Neuroscience, 9,* 257.

White, N. E., & Richards, L. M. (2016). An Integrative Approach to Optimizing Neural Function. In T. F. Collura, & J. Frederick (Eds.), *Handbook of Clinical EEG and Neurotherapy.* New York, NY: Routledge.

Womelsdorf, T., Ardid, S., Everling, S., & Valiante, T. (2014). Burst firing synchronizes prefrontal and anterior cingulate cortex during attentional control. *Current Biology, 24*(22), 2613–2621.

Zaghi, S., Acar, M., Hultgren, B., Boggio, P. S., & Fregni, F. (2009). Noninvasive brain stimulation with low-intensity electrical currents: putative mechanisms of action for direct and alternating current stimulation. *The Neuroscientist, 16*(3), 285–307.

Further Reading

Guleyupoglu, B., Schestatsky, P., Fregni, F., & Bikson, M. (2015). Methods and technologies for low-intensity transcranial electrical stimulation: Waveforms, terminology and historical notes. In H. Knotkova, & D. Rasche (Eds.), *Textbook of Neuromodulation* (pp. 7–16). New York, NY: Springer-Verlag.

Krause, B., & Kadosh, R. C. (2014). Not all brains are created equal: the relevance of individual differences in responsiveness to transcranial electrical stimulation. *Front. Syst. Neurosci. Frontiers in Systems Neuroscience, 8,* 25.

Saarimaki, H., Gotsopoulos, A., Jaaskelainen, I. P., Lampinen, J., Vuilleumier, P., Hari, R., & Nummenmaa, L. (2016). Discrete neural signatures of basic emotions. *Cerebral Cortex, 26*(6), 2563–2573.

CHAPTER 6

The Use of Music for Neuromodulation

Eric Miller*, Lynn Miller‡, Robert P. Turner**, James R. Evans†

*Montclair State University, Montclair, NJ, United States; **University of South Carolina School of Medicine, Columbia; Network Neurology LLC, Charleston, SC, United States; †Sterlingworth Center, Greenville, SC, United States; ‡Expressive Therapy Concepts, Phoenixville, PA, United States

INTRODUCTION: THE UNIVERSALITY OF MUSIC AND NEUROMODULATION

The term "universal" as used here, indicates applicability in every case or to every individual, as is true for both music and neuromodulation. Music is universal throughout: (1) time/history, (2) ethnocultural populations, and (3) in its inherent components of syntax, language, and so on. That neuromodulation occurs universally, whether or not we choose to recognize, acknowledge, and accept it, has been increasingly demonstrated in the neurosciences since the early 1990's. Our brains and bodies are always changing. Change, or "neuroplasticity", is inevitable—our nervous systems and bodies are either:

1. moving forward/positively toward healthy neuroregulation, or
2. moving backward/negatively toward diseased neurodysregulation. Fundamentally, learning can only occur because our nervous systems can change, and adapt, and that inherent capacity toward neuroplasticity lasts our entire lifetime.

In this chapter, neuromodulation generally is defined as "the ability to change the nervous system through an externally, or internally, applied modality". In the authors' opinions, music has the potential to be among the greatest neuromodulators of all.

Here, we focus on various neuro-stimulation and neuro-destimulation techniques that involve music. We will describe techniques in some detail, discuss rationale for use of each, and, when possible, cite recent, scientifically sound research which supports its use. We also will address what features, in the authors' opinions, make a particular stimulation modality unique among

159

neuro-stimulation procedures. A Glossary of Terms is located at the end of the chapter for the benefit of readers who may not be familiar with music-related terminology.

Music may be our most common complex form of rhythmic stimulation. Music has long been known to have evident neuromodulatory effects on emotions, movements, memories, attitudes, physiology, and cultures. People appear universally to relate to, and usually do not fear, music, and most people believe they understand music in some way. Seemingly adding to a close association between music and neuromodulation is the fact that music and the brain's electrical activity share physical parameters that mediate and identify patterns in each. For example,

Volume, amplitude;
Pitch, frequency (Hertz; Hz)
Harmony, coherence/phase relationships among brain regions
Melody, frequency/oscillations relations over time;
Rhythm, periodic changes in frequency and amplitude

UNIVERSAL IMPORTANCE OF MUSIC

Basic human priorities and needs may be considered to center around sustenance (sufficient food, water, air), procreation (survival of the species), concern with an afterlife, social affiliation, and self-respect/self empowerment. These, of course, are closely associated with concerns, beliefs and developments regarding food production, water and air quality, mating behaviors, and sexuality, religion, social groupings, and individual freedom. Historically, it has been the perceived, feared, and/or actual frustration of such needs (and related beliefs) that were major forces driving wars, interpersonal conflicts, cultural traditions and rules, politics, life philosophies, and many discoveries which have improved quality of life. Evidence for the primacy of such priorities and associated developments may be seen in the various selections of history's most significant persons. For example, a May, 2013 article in Time Magazine rating of the 100 Most Significant Persons in History included among the top thirty the following persons (with present authors' interpretations of related areas of relevance in parentheses): Jesus, Mohammed, and Martin Luther (religion); Adolf Hitler, Napoleon, and Alexander the Great (war and empowerment); Abraham Lincoln and George Washington (individual freedom); Aristotle and Karl Marx (political and life philosophies). Among the remaining seventy selections, not only were Christopher Columbus, Albert Einstein, Leonardo Da Vinci, Sigmund Freud, Gautama Buddha, and Plato included, but also Wolfgang Mozart, Richard Wagner, Johann Bach, Ludwig Van Beethoven, and (number 69) Elvis Presley. Perhaps music should be added to the previously mentioned list of basic human priorities and needs.

Music is, and for all of written history (or even before) has been ubiquitous among human societies (Mithin, 2005). Today it is a multibillion dollar industry, with the names and lives of the more famous singers and music groups often being more widely known than those of past and present world leaders or historical figures in science, religion, medicine, and other fields. When considering reasons for its obviously major impact one might mention its ability to arouse and enhance and modify emotions, or its social value in facilitating commonality and cooperation within groups, or the reported ability of certain music to improve cognitive function, or its healing properties. For anecdotal or scientific support for such reasoning one could site its consistent presence at church services, weddings, funerals and other emotionally charged events, its usage by military leaders and sports teams to build and maintain a "fighting spirit", the controversial research concerning the "Mozart effect" on development of aspects of visual spatial functioning, or the many successes of music therapy. Some have assigned loftier, sublime, even cosmological significance to music with notions of music being synonymous with, or inherent in, concepts of "spirit" or "soul", or even of life itself as some have interpreted T.S. Eliot's statement, "You are the music while the music lasts." Such concepts seem reflected in the words of popular songs such as Drift Away, for example, "Give me the beat boys, and free my soul, I wanna get lost in your rock n' roll and drift away", and in the labeling of an entire genre of music as "soul music". One of the most compelling counter-existentialist arguments comes from a 1931 song by Duke Ellington and Irving Mills: "It don't mean a thing (if it ain't got that swing)".

Of special relevance to this chapter, a basic feature of music is rhythm. And, certainly with today's understanding of the role of the brain and central nervous system in emotion, cognition and behavior it would be natural to expect that music's wide-ranging effects may be due in large part to its rhythms having modulating effects on neural function. This view may be considered a modern "take" on centuries of speculation on questions of the "whys and hows" of music.

The major importance of music to humans of all cultures, since earliest recorded history has been cited by many writers. Its roles in eliciting and modifying emotion, inducing altered states of consciousness, facilitating group cohesion and work activity, physical, and mental healing and providing enjoyment and recreation have been emphasized and explored extensively. A reader, therefore, may ask, "What can a relatively short chapter in yet another book add to that which already has been written"? The authors' response relates to several facts: (1) this book is about rhythmic stimulation procedures which modify neural functioning. Since a core feature of music (some would say *the* core feature) is rhythm, and since the power of music to modify emotion, and states of consciousness (which have neural underpinnings) has been demonstrated and generally accepted, it is reasonable to consider music as a neuromodulator; (2) Although the functions of music have been written about for many years, the topic of music and brain function is relatively new, and expanding very rapidly. It is safe to say that more books and scientific articles have been published on that topic since 2000 than in all previous years combined, due in very large part to rapid advances in brain imaging techniques such as functional MRI and quantified EEG; (3) Apparent similarities between music and organization of brain electrical activity is leading to innovative approaches to neurofeedback and music therapy which only recently have been discussed in the literature on music and the brain.

Following from the aforementioned facts, topics of this chapter begin with an overview of history of the uses of music, with special emphasis on its role in healing. This is followed by a brief description of the rise of, and rationale for, music therapy in the United States, its present status here and abroad, and brief descriptions of some general tenets and specific procedures presently used by music therapists and those who incorporate music into the practice of neurofeedback. Finally, topics under the general heading of the human voice as a musical instrument will be presented along with results of some related pilot research.

HISTORICAL FOUNDATIONS

McClellan (2000) suggests that the earliest music may have been a "wailing vocal chant", monotonal and rhythmic in nature, with its rhythmicity based on breath length and heart pulse. Subsequently, chanting and music became common activities of shamanism.

Shamanism dates back over 10,000 years, and is still present in many cultures. There has been a movement in the 20th century by the western culture to create modern day shamanism as a form of Spirituaity. Shamans were (and are) considered messengers between the human world and the spirit worlds, treating illness by restoring the soul (Eliade, 1972). The belief is when there is illness, there is soul loss. The Shaman goes into an altered state induced by rituals involving drums and songs to retrieve the soul. A Shaman is said to connect with helping spirits through nature. The songs often include imitations of nature sounds, calling or honoring animal and plant spirits. The animal and plant spirits are believed to have symbolic qualities to strengthen, and bring illness back to balance and wholeness. In Peru, Icaros, (healing songs) are believed to be transmitted through plants. The plant spirit gives the Shaman the song, which tells which medicinal plant to use for healing (Nakkah, 2007). Many Shamans also use hallucinogenic plants to induce altered states to message with spirits in other worlds.

The role of the drum is theorized in various ways, for example, a trance state is psycho-acoustically induced through modulating rhythms. The Shaman communicated with helping spirits through rhythms and overtones. In tribal drumming, repetitive drumming is typically in a rhythmic beat close to the frequency of EEG delta and lower theta waves (0.8–5.0 cycles per second). Some theorists consider the auditory driving of the Theta wave to be a possible entry into an altered state of consciousness (Acterberg, 1985). Tuvan Shamans believed the drum restored the rhythmic work of the heart, as well as harmonic work in the respiratory organs for someone with mental illness or possession. (Mongush and Kenin-Lopsan, 1997).

In many African tribes, the medicine man, often characterized as "witch doctor", served as priest and chief musician (Peters, 1987), using special songs and rhythms to drive away diseases. They used drums, bells, rattles and dance in their religious and healing ceremonies. American Indians used specific songs for specific ailments. According to Peters, the Navajo used song, dance and sand paintings in specific patterns to "cure specific illnesses," and the Walla Walla medicine men and their patients sang healing songs, with the patients sometimes being directed to sing for several hours a day.

One of the older and most well-known Chinese systems of music-medicine within the discipline of traditional Chinese medicine (TCM) is that of the Qi-Gong Taoist Healing Sounds. This system goes back 2,500 years, with records from the book: the QiGong book: Baopuzi (283–343), which recorded that Chinese medical physicians encouraged practicing the healing sounds. According to this system, there are specific sounds and colors that correspond with specific organs in the body to release negative emotions and bring in healthy Qi (Twicken, 2013). This system is still practiced by some today in attempts to detox the body and transform unfavorable emotions.

In India there is reference to chakras and healing sounds in Yogic texts (Rig Vedic), dating back to 1500–1200 BCE. Hypothetical energy centers referred to as chakras are said to correspond with different neural areas in the body. Sanskrit sounds are used to help balance the vibration of each center. According to Indian philosophy, Nada is the vibration and sound that comprises the Universe, and Nada Yoga is the practice of healing with sound. The Vedics practiced a science of sound, and viewed Nada as essential for healing all conditions. Ragas were used for specific ailments in ancient India, for example, Ragas Todi and Ahir Bhairav for treating headache and high blood pressure, Chandrakauns for diabetes, Darbari Kanara, Hansdhwani, Kalavati, and Durga for relaxation and easing tension, and Bihag, and Bahar for restful sleep (Gopal, 2016; Sumathy, 2005).

Healing in Ancient Egypt dates back 4,000 years. Vowel sounds were considered sacred and sung in chants, and their hieroglyphic language did not contain vowels (Reid & Reid, 2011). It appears that priests of ancient Egypt considered vibratory phenomena, presumably including the human voice and aspects of music such as specific tones, to have powers to create material forms. This is suggested by McClellan (2000, p. 112) with citations of quotes from a 1977 publication (Schwaller De Lubics, 1977): for the priests, "speaking was a process of generating sonar fields establishing an immediate vibratory identity with the essential principle that underlies any object or form", and that within the religious rites of the temple " it is the voice which seeks afar the Invisibles summoned and makes the necessary objects into reality".

For the ancient Greeks, music was intertwined with religious practice, with certain deities (Gods/Goddesses), such as the extremely influential Apollo being considered both God of music and founder of medicine. Therefore, as stated by Meinecke (in Schillian & Schoen, 1948) "music and medicine were intimately commingled in his divine nature as an integrated unity". Furthermore, music was considered to have major influence on moral/ethical development of both individuals and groups such that changes in musical mode could even "undermine basic foundations of a state".

The contributions of several ancient Greek philosophers, mathematicians and writers, familiar to anyone who has studied history of Western civilization, were instrumental in development of the music/healing connection. Aristotle, Hippocrates, Homer, Plato, and Pythagoras were prominent among these. In Greek mythology a son, Aesculapius, was born of Apollo, and came to have status regarding medicine, healing, and music. Temples referred to as Aesculapia and devoted in part to healing were established throughout the Greek (later, the Roman) world. Descriptions of healing practices were reminiscent of some involved in today's mental health treatment, for example, quiet surroundings, administration of various herbs, and use of poetry, dance, theater and music. Such practices were based largely on a view that disease is a lack of harmony among the elements of man's physical and "psychical" natures. Although terms, such as "soul" and "psychical" were used, it appears that this view was very similar to today's "psychosomatic" and "psychophysiological".

The Greek mathemetician/philosopher Pythagoras, studied the physics of sound through the use of a monochord, which determined ratios of perfect musical consonance, for example, the octave, fifth, and fourth intervals, which remain the fundamentals of today's tonal system. (Meinecke, 1948). This concept of musical consonance, being essentially synonymous with harmony, coincides with the view that health is, or involves, psychical/physical harmony, and disease a lack of harmony. It is not surprising that some ancient Greek philosophers would note this coincidence, and apply it to healing practices.

Plato described music as medicine of the soul, which could change (apparently meaning heal) a soul when it has lost its harmony, rhythm and melody. It became apparent to many that participation in and/or exposure to such rhythmic activities "render the body more harmonious". Pythagoras, himself, is said to have advocated music, poetry and dance to help insure emotional stability, and to assist in development of "ethical perfection". Homer wrote that music helps one avoid negative emotions such as anger, worry and fear, and promotes healthful recreation and morality. Aristotle and others from the era mentioned the value of music in arousing emotion as a form of catharsis (Meinecke, 1948).

The harp, flute, and lyre were the most common musical instruments of ancient Greece. Much emphasis was placed on the varying effects of different musical modes. For example, the Phrygian mode generally was described as "violently exciting and emotional", the Dorian as "more composed and manly", and the Lydian as "decorous and educative" (Meinecke, 1948). Although rhythmic activities, such as music, dance and poetry may have been primary components of healing methods of ancient Greece, "medications" also played a role. These most often may have been natural herbs chosen for their purgative powers. However, it is interesting to note that some in this era believed that it was not the herb itself, which cured, but rather the "spirit" of the herb fighting (perhaps meaning modifying) the "spirit" of the disease.

According to Meinecke (1948), many of the ancient Greek notions concerning music and medicine subsequently were adopted by the Romans. In discussing this, he mentions how the famous Roman statesman, philosopher and orator, Cicero, applied the principles of music to the writing and speaking of Latin prose. He cites sources of statements attributed to Cicero, or to historians, such as, "there is nothing so kindred to our feelings as rhythmic cadences and musical sounds", and "the basic principles of poetry, song, melody and rhythm are the very essence of his (Cicero's) speech". And, while Cicero's voice may represent the highest development of speech/music fusion, it still likely involved elements of the simple "wailing, rhythmic, vocal chants" which McClellan (2000) suggests constituted the earliest music.

Ancient civilizations had an interest in and understanding of music and healing. Some of this ancient knowledge was put to rest until the 1700's when Diogel measured physiological responses on the effects of music. He brought live musicians to his laboratory and his patients' bedsides to conduct experiments and record his findings. His published paper in the 1800's showed that music lowers BP, increases cardiac output, decreases pulse rate and, in general, assists the work of the parasympathetic system (Meymandi, 2009).

During the 19th century there was some interest in music and the brain through the study of aphasia. Since singing involves language, investigators were looking at areas of the brain involved with music (Graziano & Johnson, 2015). However, it has been only recently, in the 20th century, that there was a surge in research studies and activities concerning music and the brain.

FROM THE RELAXATION RESPONSE TO MUSIC AND THE BRAIN

Herbert Benson described an effect he called the relaxation response (Benson 1975; Wallace and Benson, 1972) that established a set of physiological changes related to relaxation, including reduced oxygen consumption, heart rate, and arterial blood pressure. He was an early adopter of using biofeedback monitoring devices in conjunction with diverse practices, such as meditation and breathing techniques. Benson, as a medical doctor, was a pioneer in the area of integrating biofeedback training with mainstream medical practice. Subsequent research by Benson and colleagues at the Deaconess Hospital and Harvard Medical School studied EEG responses in addition to the relaxation response measures (Jacobs, Benson, & Friedman, 1996). This contributed further to the increasing interest in this line of investigation, and set a stage for the study of neuromodulation as related to relaxation and meditative practices, such as those mentioned in the following paragraphs.

In treating chronic pain and depression, mindfulness meditation practice has been theorized to "optimize" a wide band (7–14 Hz) Alpha rhythm (Kerr, Sacchet, Lazar, Moore, & Jones, 2013) that helps organize sensory information, and filter inputs to the primary sensory neocortex. The mindfulness meditation exercises focus on attending to the breath and body sensations in the here and now. Zen meditators have been found to reduce activity in executive, evaluative and emotion areas (prefrontal cortex, amygdala, hippocampus) during pain compared to controls; and the meditators with the most experience showed the largest reductions (Grant, Courtemanche, Duerden, Duncan, & Rainville, 2010). The Zen meditators activated primary pain processing regions (anterior cingulate cortex, thalamus, and insula), as was strongly predicted by reductions in functional connectivity between executive and pain-related cortices. The authors suggest a mechanism of functional decoupling of the cognitive-evaluative and sensory-discriminative dimensions of pain, that may allow the Zen meditators to interpret the painful stimuli in a more neutral than negative frame of mind, thereby reducing the impact of the pain stimulus.

Studying the effects of meditation leads into the musical realm through the ancient practices of mantra meditation and chanting. In mantra meditation, a word or phrase is repeated over and over. The mantra may be repeated aloud or internally. The mantra might be a single syllable such as "OM" or a complex phrase such as "Om mani padme hum" meaning the jewel is in the heart of the lotus or "nam myoho renge kyo" conveying the idea that we all have the capacity to transform any suffering at each moment. Practitioners of Transcendental Meditation have been shown to exhibit increased frontal alpha (8–10 Hz) coherence as during mantra meditation practice as compared to an eyes-closed resting state. The shift in EEG was accompanied by physiological changes during the first minute of meditation, that is, lower sympathetic activity, higher parasympathetic activity, and a slower breathing rate (Travis, 2013).

Matthieu Ricard had originally trained as a cellular biologist before leaving France 40 some plus years ago to become a Buddhist monk in the Himalayas. Together with a team of researchers, Ricard compiled results from a number of EEG and MRI brain imaging studies to identify specific regions in the brain associated with three different kinds of meditation practices (Ricard, Lutz, & Davidson, 2014).

The three kinds of meditation studied were:

- Focused Awareness
- Mindfulness
- Compassion and Loving Kindness.

During focused awareness, the meditator concentrates attention on the inhalation and exhalation of the breath. The researchers identified four phases of a cognitive cycle related to the focused awareness meditation process:

- Episode of mind wandering
- Moment of awareness of the distraction
- Reorientation of attention
- Resumption of focused attention

Particular brain networks were associated with each of the phases. The default-mode network (DMN) that is wide-ranging (including the medial prefrontal cortex, posterior cingulate cortex, inferior parietal lobe, lateral temporal cortex, and precuneus), was activated during mind wandering. Awareness of distraction was found in the salience network that includes the anterior insula and anterior cingulate cortex. This area is thought to be involved with detecting novel events and mode switching during meditation. The third phase—reorienting attention, saw the addition of the lateral inferior parietal lobe and the dorsolateral prefrontal cortex which remained activated as the meditator returned to the meditative task in the fourth phase.

During mindfulness practice, the meditator attends to sights, sounds, self-talk, and bodily sensations, which she or he notices, without preoccupation or attachment to any single thought, and returns to detached focus whenever the mind strays. Similar to the Grant et al. (2010) study cited above, experienced meditators were less reactive to pain and exhibited less anxiety than novices.

Experienced participants meditating with compassion and loving kindness activated brain areas of the temporoparietal junction, medial prefrontal cortex, and superior temporal sulcus. When the experienced meditators heard a "distressed voice" they showed increased activation (vs. novices) in the somatosensory and insular cortices—areas reportedly involved with the experience of empathy. Of special note is that one of their MRI studies showed experienced meditators had greater increase in grey matter mass in the insula and prefrontal cortex (Brodmann areas 9 and 10) than did controls. The authors identified related functional tasks in these areas as the processing of attention and sensory information as well as internal bodily sensations.

Returning now to music, we note that Jäncke (2009) reviewed a number of studies demonstrating that musical experience changed grey and white matter densities in musicians. One that he found especially remarkable was by Schneider et al. (2002) using MEG measures comparing musicians with nonmusicians. Activity in the primary auditory cortex between 19 and 30 ms following tone onset was more than 100% larger in musicians, and grey matter volume in the anteromedial part of Herschl's gryus was 130% larger in musicians. Both of the measures correlated highly with musical aptitude (Jäncke, 2009). The take away here is that practicing music clearly increased grey matter densities in the brain. However, the effect could go both directions as grey matter gain decreased when practicing stopped! Hence the credo: "use it or lose it" seems to apply here.

RISE OF MODERN MUSIC NEUROMODULATION/THERAPY

What is the relationship between music and the brain? One of the earlier persons scientifically researching this was a leading theoretician, researcher, and scientist, the late Dr Gordon Shaw, most famous for discovering The Generalized *Mozart Effect* with Fran Rauscher (Rauscher, Shaw, & Ky, 1993). Following publication of Don Campbell's popular book The Mozart Effect (Campbell, 1997), it was not long until people were playing Mozart to babies in their cribs. In fact, however, what Shaw had shown was simply that listening to Mozart improved performance on spatiotemporal tasks for 10 min.

The ground-breaking neuroscience research and clinical applications demonstrated by Dr. Shaw quickly became overshadowed by media and marketing advances of others. Working at the University of California at Irvine, Shaw focused on cortical organization, developing his seminal, columnar-based trion model. A list of his papers is on the web site of the MIND Research Institute, the Institute Shaw founded to continue his brain research, and growing applications in elementary education, in the form of the Math + Music program. The latter combines nonlanguage based computer math games with specialized piano training.

Shaw's model for the architecture of the cortex was set forth in a paper entitled "Model of cortical organization embodying a basis for a theory of information processing and memory recall" (Shaw, Silverman, & Pearson, 1985). The abstract states: *Motivated by V. B. Mountcastle's organizational principle for neocortical function, and by M. E. Fisher's model of physical spin systems, we introduce a cooperative model of the cortical column incorporating an idealized substructure, the trion, which represents a localized group of neurons. Computer studies reveal that typical networks composed of a small number of trions (with symmetric interactions) exhibit striking behavior, e.g., hundreds to thousands of quasi-stable, periodic firing patterns, any of which can be selected out and enhanced with only small changes in interaction strengths by using a Hebb-type algorithm.*

A particularly intriguing aspect of Shaw's work is that humans love music because it resonates with the innate columnar cortical structure. Xiaodan Leng, working with Shaw, then derived music directly from these theories, yielding amazingly human-sounding, classical-like pieces of music.

Within the past two or three decades a great deal more research has been conducted and books written on the brain-music connection, with a couple of the more popular publications being: This is Your Brain on Music (Levitin, 2006), and Rhythm, Music and the Brain: Scientific Foundations and Clinical Applications (Thaut, 2008). Due to space limitations, and since this is well-covered elsewhere, topics, such as present knowledge of anatomical regions and networks of the brain mediating music appreciation and performance at different ages, and in musicians as compared to non-musicians, etc is not covered in this chapter. However, since a first step toward integrating an understanding of musical functions within a neurologic context may be to start with some conception of where musical tasks happen in the brain, the following schema (Fig. 6.1) depicts EEG electrode sites over cortical regions and Brodmann areas

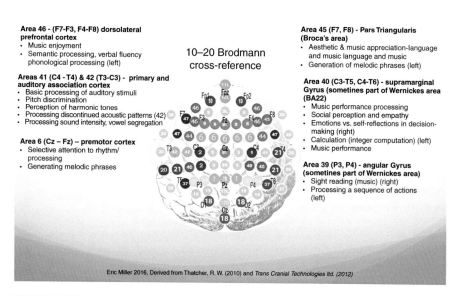

Music-related functions at 10–20 sites & Brodmann areas

Area 46 - (F7-F3, F4-F8) dorsolateral prefrontal cortex
- Music enjoyment
- Semantic processing, verbal fluency phonological processing (left)

Areas 41 (C4 - T4) & 42 (T3-C3) - primary and auditory association cortex
- Basic processing of auditory stimuli
- Pitch discrimination
- Perception of harmonic tones
- Processing discontinued acoustic patterns (42)
- Processing sound intensity, vowel segregation

Area 6 (Cz – Fz) – premotor cortex
- Selective attention to rhythm/ processing
- Generating melodic phrases

10–20 Brodmann cross-reference

Area 45 (F7, F8) - Pars Triangularis (Broca's area)
- Aesthetic & music appreciation-language and music language and music
- Generation of melodic phrases (left)

Area 40 (C3-T5, C4-T6) - supramarginal Gyrus (sometimes part of Wernickes area (BA22)
- Music performance processing
- Social perception and empathy
- Emotions vs. self-reflections in decision-making (right)
- Calculation (integer computation) (left)
- Music performance

Area 39 (P3, P4) - angular Gyrus (sometimes part of Wernickes area)
- Sight reading (music) (right)
- Processing a sequence of actions (left)

Eric Miller 2016, Derived from Thatcher, R. W. (2010) and *Trans Cranial Technologies ltd. (2012)*

FIGURE 6.1 Music-related functions at 10–20 sites and Brodmann areas.

known to be heavily involved in mediating music-related function. It is derived from several sources: (Thatcher, 2010) and *Trans Cranial Technologies Ldt. (2012)*.

SPECIFIC MUSIC-NEUROMODULATION TECHNIQUES

Drawing in part on a long history of using music for tribal ceremonies, healing rituals, and seasonal celebrations, contemporary professionals utilize an abundance of refined musical techniques to achieve clinical outcomes. Scientific evidence for these procedures does not always exist. While many anecdotal cases certainly report impressive results that imply neuromodulation, there is scant research actually monitoring neurological measures for the multitude of clinical approaches found in today's health marketplace.

In this chapter we will focus on a few practices that we hope represent some of the diversity in the field, and at the same time have at least some background-supporting neurological research. We review the following musical techniques for neuromodulation:

- Melodic intonation therapy
- Guided imagery with music
- Bio-guided music therapy
- Brain/music therapy
- Native flute biofeedback
- Sound healing with vocal toning and Tibetan bowls

MELODIC INTONATION THERAPY

When Congresswoman Gabby Giffords was shot in 2011 and left in critical condition and unable to speak, music therapy helped her regain speaking skills. Diagnosed with aphasia, the inoperative area on the left side of her brain which had controlled her speech had to be circumvented so that she could re-learn how to talk using alternate neural circuitry. For the majority of persons, speech and other verbal-related activities are mediated primarily by regions in the left hemisphere of the brain. According to ABC News reporters covering the case, "Protocols like music speech stimulation and Melodic Intonation Therapy can help patients with damage to the brain's communication center, like Giffords, learn to speak again" (Moisse, 2011).

Singing and humming are being used to help stroke survivors and people with aphasia with little or no speech, relearn speech through Melodic Intonation Therapy (MIT). Neurological researchers Sparks, Helm, and Albert developed MIT in 1973. Patients with aphasia often have damage in the left hemisphere of the brain. MIT speech processing is accessed through the right hemisphere of the brain by singing melodic patterns and rhythms of speech.

There may be variations in protocol in MIT sessions, depending on the therapist, who generally is a speech or music therapist. Once the techniques are learned, family members can reinforce the treatment. Typically, the therapist hums a few notes similar to a chant or simple rhythmic song, using accents similar to speech patterns. The patient taps the patterns with their left hand. Tapping the left hand may engage a right-hemisphere sensorimotor network that controls both hand and mouth movements (Norton, Zipse, Marchina, & Schlaug, 2009). Eventually the patient repeats the hummed patterns while tapping the rhythm. As the sessions progress, the therapist sings phrases of high probability words such as "water", and social phrases, such as "thank you" accompanied by visual cues on cards. The patient repeats the phrase. This process is repeated, requesting the patient to leave more time before repeating and trying to imagine hearing the phrase internally before repeating. This technique helps to translate MIT phrases to expressive speech and improve ability to retrieve words. For the final step, *spoken singing* replaces the constant melodic pitch, using variable pitch as involved in speech patterns. Longer phrases and sentences eventually are introduced, using normal speech patterns focusing on rhythm, tempo, and phrasing.

Further research is needed to fully understand the neural processes that underlie MIT's effect. However, studies have shown evidence for plasticity in white matter tracts, helping rebuild the connections necessary for speech. Brain areas that are likely to play a role in the speech recovery process include right-hemisphere superior temporal lobe (involved with auditory feedback control), premotor regions and posterior inferior frontal gyrus (related to planning and sequencing of motor actions and auditory-motor mapping), and the primary motor cortex (key to the execution of vocal motor actions). Some of these regions are connected by a major fiber tract, arcuate fasciculus (AF). MIT appears to increase AF fibers and volume, which may provide an explanation for sustained MIT therapy effects (Schlaug, Marchina, & Norton, 2009).

GUIDED IMAGERY WITH MUSIC

One of many guided imagery techniques was developed by Helen Bonny (1975) in which classical music serves as "cotherapist", and music therapists take the role of a guide. Bonny developed her method of GIM from a Freudian perspective where the ego resides in the center of a plethora of concentric circles of consciousness (Miller, 2011). The farther material is out from the center, the further it is from conscious awareness, as would be with the case with repressed memories. The outer reaches are said to be the realm of mystical experience. Bonny's interest in controlled manipulation of altered states stemmed from her experiences with early psychedelic drug research. In Bonny's method of Guided Imagery with Music, classical music is invoked to help elicit (unconscious) imagery with the help of the music therapist who interacts with the person throughout the process, asking such questions as: "What are you seeing? What are you feeling now? What is in front of you?" Following the imagery, there is a period of cognitive integration where the therapist and client discuss the process and content of the session and make connections with their life issues or circumstances.

Often the client or "traveler" reports profound psychological and physiological experiences during their GIM session (Hunt, 2011) that may lead to moments of catharsis and or transformation. A novel approach to looking at the significance of neural correlates with profound musical experience through the guided imagery is articulated by Hunt with a model of study called neurophenomenology (Varela, 1996). In this method, the researcher integrates EEG brainwave patterns with qualitative interview data about their experience, including sensations, thoughts, feelings, style of imagery "travel" and states of consciousness. Hunt found that her travelers exhibited activation within the same cortical areas during their imageries that would have been expected during actual engagement; for example, imagination of muscle movements within the motor region, and strong visual imagery within occipital areas. She also notes that the most vivid experiences occurred when beta and gamma activity was seen in addition to the slow wave alpha and theta increases typically seen during meditation. This supports the concept of music as a "whole-brain" activity as purported by Taylor (2010).

GUIDED IMAGERY WITH BIO-GUIDED MUSIC THERAPY

While traditional Bonny style GIM utilizes music as a "cotherapist, the bio-guided approach to music and imagery adds an additional cotherapist: the live data (Miller, 2011). In this configuration, the music therapist, or guide, is informed both by the client's answers to questions, visual cues and real time physiological data, such as Galvanic Skin Response, that provide an indicator of central nervous system modulation. This also could also be conducted using EEG Alpha as an indicator of general relaxation.

Bio-guided music and Imagery identifies five specific phases or stages of experiential process (Miller, 2011) and encourages "personalization" of each stage by identifying sensorial cues for each one that the guide may utilize. The five phases are:

- relaxation/induction
- stability/security
- transition
- transformation/change
- return

Induction is similar to many techniques geared toward deep relaxation and therapeutic process, including hypnosis, GIM, and some meditation practices. However, as used by music therapists, music which the client finds calming is involved. Additionally, when utilizing a bio-guided approach, there is an advantage of being able to monitor real-time physiological data related to the nervous system, and/or states of consciousness, and to be able to consider that data in real time as the process continues. Thus, the client's physiological measures help inform the guide in the process of induction and in further stages of the imagery. For example, if the client appears calm, yet electrodermal activity begins to elevate quickly, it is likely not an optimal time to move into a deeper stage of the process. The client identifies the images or sounds that he/she associates with calming, and the guide may utilize these in the process.

When physiological markers show reduced sympathetic nervous system activation, or increased EEG alpha (8–12 Hz) patterns, at a stable level, the client has most likely entered the "Stability/Security" stage. This is where the experience of deep relaxation is grounded and the client becomes ready for "Transformation/ Change". Images or sounds that relate to stability may vary from individual to individual. For example one who grew up on a farm in the country may feel calm and relaxed around cows and the sounds of the barnyard. This may be the inverse for the city-dweller who is not phased by taxi's honking and screechy trolley cars. The objective of this phase is to solidify the sense of stability so it may be available as a resource for future phases.

Transition is simply the shift from stability to the "work" or "transformation" phase. The object here is to create a smooth transition that does not disrupt the deep relaxation that has been established by this point. With the client supplying the sensory cues or images related to transitioning, it is not a shock and the client may be prepared and comfortable with the familiarity of his or her own transitional sound or image of choice. Physiological measures here keep the guide informed as to how smooth or seamless the transition actually is. For example, if alpha (8–12 Hz) "dives" and Beta (16–20 Hz) elevates, the traveler may be prematurely emerging from a relaxed state. Likewise, a spike in electrodermal activity could also indicate an early return to waking consciousness.

Transformation is where a beneficial change may occur. Client-supplied images are invoked here, such as replacing unwanted cigarettes with healthy coconut water, or visualizing oneself in a position of authority rather than that of victim. It is important to have the deepest relaxation possible at this stage because the potential anxiety related to the behavior change and the experiences associated with it may elevate to the point of mitigating the positive effect of the induction. Once again, real time physiological measures in this stage provide critical information as to how the body is processing the images and suggestions relative to change and if some mental event may be triggering the client away from an optimal experiential state.

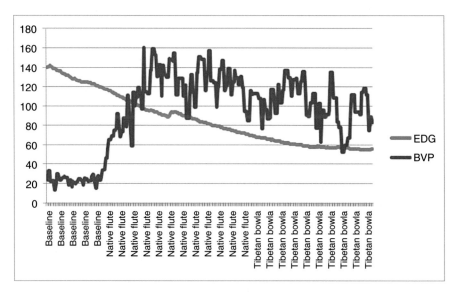

FIGURE 6.2 Graph of a bio-guided music therapy with imagery session.

Return is where the guide welcomes an increase in electrodermal activity or alpha reduction as the traveler returns to normal waking consciousness. The client's breathing pattern will also likely reflect a return with a few sharp inhalations as the body demands increased oxygen. Beta EEG rhythms may elevate as cognitive process become more active.

The physiological measures shown during the bio-guided imagery session with music in Fig. 6.2 point toward a relaxation response as indicated by the decreasing electrodermal activity and increased blood volume pulse.

BRAIN-MUSIC THERAPY

Brain-music therapy involves translating brain signals into music and playing the piece back to the client in the context of working on a clinical objective, such as improved sleep. Some researchers in this arena include Yakov Levin, Galina Mindlin, Alexander Graur, and Eric Miller, among others. There are many different methods for making the jump from brain rhythm to music, and the algorithms utilized are not always transparent. The common thread is the transformation of brain activity in some form into musical form and the subsequent presentation of that music in turn, back to the client.

Levin (1998) describes the development of his specific technique, and its use as a nonpharmacological intervention in a study involving 58 patients with insomnia. The patients listened to their brain music pieces on audiocassettes before going to bed over a period of 15 days. A control group received a placebo condition that consisted of someone else's brain music. Levin found the brain music to be effective for 80% of the insomniac patients by both subjective report and objective neuropsychological measures.

Mindlin and Evans (2009) describe a method where Mindlin works with Levin's EEG algorithms to produce both activating and relaxing musical pieces for therapeutic use. Mindlin first administers a brief questionnaire to prospective clients, as well as self-report measures such as the Beck Depression Inventory and the Subjective Sleep Inventory. She samples 5–7 min of EEG from four scalp electrode sites at central and frontal regions of the brain (F3, F4, C3, C4), and sends the data to Dr. Levin in Russia. Levin translates the EEG data into two personalized music files: one relaxing piece based on the slower EEG waves and one activating piece derived from the faster frequencies. Translation is conducted using a software sound compiler developed in the 1990s by D. G. Gavrilov (Levin, 1998). The compiler employs 18 transformation algorithms to allow for selection of one of 120 musical instruments for each EEG channel. Key EEG parameters utilized include percent power for five frequency ranges, and peak power frequency for each channel. The software is able to vary each channel's volume and tempo, modify musical parameters such as legato-staccato, octave selection and major–minor harmonization (Levin, 1998). All channels are assigned to midi no. 1, Acoustic Grand Piano and the rate of the melody playback is 120 bpm. The client listens to a CD of these pieces over a course of several weeks.

Alexander Graur concurrently developed his own form of brain music therapy in Torino Italy in the 1990's. He presents internationally and instructs at the University of Torino, Italy, Post-graduate School in the Health Psychology Graduate School in Neurosciences. Dr. Graur emphasizes the individual nature of each person's own brain wave signature and thus the necessity for creating individualized music for each patient's pathology.

Dr. Graur conducts his sessions in a three-phase format. Prior to the session, vital signs are taken including, blood pressure, pulse, and respiratory rate. The first phase of the session is introductory and utilizes music for general relaxation. When the patient's respiratory rhythm stabilizes they are ready for the second phase where music specific to their pathology is introduced. The final phase of the session lasts for 5–7 min and is intended to stabilize and integrate a new homeostatic pattern (Graur, 2006). When constructing individualized brain music for specific diseases, Graur considers the "favored frequency" in Hertz of body organs with their musical note and octave. For example: Thalamus with favored frequencies 2300, 2349, and 2400 Hz, note D and 7th octave, or Pituitary gland with favored frequencies 7700, 7800, and 7902 Hz, note B and 8th registry. He presents these data with body organs along with their frequencies and musical notes in chart form (see Graur, 2006 for easy reference).

In a Bio-guided model of brain music therapy used by the first author of this chapter the client receives a real time translation into music of one or more components of his or her EEG. The therapist can manipulate musical parameters such as key, scale and frequency range which affects the sensitivity of the pitch change function to respond to a wider or narrower window. Miller (2011) described the application of this style of brain music therapy to children with attention deficit/ hyper-activity disorder (ADHD). Originally, it was presented as "Brain-jamming for focus" in Seoul South Korea in 2006. In this procedure the child receives direct musical feedback of their EEG Theta band (4–7 Hz) in scale and in key with a selection of background music. Initially the music therapist conducts sessions choosing from a menu of diverse musical background tracks, but subsequently the children may select a background track of preference or even bring in their own.

Early pilot research in the 1990's had used this approach via an NIH mini-grant (Miller, 2007) which involved children diagnosed with ADHD and a control group. Works of such musical artists as the Talking Heads, Steely Dan, Usher, Vivaldi, Mozart, and Bach were included. Video clips played an important role in the clinical session and were attached to Theta in addition to the audio feedback signals. That is, the videos were set to pause when Theta power elevated above an autoadjusting threshold level, thus challenging the child to increase vigilance over a 5 min trial duration (Miller, 2011). Video clips that worked particularly well were fast-moving and attention-holding, such as the pod race scene or the Darth Maul/Obi Wan light saber battle from Star Wars, or sports segments such an Olympic badminton exhibition featuring US players Bob Malythong and Tony Gunawan. Sessions consisted of between 3 and 5 trials; however, on occasion a competitive child attempted to "beat" his or her own record and go for a 6th or even 7th trial. Additional trials were discouraged at any point when the child exhibited difficulty reducing Theta and/or muscle tension, the latter indicated by 25–35 Hz power elevation. With the availability of client-preferred audio and video tracks, component designs were saved for future use with such names as Vivaldi–Star Wars, Back Jack–Fractals, or Sheriff–Badminton. Children were able to engage in the process and maintain a course of training over several months or 120, 5-min trials.

Results of this study showed overall improvement on several of the Conners ADHD scales, and some individual children showed dramatic academic performance improvement, increasing reading level and reducing disruptive behavior as per teacher report. Several indices within a psychological testing battery showed significant improvement of treatment participants as compared with control group at the $P < 0.05$ level. These measures were the NEPSY Audio subtest ($P = 0.01$), and the Conners Parent Survey Cognitive ($P = 0.043$) and ADD ($P = 0.015$) subscales. Prior neurofeedback studies had shown academic improvement, yet did not necessarily show improved neurological markers. In this pilot study which had hypothesized EEG changes related to improvement in ADHD behavior, treatment participants significantly decreased their theta/beta ratios ($P = 0.004$), and significantly increased their sensori-motor rhythm (SMR) ($P = 0.012$). One child was able to move to mainstream classes following the music/neurofeedback training, something which his Mother never thought possible.

MISCELLANEOUS PRESENT AND POSSIBLE FUTURE VARIATIONS OF BRAIN MUSIC THERAPY

Researchers Miranda and Eaton at the Royal Hospital for neurodisability and Plymouth University in the United Kingdom devised a unique system of brain music creation and collaboration via EEG signals and a brain–computer interface (BCI). Four musicians are able to choose a musical sequence by concentrating on a corresponding matrix of flashing lights. Their choice is sent to another musician who plays the part. Dynamics or volume level of the part is controlled by the degree of intensity of the four musicians levels of concentration. Dr. Julian O'Kelly in the Music Therapy Department at the Royal Hospital for neuro-disability presents a demonstration of this system via video (http://blogs.discovermagazine.com/d-brief/2016/02/15/disabled-musicians-make-music-with-their-minds/). This approach to brain music therapy reportedly was able to bring back the joy of music making to musicians who had subsequently become disabled.

An interesting use of music-related stimuli in conjunction with neurofeedback is that involved in the high resolution, relational, resonance-based electroencephalic mirroring (HIRREM) procedure developed by Brain State Technologies, LLC, and being used in ongoing research in the Dept. of Neurology at Wake Forest University. As described by Gerdes, Lee, & Tegler (2013, p. 3), the HIRREM procedure initially involves sampling of EEG from several electrode sites during resting and on-task conditions to help determine existence of hemispheric asymmetries and imbalances or disproportions of power at various frequencies. Treatment then proceeds via real-time reflection of EEG activity from selected scalp electrode sites back to the brain in the form of "auditory musical tones corresponding to the individual's dominant EEG frequencies". It is believed that this allows resonance between the tones and neural oscillations. Specific details of individual training protocol selections apparently are derived using a proprietary formula. Gerdes notes that, provided this type information, the brain autocalibrates its activity, facilitating both mental and physical health without need for "energetic input to the brain, cognitive guidance or education from a clinician or referencing against population norms for the EEG".

Recent scientific research support for HIRREM is provided by Tegeler et al. (2016). In their study, six athletes in the 15–22 age range who had reported continuing postconcussion symptoms received a median of 18.5, 90 min HIRREM sessions. Pre- and post-assessment measures of self-rated insomnia and depressive symptoms, blood pressure, heart rate variability (HRV) and baroreflex sensitivity were calculated. Statistically significant improvement was found for the measures of insomnia, depressive symptoms and baroreflex sensitivity, with all participants returning to athletic activity.

Another interesting possible variation on Brain Music Therapy was proposed by Evans (2015), and might appropriately be termed "dueling brains". This neurologically based call and response type procedure would deliver feedback in the form of EEG-transformed music to a client reflecting some essential, but compromised EEG parameters as evidenced by being significant degrees away from the norm. An "ideal" or normative data-based version of the music then immediately would be played back to the client in "conversation" style for comparison. Another short sample of the client's brain music then would be taken and presented to him or her as feedback music, followed immediately by another presentation to the client of "ideal" brain music. Iterations of this would continue until the client's brain music began to approach that of the "ideal". This model came to Evans while watching the movie Deliverance during the famous "Dueling Banjos" scene. The following is a step by step sequence envisioned for this process:

- Compose "brain music" based on "ideal for age" very well-artifacted EEG data. Perhaps such data could be gathered from persons scoring within + or −.5 SD on all measures of a large normative database such as Neuroguide. Development of algorithms for EEG-to-music has some history, and could be a place to begin.
- Gather EEG data from a client/patient and translate it in real time to music, using the same EEG-to-music algorithm. Initially this might be only one or a very few sequential notes or chords, but they would be heard by the client (in the form of feedback). Immediately the "music" would cease, and the client would hear a similarly sized, but "ideal" sequence of notes (based on the normal-for-age EEG data).
- This process would continue with increasingly longer samples of client EEG—ideal EEG "music", being continuously adjusted up or down in length and/or other degree of complexity as the two forms of music began to match in regard to various parameters.
- After some variable time period with "close matching" occurring, it might be noted that the client's "music" begins to take on a

personalized character or style, still coherent (or perhaps whatever makes music "musical"), but less "matched" and uniquely reflective of the individual who, by now, hopefully is well regulated and symptom-free and can cease treatment.

Of course there likely would be technical difficulties implementing this type procedure, but given the speed and memory capacities of today's computers, it does not seem impossible. One might, however, question how this might be an improvement over other stimulation procedures, or neurofeedback in general? It might not be, of course, but it would have an appeal to many clients (after all it would be their "music"), and it could be automated, thus making it more amenable to double blind research. This "ideal" musical sound stimulation could involve EEG-related components such as varying tone, harmonics, harmony and melody which make music much "richer", and, therefore, perhaps more broadly and effectively entraining than simpler forms of stimulation. Additionally, it would be EEG biofeedback, not simply stimulation.

In a promising innovation in auditory biofeedback, Miller replaced synthesized midi feedback tones with (real audio) short Native Flute melody phrases played by Dr. Clint Goss. The latter played brief melodic phrases at 17 tonal centers, 8 levels below the mean and 8 levels above the mean, and then used a distinctive drone flute for phrases at the mean level itself. Miller programmed BioDisplay educational software to randomly trigger one of four analogous phrases as a subject's physiological measure of valence (direction of change) passed through its corresponding level above or below the mean previously acquired. Using a simple algorithm at each measurement reading in time (valence = valence + 1 where valence increases by 1 if the physiological measure x is higher than it's last prior reading, and valence = valence − S1 if the physiological measure is lower than its last prior reading) to roughly determine the strength of the directional trend, the short phrases produce a flute melody that descends in pitch as valence drops (for example as electrodermal activity decreases) and the melody, again comprised of the short phrases, ascends in pitch as the physiological measure or valence assigned to it increases. The melody could be assigned to such measures as electrodermal activity, heart rate, blood volume pulse, temperature and EEG bandwidths. The direction of the melody could be reversed, as for instance in the case of HRV or EEG Alpha, if increase is desired, such that the generally more relaxing lower pitched phrases are accessed as the measure increases rather than when it decreases. The Bio-Display software is currently in use at Music Therapy Departments at Montclair State University, Immaculata University and Marylhurst University. The aesthetic experience is enhanced by placing a complementary synthesizer drone track underneath the Native flute feedback melody, thereby creating a relaxing and meditative musical space while the subject is learning neuromodulation through the flute feedback.

TONING: THE HUMAN VOICE AS A MUSICAL INSTRUMENT FOR NEUROMODULATION

Toning can be considered one of the most natural and direct tools for neuromodulation since it uses the voice, an instrument of the body. Toning is intentional vocal sounding for the purposes of relaxation, meditation, prayer, self-expression, emotional release, and/or pain reduction. When we tone, the vibration from the sound physically resonates through the body creating a sensation resembling an internal massage. Unexpressed emotions may become locked or "crystallized" in our physiology (Pert, 1999), and the vibrational movement of toning may move "energy", with the possibility of unblocking congested areas in the physical, emotional and spiritual aspects of the body. Toning sometimes is considered a form of meditation, for it can quiet the mind by listening to sound.

There is evidence that both music and meditation activate and help develop the frontal cortex of the brain, accessing a psychosomatic network through peptides and receptors to effect changes and healing in the body–mind (Pert, Ruff, Weber, & Herkenham, 1985). The voice is linked to the mind through speech, emotion and communication (D'Angelo, 2000). When we hurt physically or emotionally, we often innately make a sound to release the pain, unless we have been conditioned to hold it in due to societal disapproval. As babies, we cried, babbled, or self-calmed through making sounds before we were able to communicate our wants and needs through speech. In Western culture, many people feel vulnerable about their voice as they have learned to judge it by comparing their voice to professional singers or literally have been told to "be quiet" by authority figures. A cycle may form, repressing vocal expressions of feelings and emotions that, although repressed, continue to exist subconsciously and represented within the body. However, it seems we have the capability to use our voices for self-generated healing through toning.

Healing may be thought of as the process of making or becoming "sound" or healthy again. By becoming "sound" we literally "tune up" our systems. Our body is made up of sound resonators such as the cells, muscles, organs, glands, and nervous system, much like a complex musical instrument. When we are "out of tune", the body's vibrational patterns may have lost their normal frequencies, leading to a "diseased" state (D'Angelo, 2000, p. 13). Toning with intention may bring the mind/body into more "ease," by harmonizing systems through the voice. Toning also can be administered by a practitioner for the purpose of another's well-being. Some sound healers tone near or into an area of the body to move energy. Many music therapists facilitate toning individually and as a group process with a therapeutic goal. Toning also can be self-generated, without a practitioner's assistance.

The breath is an important aspect of toning. Toning on a regular basis expands breathing capacities, and the breathing alone brings oxygen to the nervous system, usually creating a calming effect. The sympathetic nervous system can be calmed through deep breathing, which may normalize physical symptoms of stress in the cardiovascular, hormonal, immune and muscular systems (Gaynor, 1999).

Pranayama is the regulation of breath involved in the practice of yoga. Prana is a term for life force and yama is breath. Diaphragmatic breathing is a form of Pranayama, or conscious breathing. Research suggests that diaphramatic breathing lowers cortisol levels, elevates serotonin, increases blood flow to the prefrontal cortex of the brain and can change brain-wave patterns (Jerath, Edry, Barnes, & Jerath, 2006). A long deep breath followed by the release of sound (toning), creates an internal massage reported to slow heart rate and relax the sympathetic nervous system (Gaynor, 1999).

There are many methods for toning. Toning does not have to be loud to be effective. Toning is simple; one does not need singing experience. One simple and very effective way to tone is through a "Humm", where you literally can feel the vibration. The process is as follows:

> Step 1: Take three slow deep breaths in the nose and out the mouth. With the following breath, allow a hum sound with lips closed to emerge for the duration of the exhale. Repeat breathing in and exhaling with lips closed, creating a hum. Put hands over ears initially to hear the vibration. Feel the vibration. Focus on the sound and the vibration for 3–15 min. Sit in silence for at least 1 min. and notice what you feel.
> Step 2: Imagine moving the sound and vibration down the body: head, shoulders, heart, mid riff, abdomen, groin, thighs, calves, and feet. If there is an area in the body that feels discomfort, send the gentle hum to this area focusing on the sound and vibration.

The "mm" sound is one of the first sounds that babies make universally, stimulating the facial muscles for all languages. Mothers hum to their babies to calm them. We often innately hum when we are happy, make a discovery, question a thought or like the way something tastes. The "mm" is used at the end of sacred sounds in different traditions such as Amen in the West and Om in the East. Toning on a hum has been practiced for centuries by both the Tibetan Buddhists, and in India as a yogic practice to quiet the mind. Bhramari pranayama (humming) has been reported to reduce blood pressure and increase parasympathetic activity (Sujan et al., 2015).

Humming on a regular basis may protect against, and help relieve sinus infections through an increase in nitric oxide (NO), which is formed and released in sinuses through humming (Weitzberg & Lundberg, 2002). NO oxygenates the brain, regulates cerebral blood flow and acts as a neurotransmitter (Toda, Ayajiki, & Okamura, 2009). NO is broadly distributed in the central nervous system, where it influences synaptic transmission and contributes to learning and memory mechanisms. NO is made in response to incoming synaptic activity (activity generated by sound received by the ear), acting as a form of gain control (Steinert et al., 2011). There is evidence that neurodegenerative conditions such as Alzheimer's disease and Parkinson's disease, involve NO in their pathogenesis. Optimal NO levels could be beneficial in the protection against dementia and other neurodegenerative disorders (Duncan, Heales, 2004).

When NO is out of balance, less oxygen and glucose are received by brain cells. Optimal levels of NO support a "tuned" nervous system. "Due to stress, NO can be compromised and require a stimulus to bring it back to rhythmicity. Some research has shown that stimulation by a 128 cps tuning fork (tuned to a fifth) will "spike NO and enhance NO puffing rhythms" (Beaulieu, 2010, p. 57).

A study on music and NO, at the Neuroscience Research Institute, NY, found that NO signaling goes through the cochlear nerve fibers and the auditory cortex, activating emotion centers in the limbic system. Researchers concluded the NO signaling system is responsible for music to act as a relaxation mechanism (Salamon, Kim, Beaulieu, & Stefano, 2003).

Based on findings such as those cited above, it seems that further research on the relationship of sound vibration and NO production could have great potential.

Toning is believed to be capable of both quieting the mind and "opening the heart" as it induces a meditative state. And, since neuro-cardiologists have discovered that the heart sends signals to the amygdala, thalamus, and hypothalamus which are involved in regulation of perceptions and emotions (McCraty, 2015), it might be expected that toning and mediation would affect "matters of the heart". Globally, when love or compassion are experienced, the sound "Ah" often is expressed. Ah is a sacred sound in Hebrew Kabbalah, Tibetan Buddhism, and the beginning of the alphabet in Indian Sanskrit (sacred language of the Hindus), and frequently is used during certain meditation practices. Physically when Ah is toned or sung, the throat is open and the jaw relaxes, allowing maximum air to be released through exhalation. The "OM" (AUM) combines both the vowel sound Ah and the consonant mmmm. It is considered a sacred primordial sound in both Hinduism and Buddhism, and is the beginning or ending sound of many mantras used with meditation. An example of specific instructions for toning "Om" is as follows: (1) take 3 slow deep breaths, in the nose and out the mouth. (2) using the cavity of the mouth to shape the sound, slowly sound Ah–oh–mmmn in long extended sounds. (3) Breathe and repeat several cycles. (4) Feel the sound resonating in the body, listening closely for overtones or sounds in between the notes.

Om chanting research has demonstrated positive changes in physiological alertness and increased sensitivity to sensory transmission (Kumar, Nagendra, Manjunath, Naveen, & Telles, 2010). An Om chanting study, involving recording and measuring waveforms of frequency modulation, concluded toning Om promotes "stabilization of the brain" and an increase of energy (Gurjar & Thakare, 2009). Om chanting research measuring galvanic skin response found an increase in electrodermal resistance, indicating a decrease in stress level (Das, 2012).

An Om chanting pilot study using magnetic resonance imaging (MRI) demonstrated that the vibration (which involves humming) stimulated the vagus nerve through auricular branches indicating limbic deactivation (Kalyani et al., 2011). The vagus nerve (10th cranial nerve) is part of the parasympathetic nervous system and has been considered a mind/body connector. It is the longest cranial nerve, and communicates with the brain, and regulates the heart, facial muscles and the abdominal viscera.

Dr. Stephen Porges, director of the Brain-Body Center at the University of Illinois, articulated the "polyvagal theory". This theory describes the evolution of the nervous system and origins of brain structures, theorizing that most social and behavioral disorders are biological (Dykema, 2006). Through neuroception, we assess safety while processing sensory information. The polyvagal theory proposes there are neural circuits that support social engagement behaviors (parasympathetic nervous system) and defensive strategies of fight, flight, and freeze (sympathetic nervous system). The autonomic nervous system reacts to safety challenges in a hierarchal order. If a biochemical request for safety is not satisfied, the sympathetic nervous system activates. If this circuit does not promote safety, the unmyelinated vagus triggers a freeze response (Porges, 2008). According to the polyvagal theory, social interaction is an important component of well-being both physically and socially, and a two-way social engagement system exists, being both receptive and expressive. Neurologically, prosodic voices (melodic tone and rhythm) and facial expressions elicit a sense of safety in the receiver (Porges, 2009). The muscles involved signal our emotional state, and are the same muscles used in singing, blowing wind instruments, and listening to music (Porges, 2008).

Deficits in the social engagement system reportedly are found in people with autism, PTSD, anxiety disorders, borderline personality, bipolar and hyperactivity. Persons with these disorders often have flat affect, little vocal intonation, and difficulty distinguishing a voice from background sounds (Dykema, 2006). People with deficits in the social engagement system often have inner and/or middle ear difficulty. Exercising the social engagement system through singing, toning, breathing through diaphragm, or playing a wind instrument may promote neural regulation of the middle ear muscles. This, in turn, may affect both behavioral and physiological states which promote increased social engagement, a sense of safety and normal functioning of the myelinated vagus which calms the nervous system. (Porges, 2009). Collective group singing, chanting and humming stimulate the vagal nerve collectively, promoting social bonding and interaction (Vickhoff et al., 2013).

PILOT EEG RESEARCH ON EFFECTS OF TONING VERSUS OTHER STIMULATION PROCEDURES

Randomized, placebo-controlled studies are important for determining if treatments can be generalized to a population with a reasonable expectation of effect. However, much also can be learned from single-study examples of outlier cases where the results are not averaged into group statistics. Examples might be if one is curious how the Dalai Lama's brain responds when in deep meditation, or when a Sufi master pierces himself with a sword. Based on such reasoning, two of the authors, Miller and Miller, investigated differences in QEEG responses of a master in the art of vocal toning to three different conditions: vocal toning (Fig. 6.6); while absorbing vibrations through the body emanating from a 50-string sound bed (Fig. 6.4); and while listening to the sounds of Tibetan Bowls (see Fig. 6.5) as compared to silent baseline (Fig. 6.3).

QEEG results from baseline and each of the three conditions are shown in Figs. 6.3–6.6.

All three types of stimulation were associated with movement toward normalization in some or all frequency bands as evidenced by changes in normative Z-score measures. The 50-string sound bed normalized alpha, beta and high beta absolute power by decreasing it, while also decreasing absolute power in delta and theta, but to abnormally low levels. This then made the *relative* power in alpha very abnormally high at all sites, and to a lesser extent in beta and high beta as well. Frontal alpha coherence also was greatly increased to an abnormal level, and there were considerably more phase abnormalities. During the 50-string sound bed condition, the subject reported feeling the presence of deceased family members and felt held by them.

The toning also seems to have normalized alpha at most all sites by decreasing the absolute power in that band, and to a lesser degree in beta and high beta as well, but without excessively lowering delta and theta absolute power. Coherence was increased among frontal sites at all frequencies except theta (although this was not as extreme in the alpha band as it was for 50-string). Phase abnormalities were fewer than at baseline and during 50-string stimulation. Interestingly, relative power was more often normal in this condition than in any other. While toning, the subject reported seeing "light around the top of the head" while eyes were closed. Measures during the Tibetan Bowls condition also seemed to indicate a normalizing effect on absolute power relative to baseline. This also was true for most frequency bands. Relative power in alpha did not

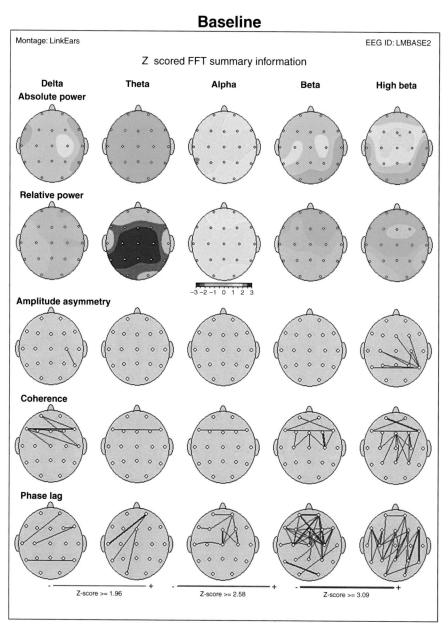

FIGURE 6.3 Silent baseline qEEG.

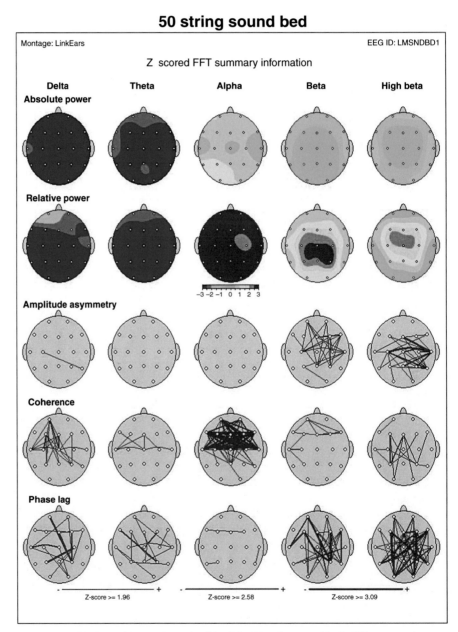

FIGURE 6.4 Playing strings on the underside of the 50 string-sound bed qEEG.

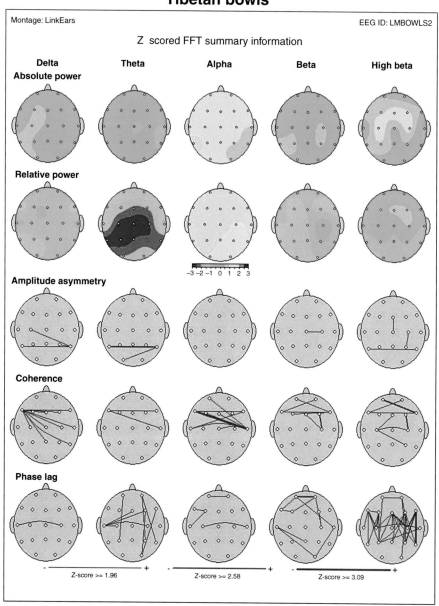

FIGURE 6.5 Tibetan bowls qEEG.

Toning aahs (variable pitch)

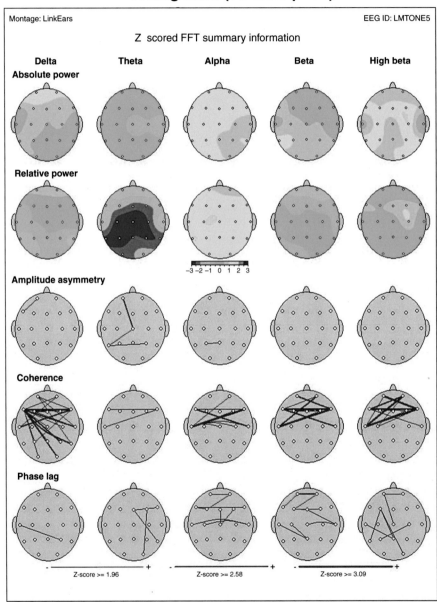

FIGURE 6.6 Vocal toning qEEG.

change much from baseline. There did not seem to be much difference in instances of coherence or phase abnormalities relative to baseline (except slightly fewer phase abnormalities in beta). During the Tibetan bowls, the subject reported feeling connected to "universal energy," something "more than herself".

These are interesting data, and call for further investigation. Overall, the 50-string sound bed stimulation seemed to have the most powerful effects with its intense relative alpha elevation. The toning and bowls stimulation generally seemed to have the most "normalizing" effects in that they were associated with more normalized (Z-scores closer to 0) absolute power scores at nearly all sites and in all frequencies. However, in all conditions there was increased coherence, with this being greatest for the 50-string sound bed and the toning.

SUMMARY

This chapter recognized the universality of both, music and neuromodulation, the associations between music, and medicine since earliest written history, and the similarities among concepts used to describe the nature of both music and the EEG, such as frequency and rhythm. Considered together, these suggest that the stimulation music provides may give it potential to be among the most powerful and effective of the many stimulation procedures used for neuromodulatuon. Following a brief mention of brain regions known to mediate various music-related processes, some specific techniques used by music therapists to help alleviate symptoms of brain damage or dysfunction were described, along with indications of scientific support where available. Some new procedures either currently being experimented with, or speculated upon as promising for future use, also were discussed. Toning and the concept of the human voice as a musical instrument useful for neuromodulation and healing purposes was covered in some detail, including use in conjunction with meditation and for stimulation of healthy neurochemical activity. Finally, results of some of the authors' preliminary research concerning effects on the EEG of toning compared to that of other stimulation procedures was presented.

References

Beaulieu, J. (2010). *Human Tuning: Sound Healing with Tuning Forks*. High Falls, NY: BioSonic Enterprises, Ltd.

Benson, H. (1975). *The Relaxation Response*. New York: Avon.

Bonny, H. L. (1975). The archives for guided imagery and music. www.temple.edu/musictherapy/home/dbs/TheArchivesforGIM.doc.

Campbell, D. (1997). *The Mozart effect*. New York: Avon.

D'Angelo (2000). *The healing power of the human voice: Mantras, chants, and seed sounds for health and harmony*. Rochester, Vermont: Healing Arts Press.

Das, A. (2012). Effect of Prayer and "Om" Meditation in Enhancing Galvanic Skin Response. *Psychological Thought, 5*(2), 141–149.

Duncan, A., & Heales, S. (2004). *Nitric oxide and neurological disorders. division of neurochemistry and neurometabolic unit.* London: Institute of Neurology, Queen Square, London WC1N 3BG, U.K. Elsevier Ltd.

Dykema, R. (2006). Don't Talk to me now, I'm scanning for danger: How your nervous system sabotages your ability to relate. An interview with Stephen Porges about the Polyvagal theory. Nexus, March–April.

Eliade, M. (1972). *Shamanism, Archaic Techniques of Ecstasy*, Bollingen Series LXXVI, Princeton University Press, Princeton, NJ, pp. 3–7.

Evans, J. (2015). Personal Communication, July.

Gaynor, M. (1999). *Sounds of healing: A physician reveals the therapeutic power of sound, voice and music.* NY: Random House.

Gerdes, L., Gerdes, P., Lee, S. W., & Tegler, C. H. (2013). HIRREM: A non- invasive, allostatic methodology for relaxation and auto-calibration of neural oscillations. *Brain and Behavior, 3*(2), 193–205.

Gopal, L. (2016). Ragas that Heal. The Medindia Medical Review: July. Available from: http://www.medindia.net/patients/patientinfo/raga-therapy-for-healing-mind-and-body-healing-ragas.htm

Grant, J. A., Courtemanche, J., Duerden, E. G., Duncan, G. H., & Rainville, P. (2010). Cortical thickness and pain sensitivity in zen meditators. *Emotion, 10*(1), 43–53.

Graur, A. (2006). *Music integrative neurotherapy.* Torino, Italy: Medicamus Intaliana.

Graziano, Amy B., & Johnson, Julene K. (2015). Music, neurology, and psychology in the nineteenth century. *Progress in Brain Research, 216*, 33–49, UC Office of the President: Multicampus Research Programs and Initiatives (MRPI); a funding opportunity through UC Research Initiatives (UCRI). Available from: http://escholarship.org/uc/item/47x082v7.

Gurjar, S., Ladhake, T. (2009) Analysis Of Acoustic "Om" Chant to study effect on Nervous system. International Journal of Computer Science and Network Security, VOL.9 No.1.India.January 2009.

Hunt, A. M. (2011). *A neurophenomenological description of the guided imagery and music experience.* Philadelphia, PA: Temple University.

Jacobs, G. D., Benson, H., & Friedman, R. (1996). Topographic EEG mapping of the relaxation response. *Biofeedback and Self Regulation, 21*(2), 121–129.

Jäncke, L. (2009). The plastic human brain. *Restorative Neurology and Neuroscience, 2*, 521–538.

Jerath, R., Edry, J., Barnes, V., & Jerath, V. (2006). Physiology of long pranayamic breathing: Neural respiratory elements may provide a mechanism that explains how slow deep breathing shifts the autonomic nervous system. *Medical hypotheses, 67*(3), 566–571. Published by Elsevier Inc..

Kalyani, B. G., Venkatasubramanian, G., Arasappa, R., Rao, N. P., Kalmady, S. V., Behere, R. V., & Gangadhar, B. N. (2011). Neurohemodynamic correlates of 'OM' chanting: A pilot functional magnetic resonance imaging study. *International Journal of Yoga, 4*(1), 3–6.

Kerr, C. E., Sacchet, M. D., Lazar, S. W., Moore, C. I., & Jones, S. R. (2013). Mindfulness starts with the body: somatosensory attention and top-down modulation of cortical alpha rhythms in mindfulness meditation. *Frontiers in human neuroscience, 7*, 12.

Miller, E. (2007). Getting from psy-phy (psychophysiology) to medical policy via music and neurofeedback for ADHD children. *ProQuest. Dissertation Abstracts International: Section B: The Sciences and Engineering, 68.*

Mongush, B., & Kenin-Lopsan, M. B. (1997). *Shamanic Songs and Myths of Tuva.* Los Angeles, CA: Akademiai Kiado, Budapest, Hungary and International Society for Trans-Oceanic Research.

Kumar, S., Nagendra, H., Manjunath, N., Naveen, K., & Telles, S. (2010). Mediation on Om: relevance from ancient texts and contemporary science. *International Journal of Yoga, 3*, 2–5.

Levin, Y. I. (1998). "Brain music" in the treatment of patients with insomnia. *Neuroscience and Behavioral Physiology, 28*(3), 330–335Available from: https://proxy.brynmawr.edu/login?url=http://search.proquest.com.proxy.brynmawr.edu/docview/619352004?accountid=9772.

Levitin, D. J. (2006). *This is your brain on music: The science of a human obsession*. New York: Dutton.

McClellan, R. (2000). *The healing forces of music: History, theory, and practice*. Lincoln, NE: toExcel.

McCraty, R. (2015). Heart-brain neurodynamics: the making of emotions. In M. Dahlitz, & G. Hall (Eds.), *Issues of the heart: The neuropsychotherapist special issue* (pp. 76–110). Dahlitz Media: Brisbane.

Meinecke, B. (1948). Music and medicine in classical antiquity. In D. M. Schullian, & M. Schoen (Eds.), *Music and medicine*. New York: Henry Schuman, Inc.

Meymandi, A. (2009). Music, Medicine, Healing, and the Genome Project. *Psychiatry (Edgmont)*, *6*(9), 43–45.

Miller, E. B. (2011). *Bio-guided music therapy: A practitioner's guide to the clinical integration of music and biofeedback*. London, England: Jessica Kingsley.

Mindlin, G., & Evans, J. (2009). Brain music treatment: A brain/music interface. In T. H. Budzinski, H. K. Budzinski, J. R. Evans, & A. Abarbanel (Eds.), *Introduction to Quantitative EEG and Neurofeedback* (2nd Ed.). New York: Academic Press.

Mithin, S. (2005). *The singing neanderthals: The origins of music, language, mind and body*. London: Weidenfeld and Nicholson.

Moisse, K. (2011). Music Therapy Helps Gabrielle Giffords Find Her Voice After Tucson Shooting. ABC News (Mar 8). http://abcnews.go.com/Health/Wellness/gabrielle-giffords-music-therapy-rewires-brain-tragedy-tucson/story?id=13075593.

Nakkah, S. (2007). Yoga of the Voice, Training Materials, Vox Mundi, CA 9/2007.

Norton, A., Zipse, L., Marchina, S., & Schlaug, G. (2009). Melodic intonation therapy: shared insights on how it is done and why it might help. *Annals NYAcademy Science*, *1169*, 431–436.

Pert, C. (1999). *Molecules of Emotion: The Science Behind Mind-Body Medicine*. NY: Simon & Schuster.

Pert, C., Ruff, Weber, & Herkenham (1985). Neuropeptides and their receptors: a psychosomatic-network. *Journal of Immunology*, *135*(2 Suppl), 820s–826s.

Peters, J. S. (1987). *Music Therapy: An Introduction*. Springfield, Illinois: C.C. Thomas.

Porges, S. W. (2008). Music Therapy and Trauma: Insights from the Polyvagal Theory. In K. Stewart (Ed.), *Music Therapy and Trauma: Bridging Theory and Clinical Practice*. New York: Satchnote Press.

Porges, S. W. (2009). The polyvagal theory: New insights into adaptive reactions of the autonomic nervous system. *Cleveland Clinic Journal of Medicine*, *76*(Suppl. 2), S86–S90.

Rauscher, F. H., Shaw, G. L., & Ky, K. N. (1993). Music and spatial task performance. *Nature*, *365*, 611.

Reid, J., & Reid, A. (2011) Ancient Sound Healing. Available from: http://www.tokenrock.com/sound_healing/sounds_of_the_ancients.php

Ricard, M., Lutz, A., & Davidson, R. J. (2014). Mind of the meditator. *Scientific American*, *311*, 38–45.

Salamon, E., Kim, M., Beaulieu, J., & Stefano, G. B. (2003). Sound therapy induced relaxation: down regulating stress processes and pathologies. *Medical Science Monitor*, *9*(5), RA96–RA101.

Schillian, D. M., & Schoen, M. (1948). *Music and medicine*. New York: Henry Schuman, Inc.

Schneider, P., Scherg, M., Dosch, H. G., Specht, H. J., Gutschalk, A., & Rupp, A. (2002). Morphology of Heschl's gyrus reflects enhanced activation in the auditory cortex of musicians. *Nature Neuroscience*, *5*(7), 688–694.

Schlaug, G., Marchina, S., & Norton, A. (2009). Evidence for plasticity in white-matter tracts of patients with chronic boca's aphasia undergoing intense intonation-based speech therapy. *Annals of the New York Academy of Sciences*, *1169*, 385–394.

Schwaller De Lubics, R. A. (1977). *The temple in man*. Brookline, MA: Autumn Press (p.19).

Shaw, G. L., Silverman, D. J., & Pearson, J. C. (1985). Model of cortical organization embodying a basis for a theory of information processing and memory recall. *Proceedings of the National Academy of Sciences of the United States of America.*, *82*(8), 2364–2368.

Steinert, J. R., Robinson, S. W., Tong, H., Haustein, M. D., Kopp-Scheinpflug, C., & Forsythe, I. D. (2011). Nitric oxide is an activity-dependent regulator of target neuron intrinsic excitability. *Neuron, 71*(2), 291–305.

Sujan, Deepika, & Malakur (2015). Effect of Bhramari pranayama (humming bee breath) on heart rate variability and hemodynamic—a pilot study. *Autonomic Neuroscience: Basic and Clinical, 192*, 82.

Sumathy, Sundar (2005) The Ancient Healing Roots of Indian Music. Voices Resources. Available from: http://testvoices.uib.no/community/?q=country/monthindia_march2005a

Taylor, D. B. (2010). *Biomedical Foundations of Music as Therapy* (2nd edition.). Eau Claire, WI: Barton Publications.

Tegeler, C. H., Tegeler, C. L., Cook, J. F., Lee, S. W., Gerdes, L., Shaltout, H. A., Miles, C. M., & Simpson, S. L. (2016). A preliminary study of the effectiveness of an allostatic, closed loop, acoustic stimulation neurotechnology in the treatment of athletes with persisting post-concussion symptoms. *Sports Medicine-Open Dec, 2*, 39.

Toda, N., Ayajiki, K., & Okamura, T. (2009 Aug). Cerebral blood flow regulation by nitric oxide in neurological disorders. *Canadian Journal of Physiology and Pharmacology., 87*(8), 581–594.

Thatcher, R. W. (2010). Symptom check list and functional specialization in the brain link between structure and function. Applied Neuroscience Inc. http://www.appliedneuroscience.com/SymptomCheckListdoc.pdf.

Thaut, M. H. (2008). *Rhythm, music, and the brain: scientific foundations and clinical applications.* New York: Taylor and Francis.

Travis, F. (2013). Transcendental experiences during meditation practice. *Issue: Advances in Meditation Research: Neuroscience and Clinical Applications, N.Y. Acad. Sci.,* 0077-8923.

Twicken, D. (May 2013). Medical Qi Gong: The Spleen Healing Sounds. *Acupuncture Today, 14*(05).

Varela, F. (1996). Neurophenomenology: A methodological remedy for the Hard Problem. *Journal of Consciousness Studies, 3*(4), 330–349.

Vickhoff, B., Malmgren, H., Åström, R., Nyberg, G., Ekström, S. -R., Engwall, M., & Jörnsten, R. (2013). Music structure determines heart rate variability of singers. *Frontiers in Psychology, 4*, 334.

Wallace, R. K., & Benson, H. (1972). The physiology of meditation. *Scientific American, 226*, 84–90.

Weitzberg, E., & Lundberg, J. O. (2002). Humming Greatly Increases Nasal Nitric Oxide. *American Journal of Respiratory and Critical Care Medicine, 166*(No. 2), 144–145.

Further Reading

Jilek, W. G. (1982). Altered states of consciousness. In R. Prince (Ed.), *Trance and Possession states* (pp. 69–95). Montreal: R.M. Bucke Memorial Society, McGill University.

Miller, L. M. (2013). *Spiritarts*. Phoenixville, PA: Expressive Therapy Concepts.

Randall, J., Matthews, R., & Stiles, M. (1997). Resonant frequencies of standing humans. *Ergonomics, 40*(9), 879–886.

Rossiter, T. (2004). The effectiveness of neurofeedback and stimulant drugs in treating AD/HD: Part II. Replication. *Applied Psychophysiology and Biofeedback, 29*(4), 233–243.

Rossiter, T., & LaVaque, T. (1995). A comparison of EEG biofeedback and psychostimulants in treating attention deficit/hyperactivity disorders. *Journal of Neurotherapy, 1*(1), 48–59.

CHAPTER 7

Noninvasive Transcranial Magnetic and Electrical Stimulation: Working Mechanisms

Dirk De Ridder*, Theresia Stöckl, Wing Tin To†,
Berthold Langguth‡, Sven Vanneste¶**

**University of Otago, New Zealand; **Neurodevelopmental Clinic, Gauting, Germany;
†Center for Brain Health, The University of Texas, Dallas, TX, United States;
‡University of Regensburg, Regensburg, Baveria, Germany; ¶The University of Texas,
Dallas, TX, United States*

THE BRAIN AS A COMPLEX ADAPTIVE SYSTEM

What is a Complex Adaptive System?

In order to qualify as a complex adaptive system, a system, whether internet, economy, ant society or brain, has to fulfill two criteria (Amaral, Diaz-Guilera, Moreira, Goldberger, & Lipsitz, 2004). It needs to have a small world topology and has to embed noise (Amaral et al., 2004). The reason for these criteria relates to the adaptiveness of the system. Systems can basically be topologically structured in three ways (Bullmore & Sporns, 2012) (Fig. 7.1): at one extreme the structure can have a lattice or regular topology, which means that every stimulus will always result in exactly the same processing, as the structure of the network is regular. At the other extreme a system can be random, which is inefficient as every stimulus will have a completely random outcome, which is not beneficial. An intermediate structure has a small world topology, which permits flexibility and adaptation to changing environments through variability. The brain is noisy, but the noise is structured, generally following a power law distribution (Buzsaki & Mizuseki, 2014). This means that it has memory, and can carry information, in contrast to white noise (Keshner, 1982). Thus, such a system can learn, while still maintaining stability. Importantly, one of the fundamental characteristics of complex adaptive systems is emergence, meaning that the whole is more than the sum of components and the very specific connectivity creates a new property.

193

Rhythmic Stimulation Procedures in Neuromodulation. http://dx.doi.org/10.1016/B978-0-12-803726-3.00007-9

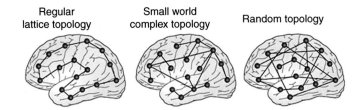

	Regular lattice	Small world complex brain	Random
Freedom	Determined	Flexible	Completely free
Predictability	Complete	Partial	Inpredictable
Noise	No	Pink/red $1/f^{\beta}$	White $1/f^{0}$
Memory	Only	Short to long	No
Learning capacity	None	Yes	None

FIGURE 7.1 Based on (Bullmore & Sporns, 2012): complex adaptive systems (CAS) are characterized by a small world topology and the presence of noise ($1/f^{\beta}$). The brain's small world topology consists of a hierarchical modular scale-free network, being flexible, containing memory, and learning capacity typical of CAS.

Characteristics of Complex Adaptive Systems

All complex adaptive systems share common characteristics, irrespective of whether the complex adaptive system is the economy, the internet, an ant colony or the brain (Holland, 2014; Johnson, 2010). First of all the system is complex, it contains many diverse and specialized agents, components or parts (>3) in an intricate arrangement, which are the building blocks of the CAS. Second, it is adaptive, it has the capacity to change under influence of feedback or memory (learn from experience) and thus evolve, giving it resilience in the face of perturbation. CAS are usually open systems, i.o.w. a system which continuously interacts with its environment. This permits feedback. The interaction can take the form of information, energy, or material transfers into or out of the system boundary. Neighboring interactions predominate in CAS, which can be both positive and or negative feedback, but some long range interactions exist, which creates a small world topology. CAS have a history or memory: they evolve and their past is coresponsible for their present behavior. The most important characteristic of CAS is that they show emergence: the whole is more than the sum of the components and the very specific connectivity creates a new property. CAS are self-organizing, that is, the complexity of the system, and thus emergence, increases without an external organizer but by making the components competitive. This results in nonlinear behavior: small causes can have large results, also known as the butterfly effect. Thus anything can emerge, depending on the feedback, and if you wait long enough it will happen. The emergence arises once a lever or tipping point has been reached. CAS have self-similarity, meaning that the whole has the same shape as one or more of the parts, that is it has a fractal nature, which is similar to structured noise. CAS operate far from equilibrium: there has to be a constant input of energy to maintain the organization of the system, and this is essential for emergence (Fig. 7.2).

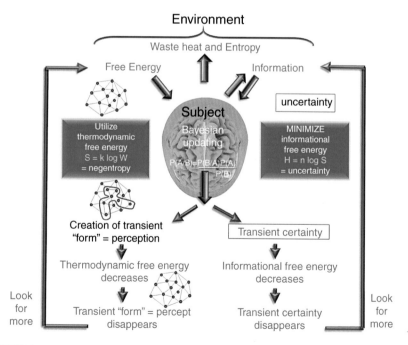

FIGURE 7.2 The brain uses Gibbs' free energy (O_2 and glucose) to transiently create structure which represents an informational percept that transiently reduces uncertainty. Due to the second law of thermodynamics this structure dissolves and thereby the informational percept disappears as well, making place for a new quantum of information to arise. *(Figure adapted from (De Ridder, D., Verplaetse, J., Vanneste, F.S. (2013). The predictive brain and the "free will" illusion.* Frontiers in Psychology, 4, 131.)

Emergence and Information

Emergence is a process whereby larger entities, patterns, and regularities arise through interactions among smaller or simpler entities that themselves do not exhibit such properties (Wikipedia, 2015). All parts of car do not make a car. Only when all those parts are put together in a very specific way, are connected in a very specific way does a riding car emerge. In a similar way it has been suggested by John Stuart Mill that chemistry is based on emergent properties (Mill, 1843). For example, it is the specific bonds between the two H atoms and the O atom that creates water as an emergent property, which has emergent properties different from the H and O atoms. In the same way emergentism in philosophy of mind supports the belief that consciousness is an emergent property of brain function.

Information is an elusive word. It is neither matter nor energy (Gleick, 2011). It requires matter for embodiment and needs energy for communication (Gleick, 2011). Plato proposed that information emerged when a form, a pattern was imposed on data points, and as such randomness can be seen as a lack of form, a lack of structure, a lack of pattern. The amount of information can be quantified by Shannon's formula: $H = n \log S$ with H = information, n = number of symbols, S = number of possible symbols in transmission. Information entropy can then be defined as missing information, analogous to thermodynamic entropy, which is a measure of disorder or a lack of structure. Thus, it is not surprising that the Boltzmann entropy formula ($S = k \log W$ with S = entropy, k = konstante, W= number of ways) is analogous to Shannon's information formula (Gleick, 2011). The brain uses energy (O_2 and glucose) to transiently create structure which represents an informational percept, as an emergent property, that transiently reduces uncertainty (Fig. 7.2). Due to the second law of thermodynamics this structure dissolves and thereby the informational percept disappears as well, making place for a new quantum of information to arise. The BOLD signal in fMRI represents this transient use of O_2 and as such reflects the creating and dissolving of a transient pattern or structure in the brain, i.o.w. an information quantum. As such the brain can be seen as to produce a cinematographic representation of the world by actively sampling the environment to update predictions (Freeman, 2006). The information gets meaning by placing it in a context (Merleau-Ponty, 1945) and relating it to the self (Llinas & Roy, 2009). Depending on the context a cold stimulus can be perceived as pleasant, for example, in a warm environment or unpleasant, for example, in a cold environment (Mower, 1976), or a painful stimulus can be perceived as unpleasant or pleasant (e.g., in specific psychiatric disorders or in sado-masochism). Thus, in summary, the entropy of the message is its amount of uncertainty, but in information theory, 'information' doesn't necessarily mean useful information; it simply describes the amount of randomness of the message.

Connectivity in the Brain

The fact that an informational quantum is represented by the transient formation of a pattern or structure, that is network, in the brain implies that any percept, thought, emotion, feeling, or action as an emergent property of that transient network can be studied by analyzing the associated connectivity of the percept, thought and so on. Three forms of connectivity can be differentiated in this regard: (1) structural, (2) functional, and (3) effective connectivity (Sporns, Chialvo, Kaiser, & Hilgetag, 2004). Structural connectivity refers to the presence of anatomical, physical fiber pathways in the nervous system, usually studied by tracer studies invasively and noninvasively by specific diffusion weighthed imaging techniques, such as diffusion tensor imaging, diffusion spectrum imaging and diffusion kurtosis imaging. The structural connectivity encompasses any anatomical local circuits to large-scale networks of interregional pathways, and these anatomical connections are relatively static at shorter time scales (seconds to minutes), but can be dynamic at longer time scales (hours to days) during learning or development (Sporns et al., 2004). In other words the anatomical connections are not hard-wired but change with experience or deprivation of experience.

Functional connectivity is fundamentally a statistical and not an anatomical concept. It looks at correlated activity between different parts of the brain by measuring correlation or covariance, spectral coherence or phase-locking (Sporns et al., 2004). Functional connectivity is often calculated between all elements of a system, regardless of whether these elements are connected by direct structural links (Sporns et al., 2004). In contrast to structural connectivity, functional connectivity is highly time-dependent, measuring correlated activity at millisecond to second scales. A further elaboration of functional connectivity looks at cross frequency coupling, also known as nesting between different oscillatory frequencies. The underlying idea is that low frequency oscillations (infraslow, delta, theta, and alpha) fundamentally function as carrier waves, on which information waves are nested, which are spatially more restricted and essentially convey either a prediction (beta1) (Arnal, Doelling, & Poeppel, 2015; Arnal, Wyart, & Giraud, 2011) or a prediction error (gamma) (Arnal et al., 2011; De Ridder, Vanneste, Langguth, & Llinas, 2015a) (Fig. 7.3). The beta2 (20–30 Hz) waves would represent contextual salience related information (Zaehle, Bauch, & Hinrichs, 2013; Bunzeck, Doeller, Dolan, & Duzel, 2012; Bunzeck, Guitart-Masip, Dolan, & Duzel, 2011) permitting motor preparation (Roopun, Middleton, & Cunningham, 2006), in other words information detailing whether the prediction is contextually salient and worthwhile to respond to.

Functional connectivity does not assume any directional influence of the elements that are correlated in activity. This is implicitly calculated by effective connectivity. Effective connectivity can be considered directional functional connectivity, and is often based on time series, where the underlying idea is that causes

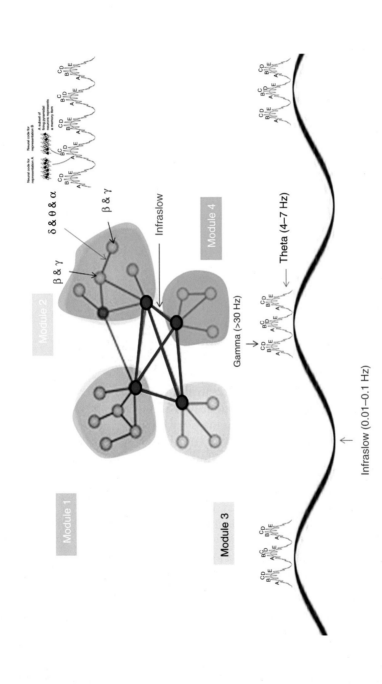

FIGURE 7.3 Within modules focal information processing is proposed to operate at high frequencies (beta and gamma) and these high frequencies are nested on lower frequencies (delta, theta, alpha), which function as carrier waves. Between modules infraslow activity integrates information between the modules. As such a multilevel hierarchy exists in which different frequency bands are coupled, with higher frequencies nested on lower frequencies.

predate effects. Thus information flow in a network follows a time sequence which is calculated by measures, such as Granger causality or transfer entropy.

Structural, functional and effective connectivity are interrelated. Functional and effective connectivity is constrained by the presence of anatomical connections and correlated activity can change structural connectivity via Hebbian mechanisms (cells that fire together wire together). These dynamical changes in structural, functional, and effective connectivity are the basis of the concept of neuroplasticity.

NEUROPLASTICITY AS AN ADAPTIVE MECHANISM

The concept of neuroplasticity is not new. Already in 1783 Charles Bonnet and Vincenzo Malacarne discussed whether experience could induce nerve growth (Bonnet, 1783), later restated by William James (1890), and Donald Hebb (1949) and finally demonstrated in the 1960's (Rosenzweig, 2007; Wiesel & Hubel, 1963). Even though no universally accepted definition exists, there are nearly as many definitions as there are review articles on the topic (Warraich & Kleim, 2010). Neuroplasticity can be operationally defined as the capacity of the nervous system to modify its structural and functional organization, adjusting itself to a changing environment. Thus, neuroplasticity can refer to both adaptive and maladaptive changes in the brain.

A number of factors, such as stress, deprivation of input, adrenal and gonadal hormones, neurotransmitters, growth factors, certain drugs, environmental stimulation, learning, and aging change neuronal structures, and functions, in other words induce neuroplasticity (Fuchs & Flugge, 2014), resulting in alterations in brain areas, changes in neuron morphology, network alterations, including changes in neuronal connectivity, the generation of new neurons (neurogenesis), and neurobiochemical changes (Fuchs & Flugge, 2014). Thus, whereas neuroplasticity encompasses basically every possibly kind of adaptive or maladaptive change in the structure and function of the brain, neuroplasticity from a connectivity point of view refers to adaptive changes of structural, functional and effective connectivity, reorganizing the more stable structural and constantly changing functional and effective connectivity networks, resulting in changing emergent properties, in other words with changing percepts, thoughts, emotions, actions and so on.

HARVESTING NEUROPLASTICITY BY NONINVASIVE TRANSCRANIAL MAGNETIC AND ELECTRICAL STIMULATION

Neuromodulation can be defined as the induction of neuroplastic changes via local application of electrical, magnetic, sound, pharmacological, or optic stimuli. This is a broader definition than the one used by the International

Neuromodulation Society: 'Neuromodulation is technology that acts directly upon nerves. It is the alteration—or modulation—of nerve activity by delivering electrical or pharmaceutical agents directly to a target area' (http://www.neuromodulation.com/about-neuromodulation).

Based on the previously mentioned discussion it can theoretically be expected that all neuromodulation techniques, whether transcranial magnetic stimulation (TMS), transcranial direct current stimulation (tDCS), transcranial alternating current stimulation (tACS), transcranial random noise stimulation (tRNS), neurofeedback, electroconvulsive therapy (ECT), or transcutaneous electrical nerve stimulation (TENS) ultimately have to exert their effect by inducing changes in cell activity, which will result in changes in connectivity, and thereby cause changes in network structure and function, finally changing the emergent properties of these changed networks, clinically expressed as altered symptom presentation or enhancement of specific functions.

Transcranial Magnetic Stimulation

The intact skull and scalp have a high electrical resistance (Barker, Jalinous, & Freeston, 1985). Large electric currents have to be applied in order for a small proportion to penetrate into the brain, leading to painful contractions of scalp muscles and activation of sensory receptors in the skin (Ridding & Rothwell, 2007). Magnetic stimulation of the cortex is particularly effective because of the ability of the field to pass through high-resistance structures. TMS produces a magnetic field of a comparable size as that of an MRI scanner, that is, 1–3 Tesla, but that lasts for only about a millisecond (Walsh & Rushworth, 1999). Because the magnetic field changes very rapidly (from zero to a very high value, then back to zero again in less than1 ms (monophasic pulse)), based on Faraday's law, it induces electrical currents in the area of the brain beneath the coil. Effectively the magnetic field 'carries" the electrical stimulus across the barrier of the skull and scalp into the brain (Ridding & Rothwell, 2007). The induced current pulse lasts for about 200 µs and is similar in amplitude to that produced by a conventional electrical stimulator applied directly to the surface of the brain (Ridding & Rothwell, 2007). It is thought to activate the axons of neurons in the cortex and subcortical white matter, rather than the cell bodies of cortical neurons (which have a much higher threshold) (Ridding & Rothwell, 2007) for which a longer pulse width ($>$1000 µs) is required (Ranck, 1975).

There are two important limitations of TMS: (1) the magnetic field falls off rapidly with distance from the coil surface, limiting direct stimulation to the outer parts of the cerebral cortex under the skull, and (2) the site of stimulation is about 2 $\times$ 3 cm and thus not highly focal (Ridding & Rothwell, 2007).

Based on TMS studies of the motor cortex it has been shown through electromyographic recordings of the activated muscles that TMS has a double effect. A single TMS stimulus evokes a burst of activity that can last for 5–10 ms after the pulse (Day, Rothwell, & Thompson, 1987), which is followed by a period lasting 100 to 200 ms during which activity is suppressed (Ridding & Rothwell, 2007). The effect of the TMS pulse is brain state dependent, as well as dependent on the position and orientation of the stimulation coil and the exact site of stimulation (Ridding & Rothwell, 2007). For example, if a TMS stimulus is given during sleep, anesthesia or coma, the stimulus will only exert a local effect and will not spread through the brain, in contrast to an awake state (Massimini et al., 2005). Furthermore TMS efficacy seems to be dependent on the stimulated person's genetic polymorphism. Certain genes, such as BDNF and 5-HT(1A) influence the sensitivity to noninvasive stimulation, both TMS and tDCS (Malaguti, Rossini, & Lucca, 2011; Cheeran, Talelli, & Mori, 2008). In view of the interindividual variability of the brain, it has been suggested that the efficacy can be improved by using neuronavigated TMS based on the individual's brain structure as demonstrated by structural or functional imaging (Fleming, Sorinola, Newham, Roberts-Lewis, & Bergmann, 2012), although it is still debated whether any resulting increase in spatial accuracy is clinically relevant (Jung et al., 2010).

Concomitant intake of medication, such as benzodiazepins or antiepileptics (Lang, Sueske, Hasan, Paulus, & Tergau, 2006; Lang, Rothkegel, Peckolt, & Deuschl, 2013) can influence the effect of TMS, as demonstrated by a change in motor cortex excitability, but other medication, such as antimigraine medication does not seem to influence the treatment effect of TMS (Almaraz, Dilli, & Dodick, 2010).

Device-related related parameters also determine the effects and side effects of TMS. The type of coil used has a strong influence on the spatial configuration of the magnetic field induced by TMS (Fleming et al., 2012). If TMS pulses are administered repeatedly in a rhythmic pattern, one speaks about repetitive TMS (rTMS). A train of rTMS induces changes in cortical excitability that outlast the stimulation period. The direction of these after-effects depends on a complex interaction of various patient-related and device- or stimulation-related parameters. Among these parameters are pulse configuration, stimulation frequency and amplitude, stimulation pattern, intertrain- and intersession interval (Speer, Kimbrell, & Wassermann, 2000) (Table 7.1).

In general high frequency stimulation (HFS) and low frequency stimulation (LFS) have opposite after-effects (Table 7.2), as demonstrated by functional imaging, with HFS exerting an increased metabolism and LFS a decreased metabolism (Speer et al., 2000; Kimbrell, Little, & Dunn, 1999). The effects on excitability and plasticity are opposite as well: whereas HFS seems to exert

Table 7.1 Patient and Device Related Factors That can Influence the Effect of TMS and tES (tDCS, tACS, tRNS)

TMS Influencing Factors		tES Influencing Factors
Person/patient dependent	Brain structure	Brain structure
	Brain state/function	Brain state/function
	History of activity in the stimulated area	History of activity in the Stimulated area
	Brain area	Brain area
	Genetic polymorphism	Genetic polymorphism
	Medication	Medication
	Scull-cortex distance	Soft tissue and bone structure
Device/protocol dependent	Coil type	Electrodes size
	Coil orientation	Electrodes positions
	Frequency	Frequency in tACS/tRNS
	Intensity	Intensity
	Stimulation pattern (burst/tonic)	Electrode polarity
	Duration	Duration
	Intertrain interval (intermittent/continuous)	
	Number of sessions	Number of sessions
	Interval between sessions	Interval between sessions
	Pulse form	

an increase in excitability (Pascual-Leone, Grafman, & Hallett, 1994) and a long term potentiation-like (LTP, i.e., an activity-dependent strengthening of synapses) effect (Wang, Wang, & Scheich, 1996; Wang, Wang, Wetzel, & Scheich, 1999), LFS seems to generate a decrease in excitability (Chen, Classen, & Gerloff, 1997), and a long term depression -like (LTD, i.e., and activity-dependent weakening of synapses) effect (Wang et al., 1996; Wang et al., 1999). Furthermore, the effect on oscillatory activity, such as measured by EEG and MEG, is different (Fuggetta, Pavone, Fiaschi, & Manganotti, 2008; Paus, Sipila, & Strafella, 2001; Brignani, Manganotti, Rossini, & Miniussi, 2008), as is the effect on neurotransmitter release (Keck, Welt, & Muller, 2002; Keck, Welt, & Post, 2001; Yue, Xiao-lin, & Tao, 2009) (Table 7.2).

It has long been recognized that the effect of TMS is widespread and not limited to the area of stimulation (Kimbrell, Dunn, & George, 2002). This might be mediated by transmission of the stimulus via structural connectivity thereby influencing functional and effective connectivity. It has indeed been shown that low and HFS exert different effects on functional connectivity. Low frequency rTMS increases functional connectivity in contrast to high frequency rTMS, which decreases functional connectivity (Fox, Buckner, White, Greicius, & Pascual-Leone, 2012). Furthermore, TMS also changes directional functional

Table 7.2 Different Effects of High (>5 Hz) and Low (1 Hz) Frequency Transcranial Magnetic Stimulation

	HFS	LFS
Excitability	Increased (Pascual-Leone, Grafman, & Hallett, 1994; Berardelli et al., 1998)	Decreased (Chen et al., 1997)
PET	Increased metabolism (Speer et al., 2000; Kimbrell et al., 1999)	Decreased metabolism (Speer 2000; Kimbrell et al., 1999)
EEG	Upper alpha and beta synchronisation (Fuggetta et al., 2008)	Lower alpha and beta synchronisation (Paus et al., 2001; Brignani et al., 2008)
Molecular	GABA and glutamate unchanged (Keck et al., 2000)	GABA and glutamate increase (Yue et al., 2009)
Plasticity	LTP-like (Wang et al., 1996, 1999)	LTD-like (Wang et al., 1996, 1999)

connectivity, in other words effective connectivity (Grefkes et al., 2010). By altering the functional and effective connectivity TMS can change the emergent property of the stimulated network and thereby exert its clinical effect.

The clinical application of TMS is growing and evidence based guidelines have been proposed for the many symptoms, syndromes, and diseases that are currently investigated as indications for TMS treatment, including pain, movement disorders, stroke, amyotrophic lateral sclerosis, multiple sclerosis, epilepsy, consciousness disorders, tinnitus, depression, anxiety disorders, obsessive-compulsive disorder, schizophrenia, craving/addiction, and conversion (Lefaucheur, Andre-Obadia, & Antal, 2014). Despite unavoidable inhomogeneities, there is a sufficient body of evidence to accept with level A (definite efficacy) the analgesic effect of high-frequency (HF) rTMS of the primary motor cortex (M1) contralateral to the pain and the antidepressant effect of HF-rTMS of the left dorsolateral prefrontal cortex (DLPFC). A Level B recommendation (probable efficacy) is proposed for the antidepressant effect of low-frequency (LF) rTMS of the right DLPFC, HF-rTMS of the left DLPFC for the negative symptoms of schizophrenia, and LF-rTMS of contralesional M1 in chronic motor stroke. The effects of rTMS in a number of indications reach level C (possible efficacy), including LF-rTMS of the left temporoparietal cortex in tinnitus and auditory hallucinations. It remains to be seen how to optimize rTMS protocols and techniques for rTMS to become relevant in routine clinical practice (Lefaucheur et al., 2014).

Many studies have looked at safety of TMS. Various studies evaluating the use of rTMS in patients with epilepsy showed that rTMS is safe in this context (Bae, Schrader, & Machii, 2007): 17% of epileptic patients have side effects, and the main side effect is transient headache (responsive to simple analgesics). The second most common side effect was a nonspecific feeling of discomfort (or weakness). The most feared side effect is the occurrence of seizures during and/ or subsequent to a session of rTMS. This is very rare, with a crude risk of 1.4% (4/280) in epileptic patients. The seizures that occur during rTMS are of the same type as the spontaneous seizures in these patients, and are neither more intense nor followed by greater postictal confusion. Some patients (3/5) actually benefit from a reduction in seizure frequency in the days following the rTMS sessions.

Transcranial Electrical Stimulation

Three different versions exist of tES depending on how the current is delivered to the brain (Fig. 7.4): tDCS, tACS, and tRNS, which is a special version of tACS (Paulus, 2011; Vanneste, Fregni, & De Ridder, 2013).

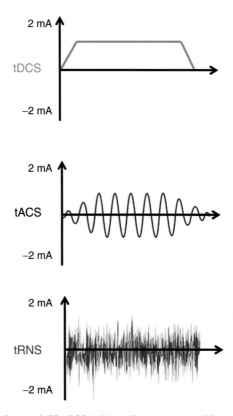

FIGURE 7.4 Different forms of tES: tDCS delivers direct current, tACS and tRNS deliver alternating current.

Transcranial Direct Current Stimulation
Conventional tDCS

Delivering direct electric current over the scalp has been reported since the times of Greeks and Romans: in 48–43 A.D., Scribonius Largus (the physician of Roman Emperor Claudius) reported on the treatment of pain by placing a live torpedo fish delivering a strong direct current - over the scalp. The amount of current was dependent on the size of the fish. In the 11th century, Ibn-Sidah, suggested the placement of a live electrical catfish on the frontal bone for the treatment of patients suffering from epilepsy (Kelloway, 1946). In the 18th century, with the introduction of the electrical battery, Galvani (Galvani, 1791; Galvani, 1797) recognized that electrical stimuli of varying duration can evoke different physiological effects (Zago, Ferrucci, Fregni, & Priori, 2008). In honor of Galvani, direct current stimulation is often called Galvanic stimulation. One of the first clinical applications of galvanic currents date back to the 19th century when Aldini (Aldini, 1804) (Galvani's nephew) and other researchers (Arndt, 1869) used transcranial electrical simulation to treat melancholia and depression. In the 1960's and 1970's, this method had a brief comeback (Lippold & Redfearn, 1964; Redfearn, Lippold, & Costain, 1964), with a more sustained revival at the turn of the millennium (Guleyupoglu, Schestatsky, Edwards, Fregni, & Bikson, 2013).

High Definition tDCS

HD-tDCS has been recently introduced to improve spatial accuracy, by using arrays of smaller "high-definition" electrodes, instead of the two large pad electrodes of conventional tDCS (Guleyupoglu et al., 2013; Datta et al., 2009; Dmochowski, Datta, Bikson, Su, & Parra, 2011; Heimrath, Breitling, Krauel, Heinze, & Zaehle, 2015; Shekhawat, Sundram, & Bikson, 2015; Villamar et al., 2013) (Fig. 7.5).

In conventional tDCS usually one anode electrode and one cathode electrode are applied over the scalp to modulate a particular brain area by inducing a controlled electrical current which flows from the anode to the cathode. Due to the high electrical resistance of the skull (Barker et al., 1985), only 50% of the transcranially applied direct current reaches the brain, the rest being shunted trough the extracranial soft tissues, as demonstrated by calculations on realistic head models, validated both in animal (Rush & Driscoll, 1968) and human (Dymond, Coger, & Serafetinides, 1975) experiments.

tDCS modulates the cellular membrane potential facilitating or inhibiting spontaneous neuronal activity (Nitsche, Cohen, & Wassermann, 2008; Moreno-Duarte, Gebodh, & Schestatsky, 2014). Anodal stimulation will produce inward current flow, resulting in depolarization of pyramidal cortical neurons and apical dendrite hyperpolarization, while cathodal stimulation will typically produce outward current flow resulting in somatic hyperpolarization

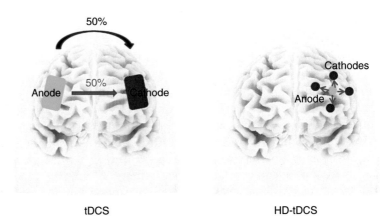

FIGURE 7.5 In conventional tDCS one large anode and cathode are used, while in high definition tDCS (HD-tDCS) multiple cathodes are used with one anode. The current flows from the anode to the cathode, and about 50% is shunted through the skin, with 50% reaching the brain.

of pyramidal cortical neurons and apical dendrite depolarization (Radman, Ramos, Brumberg, & Bikson, 2009; Zaghi, Acar, Hultgren, Boggio, & Fregni, 2010a). In contrast to TMS, tES is not directly inducing neuronal firing, but only modulates cortical excitability. As with rTMS, tES induced excitability changes outlast the stimulation period. The depolarization under the anode will result in an increase of firing and excitability under the anode, whereas the firing rate and excitability are decreased under the cathode (Bindman, Lippold, & Redfearn, 1964; Nitsche & Paulus, 2001).

However, tDCS often results in a delayed clinical effect (Fujiyama, Hyde, & Hinder, 2014; Stramaccia et al., 2015), which cannot be explained by the immediate effect of tDCS on pyramidal or interneuron cell firing. Therefore two other mechanisms have been proposed to be involved in tDCS: glial and stem cell modulation. Based on cable theory one type of glial cells, astrocytes possibly are modulated by tDCS (Ruohonen & Karhu, 2012). Astrocytes control the formation, maturation, function (and elimination) of synapses through various secreted and contact-mediated signals (Clarke & Barres, 2013) and can thereby regulate neural circuit development and function (Clarke & Barres, 2013). This could potentially explain the delayed effect of tDCS. Furthermore, another type of glial cells, microglia, which prune synapses, might also be involved. It has indeed been shown that tDCS activates microglia both under anode and cathode (Rueger, Keuters, & Walberer, 2012). Thus glial cells might be modulated by tDCS, resulting in synapse formation and/or elimination. Since this takes time, it may better explain the delayed effects of tDCS.

But, apart from modulating neurons (both pyramidal and interneurons and glial cells, and both astrocytes and microglia) tDCS could exert its delayed effects via stem cell activation. Indeed, tDCS seems to recruit proliferating neural stem cells under the cathode (Rueger et al., 2012) thereby opening the possibility of regenerative capacities for tDCS and an even more delayed clinical effect of tDCS.

The effects of tDCS depend on a lot of factors, both person/patient/client related, as well as device related factors. Among these influencing factors are the resistance of several cephalic structures including the skin, skull, blood vessels, and brain tissue (Moreno-Duarte et al., 2014; Brunoni et al., 2012; Medeiros, de Souza, & Vidor, 2012; Wagner et al., 2007). Device related factors include (1) polarity of the electrodes, (2) size of the electrodes, (3) the position of the electrodes, (4) the intensity of stimulation or the amount of current delivered (in mA), and (5) the duration of the stimulation (varies between 20–40 min in most studies) (Moreno-Duarte et al., 2014; Brunoni et al., 2012; Wagner et al., 2007; Nitsche, Kuo, Paulus, & Antal, 2015). By varying these tDCS parameters, stimulation protocols can be customized to a certain extent to achieve the desired direction, strength, focality, and duration of effects on cortical activity and excitability (Brunoni et al., 2012; Nitsche et al., 2015).

In general, no serious adverse events are seen by tDCS application, as evaluated in more than 10.000 subjects investigated in the contemporary tDCS literature (1998–2014) (Fregni, Nitsche, & Loo, 2015). The safety of tDCS depends on the strength of current, the size of the electrodes and the duration of the stimulation (Iyer et al., 2005; Poreisz, Boros, Antal, & Paulus, 2007).

For conventional tDCS, a stimulation intensity of up to 2 mA for the duration of about 20 min is considered to be safe (Iyer et al., 2005; Nitsche et al., 2003). The threshold for adverse events has been investigated in a safety study in rats, where the current density needed to induce brain damage in rats was found to be at least 100-times higher than the current density used in tDCS trials (Liebetanz et al., 2009). The most frequent side effects include a tingling sensation during stimulation, predominantly under the anode (Poreisz et al., 2007; Brunoni et al., 2011), an itching sensation (Poreisz et al., 2007; Brunoni et al., 2011; Fertonani, Ferrari, & Miniussi, 2015) directly under the electrodes, headache (Poreisz et al., 2007; Brunoni et al., 2011), moderate fatigue (Poreisz et al., 2007) and burning sensation (Poreisz et al., 2007; Brunoni et al., 2011; Fertonani et al., 2015).

For High-Definition tDCS, studies using 4 × 1 ring configuration with intensities up to 2 mA for up to 20 minutes have demonstrated its tolerability in both healthy (Nikolin, Loo, Bai, Dokos, & Martin, 2015; Kuo, Bikson, & Datta, 2013; Borckardt, Bikson, & Frohman, 2012) and patient populations (Villamar et al., 2013; Donnell, 2015). Skin irritations and skin lesions have been reported, but can largely be avoided by choosing appropriate contact media (Palm et al., 2014).

tDCS has been shown to modulate not only the areas underlying anodal and cathodal stimulation (Brunoni et al., 2012; Antal, Kincses, Nitsche, Bartfai, & Paulus, 2004; Dieckhofer et al., 2006; Matsunaga, Nitsche, Tsuji, & Rothwell, 2004; Zaehle, Beretta, Jancke, Herrmann, & Sandmann, 2011), but also functional and effective connectivity (Alon, Roys, Gullapalli, & Greenspan, 2011; Chib, Yun, Takahashi, & Shimojo, 2013; Keeser, Meindl, & Bor, 2011a; Keeser, Padberg, & Reisinger, 2011b; Meinzer, Antonenko, & Lindenberg, 2012; Meinzer, Lindenberg, Antonenko, Flaisch, & Floel, 2013; Pena-Gomez, Sala-Lonch, & Junque, 2012; Polania, Nitsche, & Paulus, 2011; Polania, Paulus, & Nitsche, 2012a; Stagg, Lin, & Mezue, 2013; Vanneste & De Ridder, 2011; Weber, Messing, Rao, Detre, & Thompson-Schill, 2014), thereby possibly changing the emergent property of the involved network (Luft, Pereda, Banissy, & Bhattacharya, 2014). The effect of tDCS primarily has been investigated on motor cortex physiology (Brunoni et al., 2012). But the visual cortex (Antal et al., 2004), the somatosensory cortex (Dieckhofer et al., 2006; Matsunaga et al., 2004), and the auditory cortex (Zaehle et al., 2011) also have been investigated. Beyond the regional effects of tDCS under the stimulation electrodes, more remote effects on topographically distant cortical and subcortical areas have been found (Lang, Siebner, & Ward, 2005).

Clinically, tDCS has been used and meta-analyses have been performed for the treatment of upper limb paresis, dysphasia, dysphagia, tinnitus, pain, depression, craving, epilepsy, and Alzheimer's disease. For upper limb improvement in stroke patients, meta-analysis and meta-regression results suggest superior motor recovery in active groups when compared to sham groups, as well as dose-response relationships relating to electrode size, charge density and current density (Chhatbar et al., 2016). For dysphasia after stroke a Cochrane meta-analysis found that there is no evidence of the effectiveness of tDCS (anodal tDCS, cathodal tDCS, and bihemispheric tDCS) versus sham tDCS for improving functional communication, language impairment and cognition in people with aphasia after stroke (Elsner, Kugler, Pohl, & Mehrholz, 2015). For poststroke dysphagia a significant effect size was found when stimulating the unaffected rather than the affected hemisphere (Pisegna, Kaneoka, Pearson, Kumar, & Langmore, 2016). For tinnitus, efficacy of tDCS could not be fully confirmed because of the limited number of studies, but all studies included in the systemic review and meta-analysis demonstrated significant tinnitus intensity improvement (Song, Vanneste, Van de Heyning, & De Ridder, 2012). For pain after spinal cord injury, meta-analytic results indicate a moderate effect of tDCS in reducing neuropathic pain, but the effect was not maintained at long-term follow-up (Mehta, McIntyre, Guy, Teasell, & Loh, 2015). For chronic pain globally, tDCS results indicated no significant difference between active and sham stimulation (O'Connell, Wand, Marston, Spencer, & Desouza, 2014). For depression, tDCS may be efficacious. However, the data are still preliminary and not yet sufficient to recommend tDCS for routine treatment in treatment-resistant depression or as an add-on augmentation treatment to psychopharmacology or psychotherapy (Meron, Hedger, Garner, & Baldwin, 2015). For craving in contrast, metaanalytic data do support the use of tDCS of the DLPFC to reduce craving in substance dependence (Jansen et al., 2013). Similarly, tDCS trials have demonstrated preliminary safety and efficacy in animals and patients with epilepsy (San-Juan, Morales-Quezada, & Orozco Garduno, 2015). For healthy aging and Alzheimer's disease subjects a significant effect size of 0.42 (healthy) and 1.35 (Alzheimer) was found for cognitive outcomes (Hsu, Ku, Zanto, & Gazzaley, 2015). Subgroup analyses indicated more pronounced effects for studies applying the stimulation during execution of the task compared with studies delivering the stimulation before the execution of a cognitive task?

Transcranial Alternating Current Stimulation

The main mechanisms by which tACS influences brain physiology has been attributed to frequency specific entrainment, that is phase alignment of endogenous brain oscillations to externally applied oscillating electric currents (Thut, Schyns, & Gross, 2011; Zaehle, Rach, & Herrmann, 2010; Witkowski, Garcia-Cossio, & Chander, 2016), and modulation of spike-timing dependent plasticity (Polania, Nitsche, Korman, Batsikadze, & Paulus, 2012b; Vossen, Gross, & Thut, 2015). It has indeed been shown that alpha frequency-tACS can enhance individual alpha frequency (Zaehle et al., 2010), but the functional effect depends on background activity (Kanai, Chaieb, Antal, Walsh, & Paulus, 2008). tACS boosts motor excitability at 140 Hz, (Moliadze, Antal, & Paulus, 2010) and decreases excitability at 15 Hz (Zaghi, de Freitas Rezende, & de Oliveira, 2010b). Whereas it has been shown that tACS can influence perception (Kanai et al., 2008), memory (Marshall, Helgadottir, Molle, & Born, 2006), motor function (Brignani, Ruzzoli, Mauri, & Miniussi, 2013) and higher-order cognition (Santarnecchi, Polizzotto, & Godone, 2013), few clinical studies have been performed using tACS as a treatment, and those which were performed were either case reports (Angelakis, Liouta, & Andreadis, 2013) or did not show a clinical benefit (Vanneste et al., 2013). Therefore tACS should be seen as an experimental approach for research at this moment, rather than a tool applicable for routine clinical practice. tACS does modulate functional connectivity (Helfrich, Knepper, & Nolte, 2014), so theoretically can be used to change emergent properties of network function.

Transcranial Random Noise Stimulation

tRNS is a modification of tACS (Fig. 7.3). The tRNS device generates alternating current which follows a white noise structure, that is all frequencies (0.1–640 Hz) have equal power, with a Gaussian amplitude structure. Low frequency tRNS is defined as 0.1–100 Hz random noise stimulation, whereas high frequency tRNS is limited to frequencies between 100 and 640 Hz. tRNS has a higher perception threshold than tDCS (1200 vs. 400 μA) (Ambrus, Paulus, & Antal, 2010).

High frequency tRNS seems to increase excitability (Terney, Chaieb, Moliadze, Antal, & Paulus, 2008), and its mechanisms of action are still unknown. Theoretically it could increase excitability by a stochastic resonance effect mediated through repeated subthreshold stimulations (Terney et al., 2008) that prevent homeostasis of the system (Fertonani, Pirulli, & Miniussi, 2011). Another possible working mechanism is through desynchronization of (pathological) rhythms (Paulus, 2011), but none of the abovementioned mechanisms of action have been proven.

tRNS modulates perception (Romanska, Rezlescu, Susilo, Duchaine, & Banissy, 2015), memory (Mulquiney, Hoy, Daskalakis, & Fitzgerald, 2011), learning (Tyler, Conto, & Battelli, 2015; Herpich et al., 2015) and other cognitive functions (Cappelletti, Gessaroli, & Hithersay, 2013) possibly by NMDA-receptor independent but sodium-channel blocker and benzodiazepines sensitive plasticity (Chaieb, Antal, & Paulus, 2015).

tRNS has been clinically used for tinnitus (Vanneste et al., 2013; Joos, De Ridder, & Vanneste, 2015), myopia (Camilleri, Pavan, Ghin, Battaglini, & Campana, 2014), depression (Chan, Alonzo, & Martin, 2012), schizophrenia (Palm, Hasan, Keeser, Falkai, & Padberg, 2013), and Parkinson's disease (Stephani, Nitsche, Sommer, & Paulus, 2011), suggesting it is a promising new modality for treating functional brain disorders. In a head to head comparison of tDCS, tACS and tRNS it was shown that tRNS was the only efficacious transcranial electrical stimulation for tinnitus suppression (Vanneste et al., 2013). Interestingly, both low and high frequency tRNS were beneficial but the combined low + high frequency tRNS was inefficacious for tinnitus suppression (Joos et al., 2015). Furthermore, both for tACS and tRNS low and high amplitudes seem to have an opposite effect. Both tACS and HF-tRNS at 0.4 mA are inhibitory but switch to excitatory modulation at 1 mA (Moliadze, Atalay, Antal, & Paulus, 2012).

THE FUTURE OF NEUROMODULATION?

In the future, noninvasive stimulation could be performed in combination with external stimuli or other treatments, coined neuromodulation 2.0 (Arns & De Ridder, 2011) analogous to the seminal work in which tinnitus was treated by combining vagal nerve stimulation (VNS) with external auditory stimuli, both in animals (Engineer, Riley, & Seale, 2011) and humans (De Ridder, Kilgard, Engineer, & Vanneste, 2015b; De Ridder, Vanneste, Engineer,

& Kilgard, 2014). The idea is that neuromodulation induced a greater level of plasticity which can then be steered in a desired direction by applying other external stimuli. Nucleus basalis stimulation or VNS paired with a tone three hundred times a day for 4 weeks has been found to result in large scale reorganization of the auditory cortex that was specific to the paired frequency (Engineer et al., 2011; Kilgard & Merzenich, 1998). These results demonstrate that precisely timed release of neuromodulators generates plasticity that is specific to acoustic features of the paired sound. This concept was applied in a tinnitus study by pairing VNS with tones (except for the tinnitus matched frequency), reversing both the behavioral perception of tinnitus and the pathological plasticity in noise exposed rats (Engineer et al., 2011). The hypothesis was that by pairing randomly interleaved competing tones with VNS, the pathological expansion of the putative tinnitus frequency can be normalized, and neuronal hypersynchrony can be decreased. This technique was extremely successful in a rat model of tinnitus, (Engineer et al., 2011) and to a lesser extent also in humans (De Ridder et al., 2015b; De Ridder et al., 2014). This concept could be translated to noninvasive stimulation as well, and is akin to similar approaches being used for rehabilitation, in which noninvasive stimulation is combined with rehabilitation or speech therapy (Dammekens, Vanneste, Ost, & De Ridder, 2014).

The idea of combining neuromodulation with another therapy can be extended to psychotherapy as well. Indeed, cognitive control therapy can be enhanced by adding tDCS in the treatment of depression, resulting in what could be called neuromodulation enhanced psychotherapy. Still other approaches can be theorized. For example, PTSD is now routinely treated by EMDR (Shapiro, 1996). Eye movement desensitization and reprocessing (EMDR) is exposure based therapy plus eye movements from left to right and back, optimally at 1 Hz (van Veen et al., 2015) and it has been shown that the eye movements are relevant (Lee & Cuijpers, 2014). Meta-analyses have further shown that EMDR is slightly superior to cognitive behavioral therapy (CBT) (Chen, Zhang, Hu, & Liang, 2015), and better for decreased intrusion and arousal severity compared to CBT. However, the avoidance aspect of PTSD was not significantly different between groups. A moderate but significant effect is seen in children as well (Rodenburg, Benjamin, de Roos, Meijer, & Stams, 2009). EMDR is associated with changes in the dorsolateral prefrontal cortex (DLPFC) (Herkt et al., 2014) and temporoparietal junction (TPJ) (Pagani, Di Lorenzo, & Verardo, 2012).

One could hypothesize that alternating stimulation on the left and right DLPFC and/or TPJ at 1 Hz could electronically replace the eye movement effect and have a similar effect on concomitant exposure therapy, such as EMDR. The advantage would be a more synchronous and better timed stimulation, and it could be performed by the patient at home.

New or presently known specific forms of neuromodulation can also be explored further, such as Limoge's current (166 kHz bursts 4 sec on 8 sec off via 2 anodes on mastoid, 1 cathode frontally), Lebedev's current (similar to Limoge but in AC:DC ratio 2:1), or modifications of this approach with pink, brown and black noise stimulation or stochastic resonance stimulation. Pulsed electromagnetic field stimulation (pEMF), which uses millitesla electromagnetic stimuli already has been investigated for depression (Martiny, Lunde, & Bech, 2010), multiple sclerosis (Richards, Lappin, & Acosta-Urquidi, 1997), and tinnitus (Roland et al., 1993). One of the important questions is what the minimal magnetic field strength is to have an effect on the brain, considering that TMS stimuli, which are classically delivered at 2–3 Tesla, are 1000 times stronger than the stimuli delivered in one aforementioned study (1.9 mT), and more than a 1,000,000 times stronger than used by the μT pEMF devices. Using EEG, it has been shown that the weakest stimuli to still modulate brain activity both during and minutes after the pEMF stimulation is in the order of 200 μT (Cook, Thomas, Keenliside, & Prato, 2005; Cook, Thomas, & Prato, 2004). The mechanism by which pEMF exerts its effect is unknown, but glutaminergic based excitation has been shown, based on paired pulse TMS evaluation of the effect of pEMF (Capone, Dileone, & Profice, 2009).

CONCLUSIONS

The brain can be seen as a complex adaptive system, adjusting its activity and connectivity to adjust to the environment. All complex adaptive systems are characterized by emergence, meaning that the whole is more than the sum of the components as when very specific connectivity creates a new property. Consciousness or actions, thoughts, symptoms can be seen as emergent properties of changing connections, in other words, of changing networks.

The structural and functional brain reorganization is known as neuroplasticity, and neuromodulation can be regarded as applied neuroplasticity or the induction of neuroplastic changes in the brain through magnetic and/or electrical stimulation. All noninvasive stimulations are hypothesized to exert their effect through modulation of functional and effective connectivity (Table 7.3), and thereby altering the emergent property (symptom, percept, action, thought,..) of the modulated network. Whereas many studies exist concerning TMS and tDCS, relatively few studies have been performed for tACS and its derived tRNS. The future of neuromodulation may lie in the combination of neuromodulation with other treatments, or neuromodulation paired with external stimuli, actions, or thoughts, that is in "neuromodulation 2.0" (Arns & De Ridder, 2011).

Table 7.3 Putative Mechanism of Noninvasive and Invasive Neuromodulation Techniques for Most but Not All Techniques it has been Shown that they Modulate Functional and Effective Connectivity, Which is Postulated to be the Final Common Pathway of All Neuromodulation Techniques

	TMS	tDCS	tACS*	tRNS*	implants	ECT	NFB
Putative mechanism	A & B synchronization	Depolarize Hyperpolarize	Entrain at specified frequency	Desynchronize?	Virtual lesion	Epileptic reset	Train oscillations
	Subthreshold for AP	Subthreshold for AP	Subthreshold for AP	Subthreshold for AP	Subthreshold for AP		Subthreshold for AP
Functional connectivity	Changes	Changes	Changes	?	Changes	Changes	Changes
Effective connectivity	Changes	Changes	?	?	Changes	?	Changes

References

Aldini, G. (1804). *Essai theoretique et experimental sur le galvanisme.* Paris: Fournier Fils.

Almaraz, A. C., Dilli, E., & Dodick, D. W. (2010). The effect of prophylactic medications on TMS for migraine aura. *Headache, 50*(10), 1630–1633.

Alon, G., Roys, S. R., Gullapalli, R. P., & Greenspan, J. D. (2011). Non-invasive electrical stimulation of the brain (ESB) modifies the resting-state network connectivity of the primary motor cortex: a proof of concept fMRI study. *Brain Research, 1403*, 37–44.

Amaral, L. A., Diaz-Guilera, A., Moreira, A. A., Goldberger, A. L., & Lipsitz, L. A. (2004). Emergence of complex dynamics in a simple model of signaling networks. *Proceedings of the National Academy of Science the USA, 101*(44), 15551–15555.

Ambrus, G. G., Paulus, W., & Antal, A. (2010). Cutaneous perception thresholds of electrical stimulation methods: comparison of tDCS and tRNS. *Clinical Neurophysiology, 121*(11), 1908–1914.

Angelakis, E., Liouta, E., Andreadis, N., et al. (2013). Transcranial alternating current stimulation reduces symptoms in intractable idiopathic cervical dystonia: a case study. *Neuroscience Letter, 533*, 39–43.

Antal, A., Kincses, T. Z., Nitsche, M. A., Bartfai, O., & Paulus, W. (2004). Excitability changes induced in the human primary visual cortex by transcranial direct current stimulation: direct electrophysiological evidence. *Investigative Ophthalmology & Visual Science, 45*(2), 702–707.

Arnal, L. H., Wyart, V., & Giraud, A. L. (2011). Transitions in neural oscillations reflect prediction errors generated in audiovisual speech. *Nature Neuroscience, 14*(6), 797–801.

Arnal, L. H., Doelling, K. B., & Poeppel, D. (2015). Delta-beta coupled oscillations underlie temporal prediction accuracy. *Cerebral Cortex, 25*(9), 3077–3085.

Arndt, R. (1869). Die electricitat in der psychiatrie. *Archives of Psychiatric Nervenkrankheit, 2*, 259–261.

Arns, M., & De Ridder, D. (2011). Neurofeedback 2.0. *Journal of Neurotherapy, 15*, 91–93.

Bae, E. H., Schrader, L. M., Machii, K., et al. (2007). Safety and tolerability of repetitive transcranial magnetic stimulation in patients with epilepsy: a review of the literature. *Epilepsy Behavior, 10*(4), 521–528.

Barker, A. T., Jalinous, R., & Freeston, I. L. (1985). Non-invasive magnetic stimulation of human motor cortex. *Lancet, 1*(8437), 1106–1107.

Berardelli, A., Inghilleri, M., Rothwell, J. C., et al. (1998). Facilitation of muscle evoked responses after repetitive cortical stimulation in man. *Experimental. Brain Research, 122*(1), 79–84.

Bindman, L. J., Lippold, O. C., & Redfearn, J. W. (1964). The action of brief polarizing currents on the cerebral cortex of the rat (1) during current flow and (2) in the production of long-lasting after-effects. *Journal of Physiology, 172*, 369–382.

Bonnet, C. (1783). *Oeuvres d'Histoire Naturelle et de Philosophie.* Neuchatel: S Fauche.

Borckardt, J. J., Bikson, M., Frohman, H., et al. (2012). A pilot study of the tolerability and effects of high-definition transcranial direct current stimulation (HD-tDCS) on pain perception. *Journal of Pain, 13*(2), 112–120.

Brignani, D., Manganotti, P., Rossini, P. M., & Miniussi, C. (2008). Modulation of cortical oscillatory activity during transcranial magnetic stimulation. *Human Brain Mapping, 29*(5), 603–612.

Brignani, D., Ruzzoli, M., Mauri, P., & Miniussi, C. (2013). Is transcranial alternating current stimulation effective in modulating brain oscillations? *PLoS One, 8*(2), e56589.

Brunoni, A. R., Amadera, J., Berbel, B., Volz, M. S., Rizzerio, B. G., & Fregni, F. (2011). A systematic review on reporting and assessment of adverse effects associated with transcranial direct current stimulation. *International Journal of Neuropsychopharmacology, 14*(8), 1133–1145.

Brunoni, A. R., Nitsche, M. A., Bolognini, N., Bikson, M., Wagner, T., Merabet, L., Edwards, D. J., Valero-Cabre, A., Rotenberg, A., Pascual-Leone, A., Ferricci, R., Priori, A., Boggio, P. S., & Fregni, F. (2012). Clinical research with transcranial direct current stimulation (tDCS): Challenges and future directions. *Brain Stimulation, 5,* 175–195.

Bullmore, E., & Sporns, O. (2012). The economy of brain network organization. *Nature Reviews Neuroscience, 13*(5), 336–349.

Bunzeck, N., Guitart-Masip, M., Dolan, R. J., & Duzel, E. (2011). Contextual novelty modulates the neural dynamics of reward anticipation. *Journal of Neuroscience, 31*(36), 12816–12822.

Bunzeck, N., Doeller, C. F., Dolan, R. J., & Duzel, E. (2012). Contextual interaction between novelty and reward processing within the mesolimbic system. *Human Brain Mapping, 33*(6), 1309–1324.

Buzsaki, G., & Mizuseki, K. (2014). The log-dynamic brain: how skewed distributions affect network operations. *Nature Reviews Neuroscience, 15*(4), 264–278.

Camilleri, R., Pavan, A., Ghin, F., Battaglini, L., & Campana, G. (2014). Improvement of uncorrected visual acuity and contrast sensitivity with perceptual learning and transcranial random noise stimulation in individuals with mild myopia. *Frontier of Psychology, 5,* 1234.

Capone, F., Dileone, M., Profice, P., et al. (2009). Does exposure to extremely low frequency magnetic fields produce functional changes in human brain? *Journal of Neural Transmission (Vienna), 116*(3), 257–265.

Cappelletti, M., Gessaroli, E., Hithersay, R., et al. (2013). Transfer of cognitive training across magnitude dimensions achieved with concurrent brain stimulation of the parietal lobe. *Journal of Neuroscience, 33*(37), 14899–14907.

Chaieb, L., Antal, A., & Paulus, W. (2015). Transcranial random noise stimulation-induced plasticity is NMDA-receptor independent but sodium-channel blocker and benzodiazepines sensitive. *Frontiers of Neuroscience, 9,* 125.

Chan, H. N., Alonzo, A., Martin, D. M., et al. (2012). Treatment of major depressive disorder by transcranial random noise stimulation: case report of a novel treatment. *Biological Psychiatry, 72*(4), e9–e10.

Cheeran, B., Talelli, P., Mori, F., et al. (2008). A common polymorphism in the brain-derived neurotrophic factor gene (BDNF) modulates human cortical plasticity and the response to rTMS. *Journal of Physiology, 586*(Pt 23), 5717–5725.

Chen, R., Classen, J., Gerloff, C., et al. (1997). Depression of motor cortex excitability by low-frequency transcranial magnetic stimulation. *Neurology, 48*(5), 1398–1403.

Chen, L., Zhang, G., Hu, M., & Liang, X. (2015). Eye movement desensitization and reprocessing versus cognitive-behavioral therapy for adult posttraumatic stress disorder: systematic review and meta-analysis. *Journal of Nervous Mental Disorder, 203*(6), 443–451.

Chhatbar, P. Y., Ramakrishnan, V., Kautz, S., George, M. S., Adams, R. J., & Feng, W. (2016). Transcranial direct current stimulation post-stroke upper extremity motor recovery studies exhibit a dose-response relationship. *Brain Stimulation, 9*(1), 16–26.

Chib, V. S., Yun, K., Takahashi, H., & Shimojo, S. (2013). Noninvasive remote activation of the ventral midbrain by transcranial direct current stimulation of prefrontal cortex. *Translational Psychiatry, 3,* e268.

Clarke, L. E., & Barres, B. A. (2013). Emerging roles of astrocytes in neural circuit development. *Nature Reviews Neuroscience, 14*(5), 311–321.

Cook, C. M., Thomas, A. W., & Prato, F. S. (2004). Resting EEG is affected by exposure to a pulsed ELF magnetic field. *Bioelectromagnetics, 25*(3), 196–203.

Cook, C. M., Thomas, A. W., Keenliside, L., & Prato, F. S. (2005). Resting EEG effects during exposure to a pulsed ELF magnetic field. *Bioelectromagnetics, 26*(5), 367–376.

Dammekens, E., Vanneste, S., Ost, J., & De Ridder, D. (2014). Neural correlates of high frequency repetitive transcranial magnetic stimulation improvement in post-stroke non-fluent aphasia: a case study. *Neurocase, 20*(1), 1–9.

Datta, A., Bansal, V., Diaz, J., Patel, J., Reato, D., & Bikson, M. (2009). Gyri-precise head model of transcranial direct current stimulation: improved spatial focality using a ring electrode versus conventional rectangular pad. *Brain Stimulation, 2*(4), 201–207.

Day, B. L., Rothwell, J. C., Thompson, P. D., et al. (1987). Motor cortex stimulation in intact man. 2. Multiple descending volleys. *Brain, 110*(Pt 5), 1191–1209.

De Ridder, D., Vanneste, S., Engineer, N. D., & Kilgard, M. P. (2014). Safety and efficacy of vagus nerve stimulation paired with tones for the treatment of tinnitus: a case series. *Neuromodulation, 17*(2), 170–179.

De Ridder, D., Vanneste, S., Langguth, B., & Llinas, R. (2015a). Thalamocortical dysrhythmia: a theoretical update in tinnitus. *Frontier in Neurology, 6*, 124.

De Ridder, D., Kilgard, M., Engineer, N., & Vanneste, S. (2015b). Placebo-controlled vagus nerve stimulation paired with tones in a patient with refractory tinnitus: a case report. *Otology & Neurotology, 36*(4), 575–580.

Dieckhofer, A., Waberski, T. D., Nitsche, M., Paulus, W., Buchner, H., & Gobbele, R. (2006). Transcranial direct current stimulation applied over the somatosensory cortex—differential effect on low and high frequency SEPs. *Clinical Neurophysiology, 117*(10), 2221–2227.

Dmochowski, J. P., Datta, A., Bikson, M., Su, Y., & Parra, L. C. (2011). Optimized multi-electrode stimulation increases focality and intensity at target. *Journal of Neural Engineering, 8*(4), 046011.

Donne, A., D. Nascimento, T., Lawrence, M., Gupta, V., Zieba, T, Truong, D.Q., et al. (2015). High-Definition and Non-invasive Brain Modulation of Pain and Motor Dysfunction in Chronic TMD. *Brain Stimulation. 8*(6), 1085–1092.

Dymond, A. M., Coger, R. W., & Serafetinides, E. A. (1975). Intracerebral current levels in man during electrosleep therapy. *Biological Psychiatry, 10*(1), 101–104.

Elsner, B., Kugler, J., Pohl, M., & Mehrholz, J. (2015). Transcranial direct current stimulation (tDCS) for improving aphasia in patients with aphasia after stroke. *Cochrane Database Systems Reviews, 5*, CD009760.

Engineer, N. D., Riley, J. R., Seale, J. D., et al. (2011). Reversing pathological neural activity using targeted plasticity. *Nature, 470*(7332), 101–104.

Fertonani, A., Pirulli, C., & Miniussi, C. (2011). Random noise stimulation improves neuroplasticity in perceptual learning. *Journal of Neuroscience, 31*(43), 15416–15423.

Fertonani, A., Ferrari, C., & Miniussi, C. (2015). What do you feel if I apply transcranial electric stimulation? Safety, sensations and secondary induced effects. *Clinical Neurophysiology, 126*(11), 2181–2188.

Fleming, M. K., Sorinola, I. O., Newham, D. J., Roberts-Lewis, S. F., & Bergmann, J. H. (2012). The effect of coil type and navigation on the reliability of transcranial magnetic stimulation. *IEEE Transactions on Neural Systems and Rehabilitation Engineering, 20*(5), 617–625.

Fox, M. D., Buckner, R. L., White, M. P., Greicius, M. D., & Pascual-Leone, A. (2012). Efficacy of transcranial magnetic stimulation targets for depression is related to intrinsic functional connectivity with the subgenual cingulate. *Biological Psychiatry, 72*(7), 595–603.

Freeman, W. J. (2006). A cinematographic hypothesis of cortical dynamics in perception. *International Journal of Psychophysiology, 60*(2), 149–161.

Fregni, F., Nitsche, M. A., Loo, C. K., et al. (2015). Regulatory considerations for the clinical and research use of transcranial direct current stimulation (tDCS): review and recommendations from an expert panel. *Clinical Research and Regulatory Affairs, 32*(1), 22–35.

Fuchs, E., & Flugge, G. (2014). Adult neuroplasticity: more than 40 years of research. *Neural Plast, 2014*, 541870.

Fuggetta, G., Pavone, E. F., Fiaschi, A., & Manganotti, P. (2008). Acute modulation of cortical oscillatory activities during short trains of high-frequency repetitive transcranial magnetic stimulation of the human motor cortex: a combined EEG and TMS study. *Human Brain Mapping, 29*(1), 1–13.

Fujiyama, H., Hyde, J., Hinder, M. R., et al. (2014). Delayed plastic responses to anodal tDCS in older adults. *Frontiers in Aging Neuroscience, 6*, 115.

Galvani, L. (1791). *De viribus electricitatis in motu muscolari animalium*. Bononiae: Bologna.

Galvani, L. (1797). Memorie sull 'elettricita' animale In L. Spallanzani (Ed.), Theoria: Bologna.

Gleick, J. (2011). *The Information*. London: HarperCollins.

Grefkes, C., Nowak, D. A., Wang, L. E., Dafotakis, M., Eickhoff, S. B., & Fink, G. R. (2010). Modulating cortical connectivity in stroke patients by rTMS assessed with fMRI and dynamic causal modeling. *Neuroimage, 50*(1), 233–242.

Guleyupoglu, B., Schestatsky, P., Edwards, D., Fregni, F., & Bikson, M. (2013). Classification of methods in transcranial electrical stimulation (tES) and evolving strategy from historical approaches to contemporary innovations. *Journal of Neuroscience Methods, 219*(2), 297–311.

Hebb, D. (1949). *The organization of behavior; a neuropsychological theory*. New York: Wiley-Interscience.

Heimrath, K., Breitling, C., Krauel, K., Heinze, H. J., & Zaehle, T. (2015). Modulation of pre-attentive spectro-temporal feature processing in the human auditory system by HD-tDCS. *European Journal of Neuroscience, 41*(12), 1580–1586.

Helfrich, R. F., Knepper, H., Nolte, G., et al. (2014). Selective modulation of interhemispheric functional connectivity by HD-tACS shapes perception. *PLoS Biology, 12*(12), e1002031.

Herkt, D., Tumani, V., Gron, G., Kammer, T., Hofmann, A., & Abler, B. (2014). Facilitating access to emotions: neural signature of EMDR stimulation. *PLoS One, 9*(8), e106350.

Herpich, F., Melnick, M., Huxlin, K., Tadin, D., Agosta, S., & Battelli, L. (2015). Transcranial random noise stimulation enhances visual learning in healthy adults. *Journal of Vision, 15*(12), 40.

Holland, J. (2014). *Complexity*. Oxford: Oxford University Press.

Hsu, W. Y., Ku, Y., Zanto, T. P., & Gazzaley, A. (2015). Effects of noninvasive brain stimulation on cognitive function in healthy aging and Alzheimer's disease: a systematic review and meta-analysis. *Neurobiology of Aging, 36*(8), 2348–2359.

Iyer, M. B., Mattu, U., Grafman, J., Lomarev, M., Sato, S., & Wassermann, E. M. (2005). Safety and cognitive effect of frontal DC brain polarization in healthy individuals. *Neurology, 64*(5), 872–875.

James, W. (1890). *The Priciples of Psychology* (Paperback Dover ed.). New York: Henry Holt and Company.

Jansen, J. M., Daams, J. G., Koeter, M. W., Veltman, D. J., van den Brink, W., & Goudriaan, A. E. (2013). Effects of non-invasive neurostimulation on craving: a meta-analysis. *Neuroscience & Biobehavorial Reviews, 37*(10 Pt 2), 2472–2480.

Johnson, N. (2010). *Simply complexity: A clear guide to complexity theory* (2nd ed.). Oxford: Oneworld Publications.

Joos, K., De Ridder, D., & Vanneste, S. (2015). The differential effect of low- versus high-frequency random noise stimulation in the treatment of tinnitus. *Experimental Brain Research, 233*(5), 1433–1440.

Jung, N. H., Delvendahl, I., Kuhnke, N. G., Hauschke, D., Stolle, S., & Mall, V. (2010). Navigated transcranial magnetic stimulation does not decrease the variability of motor-evoked potentials. *Brain Stimulation, 3*(2), 87–94.

Kanai, R., Chaieb, L., Antal, A., Walsh, V., & Paulus, W. (2008). Frequency-dependent electrical stimulation of the visual cortex. *Current Biology, 18*(23), 1839–1843.

Keck, M. E., Sillaber, I., Ebner, K., et al. (2000). Acute transcranial magnetic stimulation of frontal brain regions selectively modulates the release of vasopressin, biogenic amines and amino acids in the rat brain. *European Journal of Neuroscience, 12*(10), 3713–3720.

Keck, M. E., Welt, T., Post, A., et al. (2001). Neuroendocrine and behavioral effects of repetitive transcranial magnetic stimulation in a psychopathological animal model are suggestive of antidepressant-like effects. *Neuropsychopharmacology, 24*(4), 337–349.

Keck, M. E., Welt, T., Muller, M. B., et al. (2002). Repetitive transcranial magnetic stimulation increases the release of dopamine in the mesolimbic and mesostriatal system. *Neuropharmacology, 43*(1), 101–109.

Keeser, D., Meindl, T., Bor, J., et al. (2011a). Prefrontal transcranial direct current stimulation changes connectivity of resting-state networks during fMRI. *Journal of Neuroscience, 31*(43), 15284–15293.

Keeser, D., Padberg, F., Reisinger, E., et al. (2011b). Prefrontal direct current stimulation modulates resting EEG and event-related potentials in healthy subjects: a standardized low resolution tomography (sLORETA) study. *Neuroimage, 55*(2), 644–657.

Kelloway, P. (1946). The part played by the electrical fish in the early history of bioelectricity and electrotherapy. *Bulletin of the History of Medicine, 20*, 112–137.

Keshner, M. S. (1982). 1/f Noise. *Proceedings of the IEEE, 70*(3), 212–218.

Kilgard, M. P., & Merzenich, M. M. (1998). Cortical map reorganization enabled by nucleus basalis activity. *Science, 279*(5357), 1714–1718.

Kimbrell, T. A., Little, J. T., Dunn, R. T., et al. (1999). Frequency dependence of antidepressant response to left prefrontal repetitive transcranial magnetic stimulation (rTMS) as a function of baseline cerebral glucose metabolism. *Biological Psychiatry, 46*(12), 1603–1613.

Kimbrell, T. A., Dunn, R. T., George, M. S., et al. (2002). Left prefrontal-repetitive transcranial magnetic stimulation (rTMS) and regional cerebral glucose metabolism in normal volunteers. *Psychiatry Research, 115*(3), 101–113.

Kuo, H. I., Bikson, M., Datta, A., et al. (2013). Comparing cortical plasticity induced by conventional and high-definition 4 × 1 ring tDCS: a neurophysiological study. *Brain Stimulation, 6*(4), 644–648.

Lang, N., Siebner, H. R., Ward, N. S., et al. (2005). How does transcranial DC stimulation of the primary motor cortex alter regional neuronal activity in the human brain? *European Journal of Neuroscience, 22*(2), 495–504.

Lang, N., Sueske, E., Hasan, A., Paulus, W., & Tergau, F. (2006). Pregabalin exerts oppositional effects on different inhibitory circuits in human motor cortex: a double-blind, placebo-controlled transcranial magnetic stimulation study. *Epilepsia, 47*(5), 813–819.

Lang, N., Rothkegel, H., Peckolt, H., & Deuschl, G. (2013). Effects of lacosamide and carbamazepine on human motor cortex excitability: a double-blind, placebo-controlled transcranial magnetic stimulation study. *Seizure, 22*(9), 726–730.

Lee, C. W., & Cuijpers, P. (2014). What does the data say about the importance of eye movement in EMDR? *Journal of Behavior Therapy and Experimental Psychiatry, 45*(1), 226–228.

Lefaucheur, J. P., Andre-Obadia, N., Antal, A., et al. (2014). Evidence-based guidelines on the therapeutic use of repetitive transcranial magnetic stimulation (rTMS). *Clinical Neurophysiology, 125*(11), 2150–2206.

Liebetanz, D., Koch, R., Mayenfels, S., Konig, F., Paulus, W., & Nitsche, M. A. (2009). Safety limits of cathodal transcranial direct current stimulation in rats. *Clinical Neurophysiology, 120*(6), 1161–1167.

Lippold, O. C., & Redfearn, J. W. (1964). Mental changes resulting from the passage of small direct currents through the human brain. *British Journal of Psychiatry, 110*, 768–772.

Llinas, R. R., & Roy, S. (2009). The 'prediction imperative' as the basis for self-awareness. *Philosophical transactions of the Royal Society of London. Series B, Biological Sciences, 364*(1521), 1301–1307.

Luft, C. D., Pereda, E., Banissy, M. J., & Bhattacharya, J. (2014). Best of both worlds: promise of combining brain stimulation and brain connectome. *Frontiers in Systems Neuroscience, 8*, 132.

Malaguti, A., Rossini, D., Lucca, A., et al. (2011). Role of COMT, 5-HT(1A), and SERT genetic polymorphisms on antidepressant response to Transcranial Magnetic Stimulation. *Depression & Anxiety, 28*(7), 568–573.

Marshall, L., Helgadottir, H., Molle, M., & Born, J. (2006). Boosting slow oscillations during sleep potentiates memory. *Nature, 444*(7119), 610–613.

Martiny, K., Lunde, M., & Bech, P. (2010). Transcranial low voltage pulsed electromagnetic fields in patients with treatment-resistant depression. *Biological Psychiatry, 68*(2), 163–169.

Massimini, M., Ferrarelli, F., Huber, R., Esser, S. K., Singh, H., & Tononi, G. (2005). Breakdown of cortical effective connectivity during sleep. *Science, 309*(5744), 2228–2232.

Matsunaga, K., Nitsche, M. A., Tsuji, S., & Rothwell, J. C. (2004). Effect of transcranial DC sensorimotor cortex stimulation on somatosensory evoked potentials in humans. *Clinical Neurophysiology, 115*(2), 456–460.

Medeiros, L. F., de Souza, I. C., Vidor, L. P., et al. (2012). Neurobiological effects of transcranial direct current stimulation: a review. *Frontier Psychiatry, 3*, 110.

Mehta, S., McIntyre, A., Guy, S., Teasell, R. W., & Loh, E. (2015). Effectiveness of transcranial direct current stimulation for the management of neuropathic pain after spinal cord injury: a meta-analysis. *Spinal Cord, 53*(11), 780–785.

Meinzer, M., Antonenko, D., Lindenberg, R., et al. (2012). Electrical brain stimulation improves cognitive performance by modulating functional connectivity and task-specific activation. *Journal of Neuroscience, 32*(5), 1859–1866.

Meinzer, M., Lindenberg, R., Antonenko, D., Flaisch, T., & Floel, A. (2013). Anodal transcranial direct current stimulation temporarily reverses age-associated cognitive decline and functional brain activity changes. *Journal of Neuroscience, 33*(30), 12470–12478.

Merleau-Ponty, M. (1945). *Phénomènologie de la Perception* (English edition ed.). Paris: Gallimard.

Meron, D., Hedger, N., Garner, M., & Baldwin, D. S. (2015). Transcranial direct current stimulation (tDCS) in the treatment of depression: Systematic review and meta-analysis of efficacy and tolerability. *Neuroscience & Biobehavioral Reviews, 57*, 46–62.

Mill, J. S. (1843). On The Composition Of Causes. In *A System Of Logic, Ratiocinative And Inductive* (Vol 3). New York: Harper and Brothers.

Moliadze, V., Antal, A., & Paulus, W. (2010). Boosting brain excitability by transcranial high frequency stimulation in the ripple range. *Journal of Physiology, 588*(Pt 24), 4891–4904.

Moliadze, V., Atalay, D., Antal, A., & Paulus, W. (2012). Close to threshold transcranial electrical stimulation preferentially activates inhibitory networks before switching to excitation with higher intensities. *Brain Stimulation, 5*(4), 505–511.

Moreno-Duarte, I., Gebodh, N., Schestatsky, P., et al. (2014). Transcranial electrical stimulation: transcranial direct current stimulation (tDCS), transcranial alternating current stimulation (tACS), transcranial pulsed current stimulation (tPCS), and transcranial random noise stimulation (tRNS). In R. C. Kadosh (Ed.), *The stimulated brain: Cognitive enhancement using non-invasive brain stimulation*. London: Academia Press.

Mower, G. D. (1976). Perceived intensity of peripheral thermal stimuli is independent of internal body temperature. *Journal of Comparative Psychology, 90*(12), 1152–1155.

Mulquiney, P. G., Hoy, K. E., Daskalakis, Z. J., & Fitzgerald, P. B. (2011). Improving working memory: exploring the effect of transcranial random noise stimulation and transcranial direct current stimulation on the dorsolateral prefrontal cortex. *Clinical Neurophysiology, 122*(12), 2384–2389.

Nikolin, S., Loo, C. K., Bai, S., Dokos, S., & Martin, D. M. (2015). Focalised stimulation using high definition transcranial direct current stimulation (HD-tDCS) to investigate declarative verbal learning and memory functioning. *Neuroimage, 117*, 11–19.

Nitsche, M. A., & Paulus, W. (2001). Sustained excitability elevations induced by transcranial DC motor cortex stimulation in humans. *Neurology, 57*(10), 1899–1901.

Nitsche, M. A., Liebetanz, D., Antal, A., Lang, N., Tergau, F., & Paulus, W. (2003). Modulation of cortical excitability by weak direct current stimulation—technical, safety and functional aspects. *Supplements to Clinical Neurophysiology, 56*, 255–276.

Nitsche, M. A., Cohen, L. G., Wassermann, E. M., et al. (2008). Transcranial direct current stimulation: State of the art 2008. *Brain Stimulation, 1*(3), 206–223.

Nitsche, M. A., Kuo, M., Paulus, W., & Antal, A. (2015). Transcranial direct current stimulation: protocols and physiological mechanisms of action. In H. Knotkova, & D. Rasche (Eds.), *Textbook of neuromodulation: Principles, methods and clinical application*. New York: Springer.

O'Connell, N. E., Wand, B. M., Marston, L., Spencer, S., & Desouza, L. H. (2014). Non-invasive brain stimulation techniques for chronic pain. *Cochrane Database Systems Reviews, 4*, CD008208.

Pagani, M., Di Lorenzo, G., Verardo, A. R., et al. (2012). Neurobiological correlates of EMDR monitoring - an EEG study. *PLoS One, 7*(9), e45753.

Palm, U., Feichtner, K. B., Hasan, A., et al. (2014). The role of contact media at the skin-electrode interface during transcranial direct current stimulation (tDCS). *Brain stimulation, 7*(5), 762–764.

Palm, U., Hasan, A., Keeser, D., Falkai, P., & Padberg, F. (2013). Transcranial random noise stimulation for the treatment of negative symptoms in schizophrenia. *Schizophrenia Research, 146*(1–3), 372–373.

Pascual-Leone, A., Grafman, J., & Hallett, M. (1994). Modulation of cortical motor output maps during development of implicit and explicit knowledge. *Science, 263*(5151), 1287–1289.

Paulus, W. (2011). Transcranial electrical stimulation (tES–tDCS; tRNS, tACS) methods. *Neuropsychology Rehabilitation, 21*(5), 602–617.

Paus, T., Sipila, P. K., & Strafella, A. P. (2001). Synchronization of neuronal activity in the human primary motor cortex by transcranial magnetic stimulation: an EEG study. *Journal of Neurophysiology, 86*(4), 1983–1990.

Pena-Gomez, C., Sala-Lonch, R., Junque, C., et al. (2012). Modulation of large-scale brain networks by transcranial direct current stimulation evidenced by resting-state functional MRI. *Brain Stimulation, 5*(3), 252–263.

Pisegna, J. M., Kaneoka, A., Pearson, W. G., Jr., Kumar, S., & Langmore, S. E. (2016). Effects of non-invasive brain stimulation on post-stroke dysphagia: a systematic review and meta-analysis of randomized controlled trials. *Clinical Neurophysiology, 127*(1), 956–968.

Polania, R., Nitsche, M. A., & Paulus, W. (2011). Modulating functional connectivity patterns and topological functional organization of the human brain with transcranial direct current stimulation. *Human Brain Mapping, 32*(8), 1236–1249.

Polania, R., Paulus, W., & Nitsche, M. A. (2012a). Modulating cortico-striatal and thalamo-cortical functional connectivity with transcranial direct current stimulation. *Human Brain Mapping, 33*(10), 2499–2508.

Polania, R., Nitsche, M. A., Korman, C., Batsikadze, G., & Paulus, W. (2012b). The importance of timing in segregated theta phase-coupling for cognitive performance. *Current Biology, 22*(14), 1314–1318.

Poreisz, C., Boros, K., Antal, A., & Paulus, W. (2007). Safety aspects of transcranial direct current stimulation concerning healthy subjects and patients. *Brain Research Bulletin, 72*(4–6), 208–214.

Radman, T., Ramos, R. L., Brumberg, J. C., & Bikson, M. (2009). Role of cortical cell type and morphology in subthreshold and suprathreshold uniform electric field stimulation in vitro. *Brain Stimulation, 2*(4), 215–228228 e211–213.

Ranck, J. B., Jr. (1975). Which elements are excited in electrical stimulation of mammalian central nervous system: a review. *Brain Research, 98*(3), 417–440.

Redfearn, J. W., Lippold, O. C., & Costain, R. (1964). A preliminary account of the clinical effects of polarizing the brain in certain psychiatric disorders. *British Journal of Psychiatry, 110,* 773–785.

Richards, T. L., Lappin, M. S., Acosta-Urquidi, J., et al. (1997). Double-blind study of pulsing magnetic field effects on multiple sclerosis. *The Journal of Alternative and Complementary Medicine, 3*(1), 21–29.

Ridding, M. C., & Rothwell, J. C. (2007). Is there a future for therapeutic use of transcranial magnetic stimulation? *Nature Reviews Neuroscience, 8*(7), 559–567.

Rodenburg, R., Benjamin, A., de Roos, C., Meijer, A. M., & Stams, G. J. (2009). Efficacy of EMDR in children: a meta-analysis. *Clinical Psychology Review, 29*(7), 599–606.

Roland, N. J., Hughes, J. B., Daley, M. B., Cook, J. A., Jones, A. S., & McCormick, M. S. (1993). Electromagnetic stimulation as a treatment of tinnitus: a pilot study. *Clinical Otolaryngology Allied Science, 18*(4), 278–281.

Romanska, A., Rezlescu, C., Susilo, T., Duchaine, B., & Banissy, M. J. (2015). High-Frequency Transcranial Random Noise Stimulation Enhances Perception of Facial Identity. *Cerebral Cortex, 25*(11), 4334–4340.

Roopun, A. K., Middleton, S. J., Cunningham, M. O., et al. (2006). A beta2-frequency (20–30 Hz) oscillation in nonsynaptic networks of somatosensory cortex. *Proceedings of the National Academy of Sciences of the USA, 103*(42), 15646–15650.

Rosenzweig, M. (2007). Modification of brain circuits through experience. In F. Bermúdez-Rattoni (Ed.), *Neural plasticity and memory: From genes to brain imaging.* Boca Raton, Florida, USA: CRC Press.

Rueger, M. A., Keuters, M. H., Walberer, M., et al. (2012). Multi-session transcranial direct current stimulation (tDCS) elicits inflammatory and regenerative processes in the rat brain. *PLoS One, 7*(8), e43776.

Ruohonen, J., & Karhu, J. (2012). tDCS possibly stimulates glial cells. *Clinical Neurophysiology, 123*(10), 2006–2009.

Rush, S., & Driscoll, D. A. (1968). Current distribution in the brain from surface electrodes. *Anesthesia & Analgesia, 47*(6), 717–723.

San-Juan, D., Morales-Quezada, L., Orozco Garduno, A. J., et al. (2015). Transcranial Direct Current Stimulation in Epilepsy. *Brain Stimulation, 8*(3), 455–464.

Santarnecchi, E., Polizzotto, N. R., Godone, M., et al. (2013). Frequency-dependent enhancement of fluid intelligence induced by transcranial oscillatory potentials. *Current Biology, 23*(15), 1449–1453.

Shapiro, F. (1996). Eye movement desensitization and reprocessing (EMDR): evaluation of controlled PTSD research. *Journal of Behavior Therapy and Experimental Psychiatry, 27*(3), 209–218.

Shekhawat, G. S., Sundram, F., Bikson, M., et al. (2015). Intensity, duration, and location of high-definition transcranial direct current stimulation for tinnitus relief. *Neurorehabilitation and Neural Repair, 30*(4), 349–359.

Song, J. J., Vanneste, S., Van de Heyning, P., & De Ridder, D. (2012). Transcranial direct current stimulation in tinnitus patients: a systemic review and meta-analysis. *ScientificWorld Journal, 2012,* 427941.

Speer, A. M., Kimbrell, T. A., Wassermann, E. M., et al. (2000). Opposite effects of high and low frequency rTMS on regional brain activity in depressed patients. *Biological Psychiatry, 48*(12), 1133–1141.

Sporns, O., Chialvo, D. R., Kaiser, M., & Hilgetag, C. C. (2004). Organization, development and function of complex brain networks. *Trends in Cognitive Sciences, 8*(9), 418–425.

Stagg, C. J., Lin, R. L., Mezue, M., et al. (2013). Widespread modulation of cerebral perfusion induced during and after transcranial direct current stimulation applied to the left dorsolateral prefrontal cortex. *Journal of Neuroscience, 33*(28), 11425–11431.

Stephani, C., Nitsche, M. A., Sommer, M., & Paulus, W. (2011). Impairment of motor cortex plasticity in Parkinson's disease, as revealed by theta-burst-transcranial magnetic stimulation and transcranial random noise stimulation. *Parkinsonism & Related Disorders, 17*(4), 297–298.

Stramaccia, D. F., Penolazzi, B., Sartori, G., Braga, M., Mondini, S., & Galfano, G. (2015). Assessing the effects of tDCS over a delayed response inhibition task by targeting the right inferior frontal gyrus and right dorsolateral prefrontal cortex. *Experimental Brain Research, 233*(8), 2283–2290.

Terney, D., Chaieb, L., Moliadze, V., Antal, A., & Paulus, W. (2008). Increasing human brain excitability by transcranial high-frequency random noise stimulation. *Journal of Neuroscience, 28*(52), 14147–14155.

Thut, G., Schyns, P. G., & Gross, J. (2011). Entrainment of perceptually relevant brain oscillations by non-invasive rhythmic stimulation of the human brain. *Frontiers in Psychology, 2*, 170.

Tyler, S., Conto, F., & Battelli, L. (2015). Rapid effect of high-frequency tRNS over the parietal lobe during a temporal perceptual learning task. *Journal of Vision, 15*(12), 393.

van Veen, S. C., van Schie, K., Wijngaards-de Meij, L. D., Littel, M., Engelhard, I. M., & van den Hout, M. A. (2015). Speed matters: relationship between speed of eye movements and modification of aversive autobiographical memories. *Frontiers in Psychiatry, 6*, 45.

Vanneste, S., & De Ridder, D. (2011). Bifrontal transcranial direct current stimulation modulates tinnitus intensity and tinnitus-distress-related brain activity. *The European Journal of Neuroscience, 34*(4), 605–614.

Vanneste, S., Fregni, F., & De Ridder, D. (2013). Head-to-head comparison of transcranial random noise stimulation, transcranial ac stimulation, and transcranial DC stimulation for tinnitus. *Frontier Psychiatry, 4*, 158.

Villamar, M. F., Volz, M. S., Bikson, M., Datta, A., Dasilva, A. F., & Fregni, F. (2013). Technique and considerations in the use of 4 × 1 ring high-definition transcranial direct current stimulation (HD-tDCS). *Journal of Visualized Experiments*(77), e50309.

Vossen, A., Gross, J., & Thut, G. (2015). Alpha power increase after transcranial alternating current stimulation at alpha frequency (alpha-tACS) reflects plastic changes rather than entrainment. *Brain Stimulation, 8*(3), 499–508.

Wagner, T., Fregni, F., Fecteau, S., Grodzinsky, A., Zahn, M., & Pascual-Leone, A. (2007). Transcranial direct current stimulation: a computer-based human model study. *Neuroimage, 35*(3), 1113–1124.

Walsh, V., & Rushworth, M. (1999). A primer of magnetic stimulation as a tool for neuropsychology. *Neuropsychologia, 37*(2), 125–135.

Wang, H., Wang, X., & Scheich, H. (1996). LTD and LTP induced by transcranial magnetic stimulation in auditory cortex. *Neuroreport, 7*(2), 521–525.

Wang, H., Wang, X., Wetzel, W., & Scheich, H. (1999). Rapid-rate transcranial magnetic stimulation in auditory cortex induces LTP and LTD and impairs discrimination learning of frequency-modulated tones. *Electroencephalography and Clinical Neurophysiology Supplements, 51*, 361–367.

Warraich, Z., & Kleim, J. A. (2010). Neural plasticity: the biological substrate for neurorehabilitation. *PMR, 2*(12 Suppl 2), S208–219.

Weber, M. J., Messing, S. B., Rao, H., Detre, J. A., & Thompson-Schill, S. L. (2014). Prefrontal transcranial direct current stimulation alters activation and connectivity in cortical and subcortical reward systems: a tDCS-fMRI study. *Human Brain Mapping, 35*(8), 3673–3686.

Wiesel, T. N., & Hubel, D. H. (1963). Effects of visual deprivation on morphology and physiology of cells in the cats lateral geniculate body. *Jornal of Neurophysiology, 26*, 978–993.

Wikipedia. (2015). Emergence.

Witkowski, M., Garcia-Cossio, E., Chander, B. S., et al. (2016). Mapping entrained brain oscillations during transcranial alternating current stimulation (tACS). *Neuroimage, 140*, 89–98.

Yue, L., Xiao-lin, H., & Tao, S. (2009). The effects of chronic repetitive transcranial magnetic stimulation on glutamate and gamma-aminobutyric acid in rat brain. *Brain Research, 1260*, 94–99.

Zaehle, T., Rach, S., & Herrmann, C. S. (2010). Transcranial alternating current stimulation enhances individual alpha activity in human EEG. *PLoS One, 5*(11), e13766.

Zaehle, T., Beretta, M., Jancke, L., Herrmann, C. S., & Sandmann, P. (2011). Excitability changes induced in the human auditory cortex by transcranial direct current stimulation: direct electrophysiological evidence. *Experimental Brain Research, 215*(2), 135–140.

Zaehle, T., Bauch, E. M., Hinrichs, H., et al. (2013). Nucleus accumbens activity dissociates different forms of salience: evidence from human intracranial recordings. *Journal of Neuroscience, 33*(20), 8764–8771.

Zaghi, S., Acar, M., Hultgren, B., Boggio, P. S., & Fregni, F. (2010a). Noninvasive brain stimulation with low-intensity electrical currents: putative mechanisms of action for direct and alternating current stimulation. *Neuroscientist, 16*(3), 285–307.

Zaghi, S., de Freitas Rezende, L., de Oliveira, L. M., et al. (2010b). Inhibition of motor cortex excitability with 15Hz transcranial alternating current stimulation (tACS). *Neuroscience Letter, 479*(3), 211–214.

Zago, S., Ferrucci, R., Fregni, F., & Priori, A. (2008). Bartholow, Sciamanna, Alberti: pioneers in the electrical stimulation of the exposed human cerebral cortex. *Neuroscientist, 14*(5), 521–528.

Further Reading

De Ridder, D., Verplaetse, J., & Vanneste, S. (2013). The predictive brain and the "free will" illusion. *Frontiers in Psychology, 4*, 131.

CHAPTER 8

Toward a Frequency-based Theory of Neurofeedback

Siegfried Othmer, Susan F. Othmer
The EEG Institute, Woodland Hills, CA, United States

"Rhythmical oscillations are archetypes of time-dependent behavior in nature." J.A. Scott-Kelso

This book is mainly concerned with techniques of low-level stimulation to effect neuromodulation for therapeutic purposes. Similar objectives can be achieved through the use of a learning paradigm, one that is typically based on a reinforcement strategy—although that is not essential. That general approach is now called neurofeedback, which is the topic of this chapter for a complementary perspective. Neurofeedback belongs in the domain of biofeedback, with the distinction that it utilizes the EEG as a training variable. In fact, it used to be called EEG biofeedback in its early days. It will be seen that it is just as strongly frequency-based as the stimulation technologies.

There is a great deal of methodological commonality among the different methods of biofeedback, and there is a great deal of overlap in what they accomplish clinically. On the other hand, there are also quite significant differences, and the two disciplines, biofeedback and neurofeedback, have developed somewhat independently over the last half-century. Familiarity with neurofeedback makes this unsurprising. That is because it is difficult to overestimate the advantages conferred on neurofeedback—with respect to other biofeedback techniques—by virtue of its appeal to the frequency basis of the organization of cerebral communication.

It is the frequency basis of neurofeedback that gives it its sensitivity, its specificity, its flexibility, its breadth of application, and its ability to reveal aspects of brain functional organization. This feature also gives neurofeedback an extraordinary advantage vis-à-vis the neuroscientist in the brain laboratory. By putting the brain in the feedback loop, the brain becomes an active agent in the process. It is both an exquisitely sensitive detector of frequency-based signals

Rhythmic Stimulation Procedures in Neuromodulation. http://dx.doi.org/10.1016/B978-0-12-803726-3.00008-0

and a strong responder to brain-derived, spectrally specific information. What serves us well in the task of remediation also serves us well when it comes to the scientific exploration of brain function—and of its dysfunction.

Throughout history, we have learned about function by way of dysfunction. Sometimes the dysfunction is even deliberately introduced, as in lesion studies. Sometimes we are simply opportunistic. Surgery in cases of epilepsy provided opportunities for probing cortical function in human beings that were not otherwise available. Hans Berger took advantage of trepanation of head-injured patients to observe the EEG without attenuation by the intervening skull. Sigmund Freud's core interest was the functioning organism, but it was revealed to him largely through dysfunction. Oliver Sacks relished his odd-ball cases for the light they shed on how brain function must be organized. In neurofeedback, by contrast, we get to observe the brain under the most favorable of circumstances, a state that calls for mere engagement rather than overt challenge.

THE FREQUENCY BASIS OF THE ORGANIZATION OF NATURAL SYSTEMS

As for the frequency basis of brain functional organization, matters could hardly be otherwise. Perhaps this signifies a lack of imagination, but once one obtains insight into how the brain actually manages its affairs, one wonders how it could have been done differently. Albert Einstein considered the comprehensibility of nature as nonobvious, which is surprising in view of the fact that many things were obvious to him that were not obvious to his contemporaries or to us. In any event, nature becomes comprehensible to us through the language of mathematics. The laws of nature rest upon a mathematical foundation, one that can lay the strongest claim to obviousness among all of the sciences.

Nature favors us in yet another respect, which is that the laws of nature tend to be simple—even though the actual execution may be complex. Nature exploits the simplest concepts to the fullest. Unsurprisingly it was a physicist who thought nature's laws were simple, because physicists found their bearings with simple systems. A similar sense of simple geometrical order in nature had already inspired the ancients, the Pythagoreans, more than two millennia ago.

Our understanding of nature grew with our understanding of mathematics, and was always constrained by it. Jointly these understandings even impinged on social change. We stand before such an age of rapid social change presently as we begin to exploit the potential of recovery and enhancement of brain function, a notion that has only recently come to be accepted. As we stand at this major threshold of our human experience, it is worthwhile to recapitulate how we reached this point.

With Isaac Newton, we got the law of inertia. A single body moving in an empty universe continues on its path indefinitely and without alteration. That gave us stasis, or at least immutability. Introducing a second body into the mix gives us one of two situations. Either the two pass each other in the night, never to meet again, or they go into orbit around their mutual center of gravity. This orbit closes on itself and repeats endlessly and identically. With only two bodies, we have periodicity, the endless repetition of the same pattern. We have frequency.

The orbits obey a simple mathematical or geometrical description. They are ellipses, as first shown by Johannes Kepler, and ellipses are conic sections. These are the intersections of a plane with a cone. If the plane is parallel to the base of the cone, we have a circular orbit. But for a given size there is only one way to get a circular orbit, whereas there is an infinity of ways we can have an elliptic orbit with a given axis. So, that is what we actually get to see in nature—the infinite variety rather than the singular exception. The periodicity, however, is there in either case, and it is maintained in perpetuity.

The Newtonian view of space is the one bequeathed to us by Euclid. The universe is flat. The orbit of a two-body system lies in a plane, and Descartes is given credit for placing a coordinate system on such a plane, which brought together geometry and algebra. In the orbital motion we see both movement and perpetual repetition, or stasis. However, until Newton's lifetime, the defining feature of orbital motion was its stability, its stasis, rather than its movement, and that was reflected strongly in the society of the day, one that was deemed to reflect the natural order—with the emphasis on stasis.

It was the parallel but independent development of calculus by Leibniz and Newton, following on the heels of Descartes, that first gave mathematicians the ability to handle dynamical relationships. This development was massively resisted by the power structure of the day because it loosened the moorings underpinning the static political, social, and religious hierarchy as an ideal. The reality on the ground was already in turmoil, with the Reformation and the Thirty Years' War, and now theory became accommodating. The geometers were not happy either. The story of the "dangerous infinitesimal" has been recently told by Amir Alexander (2014).

With this tool in hand, a circular orbit could now be described in terms of two sinusoidal signals, the one a quarter cycle out of phase with respect to the other. A simple description of a single-frequency waveform had been found. Two centuries later electromagnetic radiation would be discovered that had precisely this form, the sinusoidal fluctuation of orthogonal electric and magnetic components. It is in electromagnetic radiation that the mathematical ideal of a purely sinusoidal signal came closest to expression in nature.

The 19th century also saw the emergence of non-Euclidean geometry with Gauss, although he did not publish because he feared backlash from the "blockheads" that invariably beset the frontiers of human inquiry. The universe did not need to be flat. This further undermined the classical model that had stood for millennia. And we are just now celebrating the centenary of general relativity, which described our universe as non-Euclidean. Under gravitational influence, even two-body orbits migrate over time, but they do remain predictable and thus calculable.

However, with the mere introduction of just one more body into the mix, such a system is perturbed and enters into a nonrepeatable orbit. With a mere three bodies, we already have a chaotic trajectory for all three. We have already lost long-term predictability. The orbits are still bounded, but they no longer close on themselves. The mathematical simplicity of nature gets fuzzy and unpredictable in its actualization. We call these orbits limit cycles, and the whole system would be called a limit cycle attractor.

With the interaction of many-body systems, we inevitably find ourselves in a world of complexity that our traditional tools of analysis are not suited for. We were already in trouble at three. The brain confronts us with the ultimate many-body problem, in the sense of uncountable degrees of freedom, and we came to the task with analytical tools that were suited to a very different set of problems.

Wherever we see many-body problems in nature, we tend to see clustering and the emergence of patterns. Often these patterns are periodic or wave-like. We see clustering at all spatial scales in the galaxy, even out to the largest. In galaxies we may see both periodicity and the formation of density waves in the arms. We are only too familiar with hurricanes, tornados, and closer to home, the vortex in the bathtub drain. Living systems have more degrees of freedom, and they exhibit an even greater propensity to organize into patterns. For example, we see the formation of toroidal patterns in schooling fish. Even mitochondria have been observed to oscillate.

Once again progress in mathematics was needed to probe this new realm, and these methods have come along just in the last fifty or so years. Collectively, they illuminate the rules governing self-organizing systems; they find order within the general disorder; they surface general organizing principles. This demanded a new theory of networks, new ways of doing frequency-based analysis of transient states, and new thinking on how living systems dynamically organize their states. Networks in living systems are almost never actually random in their connection scheme, although that remains the mathematician's ideal.

Our attachment to the Gaussian distribution is coming to a messy dissolution—but our fondness lingers. The complex realm of self-organizing systems is populated with scale-free distributions, which are about as far removed from the Gaussian as one can get. The Fourier transform was suitable for stationary processes, not the dynamics that now must be characterized. The new frontier calls for understanding state transitions, hence dynamics more than stasis.

Additionally, the neurosciences benefited significantly from the emergence of real-time imagery of brain function with functional magnetic resonance. First and foremost, it brought the discussion back to the living brain. Many of the questions that could now be fruitfully raised and answered, however, were also issues for us in neurofeedback decades earlier, when we were still flying blind. It was in those days that the value of placing the brain in the loop was solidly established.

In particular, neurofeedback has allowed us to explore the frequency basis of neural organization in a way that could not readily be done through brain imagery. The brain's manifest competencies, as revealed through neurofeedback, take us to the very edge of believability. That story will be told in this chapter.

THE BRAIN AS A CONTROL SYSTEM

What is the brain's burden? The brain has to be organized as a control system, which means that it has to obey all the rules that apply to such a system. The prime directive for a self-regulating control system is to maintain its own stability unconditionally. In the case of the brain, the concern is firstly with conditions such as seizures, migraines, panic, narcolepsy, cataplexy, and coma. However, it must be appreciated that the brain even manages to sustain itself through such states rather consistently, so that viability is being maintained at another level. With brain stability assured, the secondary burden is to regulate its own states with the requisite subtlety. The maintenance of proper operating points is termed homeostasis, but is perhaps more appropriately called homeodynamic equilibrium. With its own housekeeping in order, the brain regulates the rest of our bodily functions. The brain's final concern is engagement with the outside world. As observers of our own brain function, we tend to view this hierarchy the other way up, in line with our personal objectives in life, but that does not match up with the brain's own priorities. Our engagement with the outside world is a mere perturbation on what the brain is managing on an ongoing basis. For the brain, there is no respite from its core duties of self-maintenance and self-regulation.

This hierarchy is dramatically confirmed for us in both evoked potential research and the current fMRI imaging. Evoked potentials are those features in the EEG that are explicitly related to input or output functions being executed by the brain. These signals tend to be either comparable to, or smaller than, the background EEG, and in fMRI we observe that the difference in signal level between an activated and a passive background state is on the order of half to 1% typically—and at most about 5%. This is just a small increment on what the brain is doing from moment to moment in the absence of a challenge.

In neuroscience research, the bias is toward investigation of the engaged brain. We've been looking at evoked potentials seriously since the mid-sixties. Both PET and SPECT imagery involve comparison of an active state with a passive baseline state, and now fMRI research has trod the same path. All are looking at metabolic activity that is associated with a particular function, which therefore must be distinguishable from baseline.

In neurofeedback, by contrast, we are interacting with the nonengaged brain, or at least as close to that as we can get. This is a preoccupation that is largely complementary to conventional neuroscience. We are concerned with the brain that is "merely"' managing itself, which happens to be its priority. We are witness to the self-regulating brain conducting its core activity. It should not surprise anyone that this has been very fertile ground indeed.

The EEG is our window into brain function, and the EEG reflects neuronal activity at cortex within the EEG spectral band from nominally 0.5–30 or 40 Hz and beyond. We, as the outside observer, get to see a spectrum that is packed densely with spindle-burst activity that is sharply delimited in frequency, variable in frequency over time, and highly dynamic in terms of amplitude. This is illustrated in Fig. 8.1. For the brain that is observing its own EEG, however, this is a new window into its own self-regulatory activity. The brain has a very different experience with this real-time signal than the outside observer. For the brain, it is a matter of recognition by virtue of the correlations that exist between the signal and its own internal activity. The distinction perhaps becomes more apparent with reference to Fig. 8.2, which shows an epoch of EEG spectral response over a small range of frequencies under three different conditions. On the left we see the spectral with low frequency resolution and high time resolution; on the right we see the same signal under conditions of high frequency resolution and low time resolution; the middle shows an intermediate condition.

The signal looks very different under the three conditions, and yet they all represent a comparable "truth". One trace is not more "correct" than the other. The underlying EEG must have at least the degree of frequency specificity that is implied in the right panel, and it must have at least the temporal dynamics that are displayed in the left panel. The actual reality, irrespective of whether we are able to illustrate it, must be a combination of both.

FIGURE 8.1 EEG spindle-burst activity fills frequency space in the range up to 20 Hz and beyond.
Each spindle-burst is narrowly defined in terms of instantaneous frequency, and the space between the spindles appears to be entirely devoid of collective neuronal activity. The time course of spindle-burst amplitude exhibits high dynamics.

Ironically, it is the Heisenberg Uncertainty Principle that prevents us from displaying the signal in its full, glorious complexity. We cannot have both high frequency resolution and high time resolution at the same time. Note, however, that the observing brain does not have the same problem. The brain merely has to get enough information to "recognize" its own agency—its controlling role—with respect to the signal, and it can do that regardless of which of the aforementioned images it is exposed to.

However, there is yet another issue. The Fourier transform that is used to display these data has fixed frequency bins, which imposes a kind of order on the EEG signal that does not correspond to reality. The frequency spindles migrate in frequency, and they experience discontinuities in phase. The Fourier transform is basically unsuitable for dealing with highly dynamic signals. This calls for different analysis schemes, including different transforms (Gabor or Hilbert), or wavelets, or other forms of time-frequency analysis.

In any event, irrespective of how the information is delivered, the brain is not confused because the signal refers to its own lived experience. The brain then

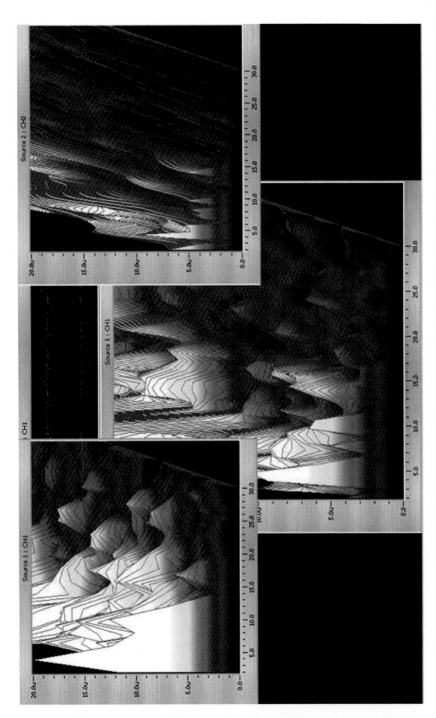

FIGURE 8.2 The same temporal window into EEG spectral activity is presented with three different choices of tradeoff between frequency resolution and time resolution.

All three representations refer to the same information subject to different signal processing. Particular aspects of the underlying "reality" are revealed in each of the three screens, and the actual reality can be intimated from a composite of the three. The brain would have no difficulty recognizing its authorship of all three if presented with the information in real time.

utilizes this information to adjust its own ongoing regulatory response. This process requires that the brain assign meaning to the signal it is observing, which follows directly from the correlations it is detecting with respect to its own internal states and state transitions. The process here described is indistinguishable from what goes on in ordinary skill learning, except for the novelty that in this case the relevant information is derived from the EEG rather than from the brain acting upon the environment and observing the response. As outsiders to this process, we now confront this same challenge as well: to extract meaning from the EEG signal that sheds light on the brain's regulatory mechanisms. We want to know what role the frequency-based ordering serves in the overall cortical regulatory schema. This question has existed ever since Hans Berger first published on the human EEG, calling it the Elektrenkephalogramm (Berger, 1929).

MASS ACTION IN THE SPATIAL AND TIMING DOMAINS

The EEG manifested the collective activity of neuronal assemblies at a time when such collective role was already being considered. As far back as 1906, Camillo Golgi was persuaded that cortical neurons functioned collectively rather than individually: "Far from being able to accept the idea of the individuality and independence of each nerve element, I have never had reason, up to now, to give up the attempt that I have always stressed, that nerve cells, instead of working individually, act together…. However opposed it may seem to the popular tendency to individualize the elements, I cannot abandon the idea of a unitary action of the nervous system…." And it was in 1906 that Sherrington published his book titled "The Integrative Action of the Nervous System." A new conception was starting to take hold (Sherrington, 1906).

The discovery of the EEG, then, gave substance to this conjecture, as the EEG clearly reflected collective behavior, and such collectives illustrated the regulatory role of the brain in organizing the neuronal assemblies into functional entities. The EEG continued to be studied and characterized, and by 1974 more than 1000 papers had been published on the alpha rhythm alone (Brown & Klug, 1974).

However, the EEG did not reveal its secrets easily. The complexity of the signal was discouraging, as was the lack of testability. A more fruitful area of study was the related question of evoked potentials, which were also a manifestation of neuronal group behavior, but in this case, it was possible to identify functional relationships more readily. For those purposes, the background EEG was seen as an irrelevancy, a mere nuisance. This was true until very recently, when it was realized that these phenomena are not independent. The main body of neuroscience, however, remained preoccupied with the study of the individual neuron as the presumed key to the understanding of brain function. This was called the "single neuron doctrine," and it succeeded in bringing about collective behavior on the part of neuroscientists: many of them chose to study the behavior of individual neurons.

Eric Kandel recalls in his memoir the instruction from his mentor Harry Grundfest during his graduate career at Columbia University: "Study the brain one cell at a time." It was of course quite necessary and appropriate that this be done, but the secret to brain function was not ultimately to be revealed there. Even worse, the rhythmic properties of neuronal firing in groups are not necessarily apparent in the individual firing streams of neurons. Rhythmicity arises out of correlations, and these may involve larger spatial scales than those under inspection. In a given neuron, the rhythmicity to which it contributes may account for only a small part of the variance in the firing stream. In consequence, there was little crosstalk between those who studied neuronal firing streams and those who were working at the level of the EEG or evoked potentials.

The study of what happens in the near neighborhood of neurons in cortex also got an early start, despite severe challenges on the instrumentation front. It was recognized early on that the most obvious anatomical feature of the interaction zone of cortex, the gray matter, was its organization as a two-dimensional (2D) surface with fairly uniform character. The essential characteristics of the neuronal system were largely conserved over the course of evolutionary development. It merely got larger, ultimately necessitating the cortical convolutions to simultaneously accommodate more cortical surface area, as well as the associated long-distance axons, the white matter. The relationship of gray matter surface area to white matter volume followed a fixed scaling law over the entire mammalian class, from the smallest shrew to the largest whale (Zhang & Sejnowski, 2000). An optimization process was at work that was driven by geometrical constraints. With increased cortical volume requiring longer axons, white matter volume increases more than the gray. The relationship is a power-law with an exponent just larger than unity: 4/3.

By now we also know that considerable evolutionary development actually did occur within the gray matter over mammalian evolutionary history. It occurred primarily within the glial system, albeit with obvious implications for the neuronal system as well. That was missed until recently in the face of our preoccupation with the neuronal system. With respect to the latter, we are right to assume that neuronal function has been serving the same purposes, and subject to the same organizational principles, throughout vertebrate development. It is in the glial system that we humans truly differentiate from our primate relatives.

THE SPATIAL MAPPING OF PATTERNS IN CORTEX

Neurons needed to be studied in their native habitat, namely in relationship to other neurons in a functioning organism. Areal mappings of neuronal activity go back to the 1940's, to John C. Lilly, who wired up a square array of 25 probes, each with its own amplifier, to drive glow tubes that would reflect the collective activity of the array. By 1955, some 50 probes were sequentially sampled with a single amplifier and the results mapped topographically onto a CRT. And by 1970 deMott had demonstrated a 400-channel topographic mapper with solid-state amplification. It was challenging work under nonideal conditions, yielding only limited findings. However, the path forward was clear (DeMott, 1970).

By 1975, Walter Freeman published his book titled "Mass Action in the Nervous System". He considered the implications of collective neuronal activity as not simply an extension of what was already known, but rather as a revolutionary new departure. In some generality, the signal being processed in cortex is not the property of individual neurons, but rather is encoded in an ensemble of neurons. Such ensembles are distributed spatially over the cortical surface layer. Their existence is also transient, serving just the immediate need and then dissipating, only to be replaced by a new pattern. One can think of this organization as cinematographic, one spatial representation successively replacing another. This sequence of "frames" may be those of a "movie", with successive frames incorporating incremental change, such as would be required to implement sequential activity. Or it could also be the frames of a slide show, with successive frames representing different processing tasks. Or it could be something involving both, like the scene change in a movie.

This kind of spatial mapping requires an organizational schema to facilitate the encoding of the signal and any processing of the signal during its brief life span. Every neuron participating in the feature representation must retain its individuality as well as express its group membership. The problem is solved if the criterion of commonality lies in the domain of timing rather than in the spatial domain. A spatial criterion is problematic as there are many activities going on simultaneously in any cortical region. It is also apparent that much of brain function is time-critical. Any cortical activity subserving time-critical processes must be highly ordered in the time domain. This manifests in the pulse waveforms of evoked potentials, for example.

THE ORGANIZATION OF FEATURE BINDING

Animal studies in the late eighties and early nineties at the Max Planck Institute in Frankfurt on visual processing led to a proposed solution to the problem of neuronal assemblies distinguishing themselves from others, the problem of feature binding. It was based on a simultaneity criterion, namely that those neuronal events occurring with a common timing signature are recognized as belonging to the relevant ensemble. The theory that simultaneity of firing specifies membership in the relevant neuronal assembly was called "Time Binding". The theory was the subject of controversy for some time. In the face of such general skepticism Christoph von der Malsburg, a participant in the ground-breaking research, said at the time: "We are in the middle of a scientific revolution, the result of which will be establishment of [time] binding as a fundamental aspect of the neural code." The term scientific revolution is used sparingly by scientists, and even more rarely by scientists in reference to their own work. After all, referring to one's own discovery as a scientific revolution is the equivalent of Napoleon crowning his own head. Most scientists are more humble than Napoleon. However, just as these scientists had the sense of a radical new departure toward a new organizing principle, so did everyone else. Eventually, a theory such as this enters one's comfort zone, and a while later one wonders just how things could have been otherwise.

The critical experiment that was done at the Max Planck Institute was a determination of the correlation of firing events in cat visual cortex as an object was moved across the visual field (Engel, Konig, Kreiter, & Singer, 1991). When neurons in striate cortex were being "illuminated" by the passing object, a correlation in firing events became apparent even across the hemispheric fissure, a correlation that was not apparent otherwise. One other thing: the cat had to be electrically stimulated in the mesencephalic reticular formation in order to be suddenly interested in that bland, featureless visual presentation. This tells us that the correlation was not due to passive visual processing of the object in its narrow sense, but rather was a feature of the visually attentive brain. The object in the scene had been stamped with significance. Visual processing occurs at a nominal rate of 40 Hz, or a period of 25 msec, and within that window, the correlations could be detected down to the millisecond level. Here was a basis for feature binding.

The finding was all the more remarkable when one considers that it is quite a task to organize timing down to the millisecond level across the hemispheric fissure with a brain architecture that is largely lateralized all the way down to the brainstem. The interhemispheric connections that do exist in cortex inevitably involve transport delay in their communications. How, then, is simultaneity in neuronal firing to be organized across the hemispheric divide? Here we must fall back upon what we know about periodic systems. They can entrain each other, and synchronous operation is achievable even if there are transport delays in the communication between them. Alternatively, bilateral synchrony could be organized at the brainstem via the bilateral connections that exist there. This can explain the substantial interhemispheric coherence that exists even in cases of agenesis of the corpus callosum.

What we have, then, is a basic frequency (40 Hz) at which visual information is processed in packets, and this periodicity is tightly coordinated between the hemispheres. Within this general periodicity, specific timing relationships also matter. These are organized with respect to the common 40 Hz periodicity serving as a timing reference.

The critical importance of timing relationships in the brain traces back to a fundamental characteristic of the action potential mechanism. This is the observation that a single excitatory input to a neuron is not capable of generating an action potential in the target neuron. Instead, the target neuron must be "primed"' through the arrival of other excitatory inputs within the relevant window of time. This time interval is given roughly by the width of the excitatory postsynaptic potential (EPSP). In the simplest possible realization, there must be a second EPSP within the window of opportunity, or about 10 ms, in order for the combined signal to cross the threshold for the generation of an action potential.

A mathematician regarding such an arrangement would say that the critical function performed by the neuron in this connection is that of coincidence detection. Since the function of an action potential is its contribution to the generation of yet another action potential downstream, we observe that the criticality of timing at the level of the individual neuron makes brain timing a critical consideration for the quality of brain function in general. It follows that the integrity of brain timing relationships must be maintained globally within the nominal 10 ms benchmark or function degrades.

The critical findings by Wolf Singer and his research group place the individual neuronal firing event into its larger context of ensemble collective activity. Coincidence at the level of the neuron translates to simultaneity of firing at the level of the ensemble, which is then observable as local synchrony at the level of the EEG. We find, then, that the EEG at a given frequency provides us with exquisitely detailed and specific information on the behavior of neuronal assemblies with respect to timing and frequency.

By virtue of having regulatory function managed via neuronal assemblies, the brain has minimized the risk of single-point failures. As John Eccles has pointed out, "the firing or nonfiring a single pyramidal cell cannot have any consequence for the brain." The ensemble basis of state regulation turns a hard failure at the single neuron level into a soft failure at the ensemble level. The integrity of ensemble activity is maintained intact despite numerous dropouts. One is rarely confronted with a complete abolition of functional capacity. Rather, we are confronted with partial functional loss. The task of neurofeedback, then, is to restore functional integrity to a mechanism that is still functional at some level. The process always builds on what already works. We do not expect the neurofeedback technique to enable function de novo.

The action potential mechanism can only have been a late feature of neuronal evolution. At the outset, neurons must have developed to implement direct chemical communication. Particularly if one views the neuronal/glial system jointly, direct chemical communication remains the dominant feature of neuronal existence, the major mechanism of information transport. And action potential formation is itself orchestrated entirely by means of chemistry in the analog domain. In this regard, the neuron can be understood as an analog-to-digital converter. In the analog domain we have the benefit of continuous variation of the parameter so that regulation can be conducted with great subtlety and precision. Digital signal transport is subject to graininess, and to limitations on the rate at which information can be communicated. On the other hand, the process is robust and relatively noise-immune—ideal for long-distance transport of information in cortex and beyond. And it is capable of great precision in the timing domain.

The action potential mechanism initially served the purpose of rapid responding to environmental demands by way of a motor response. This is perhaps best illustrated with reference to the sea squirt, which possesses a modest nervous system while it remains in a larval stage, seeking a place to settle on the ocean floor. Once it is sessile, it has no further need for its nervous system, and proceeds to resorb it. Significantly, it does not take advantage of its nervous system to retain broader sensory awareness, or to luxuriate in the existence of life itself just because it can. The higher form of sentience has lost its reason for being. If one wants to make the case that life is basically organized to permit genes to replicate themselves, then the sea squirt presents a good argument. Nature is a minimalist composer, despite appearances.

The neural system served to mediate between the sensory and the motor realms, where timely responsiveness was of the essence. In the nematode *Caenorhabditis elegans* we already see an elaborated neural system of some 302 neurons that function with an action potential mechanism. Every neuron is differentiable from all others. And every such nematode ends up with the same complement of 302 neurons, all similarly organized into functional networks. These already display some of the features of more fully developed nervous systems, showing signs of the emergence of hierarchical order, the hallmark of complex biological networks.

The complete genomic prescription of the neuronal network that we have in the nematode is out of the question for the human brain, and indeed, our human genome is substantially smaller than that of *C. elegans*. We also get along with a smaller number of neuronal types. Our genome manages with more general prescriptions, and it is able to do so because of greater hierarchical organization.

THE HUMAN VISUAL SYSTEM

Whereas the problem of limited digital signal bandwidth is not an issue with the nematode, by the time we get to the human visual system we do confront that limitation, and yet the brain seems to surmount the challenge skillfully. Here's the problem: When we gaze upon a 4 K resolution large-screen TV displaying a high-resolution image, we recognize instantly that this resolution exceeds that of "ordinary HD". And yet we possess "only" about 128 M optical sensors in our retina (to the nearest power of two), which are serviced by an optic nerve of only a million axons. The data rate that leaves the retina is only about 6 megabits/s, and by the time the signal reaches layer IV of V1 (striate cortex), the data rate is down to 10,000/s. This is clearly inadequate to represent the high spatial frequencies present in a 4 K image in real time. In the words of Marcus Raichle, "These data leave the clear impression that visual cortex receives an impoverished representation of the world" (Raichle, 2010).

We have tried to understand perception as an extrapolation of sensation, but that project falters immediately as one confronts the particulars. The paucity of the signal stream means that only a fraction of the cortical neurons receive input on each exposure, leading to high variability over the receptor field even with an invariant stimulus. And yet we perceive a stable image that represents the stimulus with startling fidelity. Increasingly it has become evident that visual processing is dominated by the brain's internal representations, as informed by sensory inputs. The evidence for this proposition is collectively compelling. As a matter of fact, even William James was persuaded of this proposition just based on the evidence available to him late in the 19th century. Striate cortex is the last stop in the chain of events in visual signal processing at which the information is mapped topographically. Beyond V1, tracing the circuitry reveals that visual information is delivered to some 40 different loci in posterior cortex.

Even at the lateral geniculate, the thalamic way station for information from the retina to visual cortex, a mere ten percent of synapses are recipients of visual information. Already at this juncture, it appears that context dominates the processing that occurs. A similar ratio applies at layer IV of V1, which drives one to the same conclusion.

Traditional experimental approaches to sensory processing have relied majorly on evoked potential studies, rendered more discriminating lately via independent component analysis. These studies share an implicit bias toward the sensory input-dominated perspective. The complementary view is that of top-down influences, by which are meant those that involve frontal-lobe directed functions, such as selective attention.

It is becoming apparent that the largest burden of sensory signal processing is borne by intrinsic brain activity that is modulated via both bottom-up and top-down pathways. In fact, this proposition is not new. The initial foray into a more realistic model of sensory processing in the mammalian brain was presented by Walter Freeman in 1975 in the book already referred to, Mass Action in the Nervous System. Freeman chose to study the olfactory system of the rabbit as a paradigm for sensory processing in vertebrates, given the primacy of chemoreception and hence the likelihood that this had precedence in our early evolutionary development.

THE OLFACTORY SYSTEM

Odorant receptors are responsive to specific odorants, and one estimates some 100,000 receptors per odorant, some fraction of which will be excited on each nasal inhale. The action potentials of the sensory receptors are projected to the olfactory bulb, which then directly reflects the variability of the original signal. Nevertheless, the emergence of a stable spatial pattern of excitation is observed over the olfactory bulb, one that is specific to the odorant being detected. Significantly, every neuron in the bulb participates in this pattern, irrespective of whether it had received an input pulse relevant to the odorant.

The original signal representing the odorant had short life expectancy within the processing sequence. It lost its identity at the bulb, where the sensory neuron first encounters brain. However, the brain had already made the signal its own. It had established a pattern of firing that represented the odorant uniquely and also persisted over time. This periodic repetition of the firing pattern takes place at a characteristic frequency, by analogy to the refresh cycle of dynamic RAMs. This frequency falls into the gamma range of nominally 40 Hz (but it can range widely). Significantly, this rhythmic pattern had to have been self-organized by the bulbar neurons themselves. The pattern is synchronous over the entire bulb, even as it varies over time or from one inhale to the next. The amplitude distribution of the gamma-band signal over the bulbar surface is the unique identifier of the odorant. It is also unique to the particular rabbit, having arisen out of the rabbit's life experience. It is also subject to change with subsequent life experience, on top of an intrinsic variability.

With each inhale, the olfactory system undergoes a state transition from an initial input-dominated mode to the brain-dominated pattern characteristic of the odorant. In between inhales, the system prevails in a state of high variability. Cortex is informed of both the stable, recursive pattern and of the input-dependent signals, but the system response is limited to the recursive pattern. What the brain pays attention to with regard to olfaction is exclusively the pattern it has itself created out of the sparse and highly variable input stream. This turns out to be a general property of our primary sensory systems. With this background, we return to consideration of the visual system.

According to Fiser, Chiu, and Weliky (2004), who tracked the visual system response pattern of ferrets through their entire course of development from eye opening to maturity, visual cortical neurons fire with a large degree of variability, even with presentation of a stable image. However, on the larger scale, a rather stable spatiotemporal pattern may be discerned. At all ages, the observed correlations in neuronal firings were only slightly affected by visual stimulation. Once again one has the impression that input signals serve to modulate established patterns that integrate over the variations in the signal stream, thus rendering a stable pattern that informs our experience of the visual field.

For our present purposes, the salient observation is that this recursive pattern occurs at a nominally 40-Hz repetition rate. It is temporally coherent with the incoming signal stream, and is colocated in the same cortical real estate. Significantly, visual processing cannot be merely a matter of coming to terms with incoming information as it happens. Because of signal transport and processing delays, relevant visual experience may well be shaped after a critical event such as watching a fastball from the batter's box. The brain manages to give us the experience of living in real time despite such processing delays, and batters do manage to hit fastballs. This means that the brain is actually organizing a prediction model on the basis of the meager visual information stream that is available to it. In order for us to live successfully in real time, the brain has to anticipate the likely trajectory of events.

In sum, the incoming information shapes the visual experience that the brain organizes cumulatively on the basis of the flow of input, combined with expectancy factors for the likely scope of subsequent inputs, predictions for the probable trajectory of events in the scene, and projections forward of our own motor responses. Visual processing can therefore only be understood as a system response in which the brain itself is the principal generator of the scene we get to observe, placed within the context of the unfolding pageantry of our lives. Further, the arguments that have just been made also suffice to make the case that our vaunted executive control system cannot be playing much of a role in this process because of the delays involved and because of the distributed nature of the processing involved. Top-down control is not an option except at the margins. Visual processing must be actualized and governed by control schema that are largely self-organized and largely buffered from explicit top-down control.

ON THE SENSE OF HEARING AND ON OUR SENSE OF PLACE

Given the centrality of our concern with frequency-based organization of brain function, we cannot overlook the sense of hearing. This too has an early evolutionary origin, and it is the sensory modality that gives us our most immediate awareness of the environment. The time delay from brainstem to cortex is a mere millisecond or so. At low audio frequencies, up to about 300 Hz, frequency is mapped as actual waveforms, and at higher frequencies, frequency is mapped tonotopically in primary auditory cortex. Our exquisite frequency discrimination bears testimony to the brain's ability to organize such fine distinctions. This makes it easier to accept that the brain takes advantage of such a capability in other ways, and this may even be directly relevant to neurofeedback.

The sense of hearing also exhibits the brain's limiting performance in terms of temporal discrimination. Determination of the direction of sound requires a comparison of relative arrival time, or equivalently, relative phase of the auditory signal at the two ears. This comparison is done in the digital domain in the midbrain, and the brain has been shown capable of discriminating time differences of less than a millisecond—a time interval that is smaller than the width of an action potential.

Precise timing is also an issue in the maintenance of our sense of space. Take, for example, the case of a runner on an oval track. A mental representation of where the person is on the track is maintained by place cells in the hippocampus. A basic theta rhythm sets the pace, and the place cells map the space in terms of the phase within the theta periodicity at which they fire. It is as if the brain occupies two worlds, and is equally at home in the frequency domain and in the time domain. Given the tight timing constraints, and the large-scale organization involved here, it is perhaps no surprise that the sense of where we are in space is the first to be lost in Alzheimer's dementia. By the same token, one can understand the visual disturbances and the loss of the sense of smell after minor head injury in terms of disturbance of the frequency-based organization of sensory perception, a mechanism that need not presuppose any structural injury.

ORGANIZATION OF THE MOTOR RESPONSE

With respect to the visual system the brain retains its greatest secret, namely how it is that we actually get to "see" the visual imagery that we do, when all the brain has to work with is neural assemblies and their distributed firing patterns. When it comes to the motor system we at least get to see the output directly. The organization of motor responses is a significant preoccupation of the brain. Charles Sherrington once put it this way: "The motor act is the cradle of the mind."

Johann Wolfgang von Goethe put it poetically: "In the beginning was the act" (Am Anfang war die Tat). With respect to the allocation of cortical resources, output refers to motor function almost exclusively. The entire frontal lobe can be viewed in terms of the hierarchy of motor control. If one includes the somatosensory system, which is so strongly interwoven with the somatomotor system, the dominance of motor control in cortical real estate is impressive.

In fact, it is helpful in this context to recognize the primacy of the somatosensory system among our primary sensory modalities. Sigmund Freud recognized that "The self is first and foremost a body self." And Oliver Sacks has shed light on the catastrophic consequences of the loss of somatosensory awareness, which entails a substantial loss of the sense of self. The somatosensory system is the only one that presents us with such a hazard. This condition tends to afflict highly intellectual people preferentially, and it can subside as readily as it arrived. A functional mechanism is therefore indicated. This places it in the class of conditions that are expected to respond to a functionally based intervention such as neurofeedback.

This part of the discussion is particularly relevant to neurofeedback, because that process is best understood by analogy to the learning of a motor skill. Historically there has been very little interest among psychologists in motor skill learning. In the behaviorist era the attempt was made to explain motor skill learning in terms of stimulus-response models, and that effort was not very productive. Similarly, operant conditioning models have served as the principal explanatory model of neurofeedback, and we will see that that is not entirely satisfactory either.

On the other hand, thinking of the problem in terms of a conventional control loop, one in which comparison is made between the desired and the present state, and an error correction scheme is mobilized, turns out to be similarly unavailing as a comprehensive description. The execution of a golf swing is best accomplished without interference from the executive control system. The playing of Rachmaninoff's Second Piano Concerto offers no opportunity for error correction on the relevant timescales. On the other hand, this error–correction model does have its zone of applicability, and it will also be relevant to our understanding of neurofeedback.

The two counter-examples offered illustrate the importance of skill learning. What is being learned is a sequential process that involves mostly "local" control, with little input from the executive control networks. Pianists may still prefer to have notes in front of them during a performance, but they serve a purpose of general cueing rather than moment-to-moment instructions. In the previous examples, one can readily talk in terms of training to mastery, because the skill is exercised in relative isolation, that is, without environmental interference.

More generally, skill learning must involve acquisition of a response capability that integrates incoming information with motor output. Life is more like tennis than golf. Once again, however, the pace of life allows for little more than general oversight by executive control functions. As in the case of sensory processing, the brain is called upon to organize a system response that is minimally dependent on top-level steering.

THE SMALL-WORLD MODEL OF THE CEREBRUM

The limited role of top-level executive control in motor activity does not imply the absence of hierarchical control of movement. It's just that the most relevant hierarchy begins at the brainstem rather than in our prefrontal cortex. Control is implemented through a hierarchical network structure with "small-world" character. That is to say, there is sufficient global interconnectivity to draw the whole network into intimate, efficient communication. Once such interconnectivity exists, there is a natural tendency toward the emergence of hierarchy. This tendency is exploited wherever possible. On the large scale, hierarchy emerges over the course of evolutionary development and becomes obvious in the cerebral architecture. This gives brain function the capacity for unitary operation. The gross hierarchy of control is nicely illustrated in a study of structural connectivity in the Macaque monkey by Modha and Singh (2010). At the very highest levels of connectivity between hubs, we have top-down control emanating from the brainstem. This is seen in Fig. 8.3. Even at the next lower level of connectivity, all of the control linkages are still top-down, and the appearance is still very modular. This is shown in Fig. 8.4. One has to drop to yet lower levels of connectivity in order to see cortical–cortical and other linkages enter the picture to facilitate global integration and feedback to the brainstem.

However, small-world character of the networks also prevails within the brainstem itself, within the thalamus, and within cortex itself. Within cortex, hierarchy emerges over the course of development as neurons differentiate in terms of connectivity (Hagmann et al., 2008). The driver here is the principle of preferential attachment (the rich get richer), which drives connectivity to a broad, scale-free distribution (Barabási, 2002). The two salient characteristics of small-world models are high local connectivity and efficient long-distance communication. It is in cortex that the small-world model is taken to extremes. The dendritic tree (of a 1,000–10,000 dendritic branches, and between 5,000 and 60,000 synaptic inputs), together with large-scale axonal branching, assures high levels of local connectivity. At the same time, global connectivity is maximized by having every pyramidal cell also participate in distal communication. The result is that essentially every pyramidal cell is accessible to every other within three synaptic linkages, a simply staggering level of large-scale interconnectivity. In fact, it can be readily argued that this number represents a biological limit. It cannot be lower than three. Large-scale connectivity has

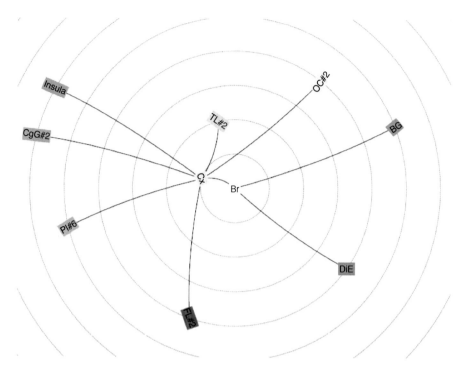

FIGURE 8.3 Linkages with the highest levels of connectivity between regions are illustrated here for the brain of a macaque monkey.
The brainstem is seen as the highest level of the regulatory hierarchy. The next level of connectivity includes cortex, the diencephalon (thalamus and hypothalamus), and the basal ganglia. Primary cortical linkages are to the temporal lobe, to frontal cortex, to parietal cortex, to the cingulate gyrus, and to the insula. *Taken from Modha, D. S., & Singh, R. (2010). Network architecture of the long-distance pathways in the macaque brain.* Proceedings of the National Academy of Sciences, 107*(30)*, 13485–13490.

been taken to its limit in cortex. Since this arrangement is energetically cost-ly, local and global connectivity must have been key drivers in evolutionary development.

With this as a background, the question to be asked is whether the aforemen-tioned hierarchy of control manifests itself in motor function in particular. If so, then that may be considered paradigmatic for brain function in general. One fruitful area of inquiry is into the coordination of movements under con-ditions where the two hands are asked to manage tasks that differ in level of difficulty—under time pressure. It is found that the brain choreographs the two independent activities so that both objectives are attained at about the same time (Kelso, 1982). This all occurs beneath the level of awareness, so it is cer-tainly not the outcome of any intention. This must be strictly a matter of the brain optimizing its own performance. By arranging the trajectories to follow

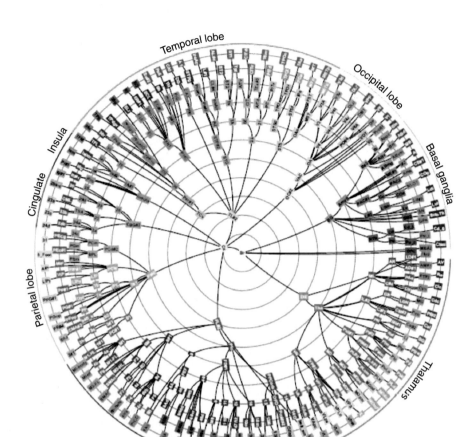

FIGURE 8.4 At the next lower level of connectivity illustrated here, all of the linkages shown still represent top-down regulatory pathways traceable back to the brainstem.
The connectivity tree still appears very modular. Interestingly, the temporal lobe commands as much real estate on this plot as the thalamus and the frontal lobe. One has to drop down to yet a lower level of connectivity to bring interregional connections and linkages back to the brainstem into the picture.
Taken from Modha, D. S., & Singh, R. (2010). Network architecture of the long-distance pathways in the macaque brain. Proceedings of the National Academy of Sciences, 107*(30), 13485–13490.*

a common time course to completion, the brain is limiting the degrees of freedom that it has to manage independently. Significantly, this simplification falls into the domain of timing.

This is such a foundational concept that perhaps another illustration is in order. Children often challenge each other to simultaneously pat themselves on the head with one hand while making a rotating motion over the stomach with the other. This is difficult to do simultaneously right out of the starting gate. When the task is achieved, however, it will be noted that most likely both motions are embedded in a common periodicity. Specifically, an integral number of pats on the head will go with a single rotation of the other hand. An overarching order is imposed into which both activities can be enfolded and jointly optimized. This degree of order emerges, however, out of a self-organizing process, without any top-down guidance, or even awareness.

Another useful probe of the underpinnings of motor control is to challenge performance near its limits. J.A. Scott Kelso famously performed a simple experiment that illuminated yet another core concept. Here's the challenge: Place both hands before you and extend both index fingers upward, folding the other fingers. Then move the index fingers toward and away from each other synchronously at a comfortable frequency. Imagine a metronome synchronized to the frequency. Now imagine a malevolent agent gradually increasing the frequency of the metronome while you try to keep up with the identical, antisymmetric movement. It will not be long before the fingers undergo a natural transition to moving in parallel rather than retaining the mirror image pattern. The brain will have migrated from one pattern to the other via a phase transition, which took it from one "basin of attraction", its preferred operating space, to one that was easier to implement and thus more suitable to the higher frequencies.

Just to add to the mystery of how the brain slips so comfortably into symmetric and antisymmetric movement, the very same results are obtained when this experiment is performed with someone whose interhemispheric connections have been severed to eliminate seizures. The basic principles that govern the self-organization of patterns of brain function override even major hardware constraints.

All of this transpires, of course, beneath the level of voluntary control and even of awareness. One must conclude that movement is organized according to basic patterns that arise out of the brain's own optimization schemes in the domain of timing and frequency.

SYNERGETICS: THE SCIENTIFIC PRINCIPLES UNDERLYING SELF-ORGANIZATION OF NATURAL SYSTEMS

This brings us then, finally, to the core issue of the principles underlying self-organization. For this discussion, we turn to a physicist, Hermann Haken, who has been engaged on the topic since the 1960's, when he concerned himself with the properties of the laser that had just been invented. The core principles, then, are already on view in inanimate systems. In the laser, atoms in a particular excited state can be stimulated to emit a photon, with the result that both the stimulating and emitted photons now possess a common phase. By this process, a large number of photons can be brought to a state of common phase—and, in the case of this quantum-mechanical system, to a common identity. One such photon is no longer distinguishable from another. They have all effectively become enslaved, each to the others. So we have slavery, yet we have no master. This theory, along with its elaboration into living self-organizing systems, is called Synergetics (Haken & Stadler, 1990).

In the two-finger experiment just described, what one finger is doing is highly predictive of what the other one is doing, even if one cannot see it. Phrased mathematically, the phase relationship between the two fingers is very stable, undergoing only small fluctuations, within the two comfort zones of the low frequency and the high. (This is similar to what is observed in the laser. All participating photons have identical phase.) This relatively stable measure can, therefore, be used to specify the degree of order in the system, the degree of similarity among the elements of the system. As such, it is termed an order parameter, which is simply a measure of the degree of prevailing order in the system.

In the finger experiment, we are witness to the behavior of the system under the forcing function of the metronome. The periodicity of the metronome is termed a control parameter in the parlance of synergetics. The experiment allows us to say that the order parameter is stable over broad ranges of the control parameter, with the exception of the transition zone between them, the region of phase change. All of the stimulation procedures discussed in this book are frequency based, and within the framework of synergetics these frequencies would be regarded as control parameters with the objective of enslaving the neural populations that are available for such recruitment.

THE BRAIN AS A NONEQUILIBRIUM SYSTEM IN A STATE OF CRITICALITY

The behavioral invariance and stability demonstrated by the finger experiment stand in stark contrast to the dynamics that are displayed in the real-time EEG, irrespective of whether we sample the EEG at sensorimotor cortex or anywhere else. One observes a densely packed array of brief spindle-burst activity that covers the entire spectral range of the EEG, as shown in Fig. 8.1. Persistence varies inversely with frequency, with lower-frequency spindle-bursts lingering longer than higher-frequency ones. It is not clear how behavioral stability emerges out of such apparent cacophony. Matters are even worse than they appear.

There is yet one more key organizing principle to be discussed before we try to fit both neurofeedback and stimulation-based methods into this framework. It is that the brain operates far from equilibrium under all circumstances. There is, in fact, no such thing as a resting state as far as the brain is concerned. The term is in common usage, to be sure, but it refers to yet another highly active state—the state of mere nonengagement. Even worse, the brain is driven to the very edge of microscopic instability. What is at issue here is a bounded instability rather than a runaway condition, such as a seizure.

This state is difficult to describe, but the general principles operative here are exhibited in the sandpile (Bak, 1997). As one adds grains of sand to the top of the sandpile, the entire conical surface will gradually approach what is called the angle of repose. Adding sand beyond that point will trigger the formation of small avalanches that will restore the surface to the quasistable angle of repose. If sand continues to be added, the sandpile continues to "live" at the edge of stability thus defined. A similar situation prevails for the brain.

Cortex prevails in a state that is perpetually ready for macroscopic state change (Plenz & Niebur, 2014). An analogy exists between this process and phase transitions in inanimate systems. When such systems are poised at the threshold of a phase transition, they are deemed to be in a critical state. In physical systems, such critical states occupy a very small part of state space. By contrast, that is where the waking brain lives perpetually, thus occupying a much larger state space. What is mere happenstance in the case of the sandpile is under active management in cortex, so that those phase transitions that do occur are ones occasioned by functional demands—of either internal or external origin. In physical systems, the phase transition is often between ordered and disordered states. In the brain, the phase transition is between one state of local order and another.

The brain faces the complementary challenges of maintaining its own macroscopic stability while also remaining poised for nearly instantaneous state change locally as circumstances may demand. To solve this problem, the brain takes full advantage of the entire frequency spectrum. It arranges for stability and continuity of state at low frequency, and for agile responsiveness at intermediate frequencies. Transient cognitive activity is managed at higher frequencies. In this manner, both stability and agility of responsiveness can be accommodated within this frequency-based schema.

The state transitions that are managed by means of the intermediate frequency range (below the gamma range of frequency) resemble phase transitions, as already indicated. They are macroscopic shifts that rapidly encompass the entire neuronal pool that is susceptible to such a shift. Large cortical regions suddenly shift from one pattern of functioning to another very different pattern, and the transition zone is very brief—just milliseconds. The stable period between transitions can be fairly brief as well, that is, fractions of a second. In the limit, brain function is organized in terms of a sequence of four microstates that toggle between giving priority to different brain regions (Lehmann, Ozaki, & Pal, 1987).

Once we have taken things apart like this, we also have to put them together again. Every initiative by the brain involves all of the frequencies, each playing its assigned role but ultimately being part of one orchestration. It is useful to think of this in terms of a kind of nesting, in which the lower frequencies set the context for the higher ones. The problem is that once one has that idea in one's head, it is easy to think of the higher frequency activity as being largely prescribed by the lower (i.e., that one is "locked" to the other), when in fact it is typically more a matter of shifting probabilities of occurrence. It is perhaps more realistic to view the lower frequencies as context-setting, as being permissive rather than prescriptive. It is the demands of life that are prescriptive for the brain's response, and it is the brain's burden to be poised for whatever response is called for. Under other conditions, the higher-frequency activity is indeed phase-locked to the lower.

NEUROFEEDBACK

The above discussion sets the table for how we should regard the potential contribution of neurofeedback to the enhancement of the brain's functional competences, and to their recovery from dysfunction. Our tool is frequency-based in that it depends principally on the trainee's response to information derived from a narrow portion of the frequency spectrum. This allows one to target different parts of the EEG spectrum for particular objectives, and it makes it appropriate to discuss the issues to a large extent within the framework of the frequency spectrum.

The principal applications of neurofeedback to date have been to the matter of state regulation. However, the issue has not always been framed in this way, and it has not always even been apparent. In the beginning, there was alpha training, and the driving objective here was its relationship to psychological states and the opening it provided to enhanced states of awareness. Sterman's SMR training was aiming narrowly toward the control of motor seizures at the outset; and Lubar's elaboration of that method initially targeted the management of hyperkinesis. There was very little engagement with issues of arousal regulation per se in the early days, as arousal-based models were falling into disfavor.

What engaged attentions in those early days were those functions in which alpha band activity and SMR band activity was explicitly involved. Their implicit involvement in matters of core state regulation remained in the background. It is difficult for the tutored mind to appreciate just how rigidly these lines were being drawn at the time. An anecdote might be helpful here. When we were first observing the beneficial effects of SMR training on anxiety states in 1990, another leader of the field was nonplussed. "But you should do temperature training for that," he declared. Autonomic regulation was a matter for traditional biofeedback, not neurofeedback. Similarly, the suggestion that beta training might be helpful for depression was categorically rejected. And when the suggestion was made that SMR-beta training was very helpful for the elimination of PMS, many people in the field were simply apoplectic. That wasn't even a recognized disorder. Critics could hardly picture a better way to get neurofeedback dismissed from learned discussion. And yet PMS was really the paradigm for disorders of dysregulation, encompassing a whole host of symptoms with a broad range of symptom presentations. They were all functional in character, and therefore susceptible to a functional remedy.

The field has largely overcome its blinkered history. The main target of neurofeedback is the quality of brain self-regulation in general rather than specific disorders or dysfunctions. The neurofeedback challenge typically evokes a more general reordering of network relations than we felt entitled to anticipate at the time. Typically it is these general effects, rather than the specific ones, that are of primary interest in clinical work. The generality and universality of the impact of neurofeedback tended to be obscured for two reasons. First, there was the obligation on the scientist to be as specific as possible with respect to findings, and secondly there was the problem that core regulatory function is not readily quantifiable.

Since neurofeedback has thus far been a clinically driven field, it has not been emphasized sufficiently that it is also an excellent probe of brain function, and more specifically a probe into its frequency-based organization. Investigating brain function typically involves the comparison of a challenge state with a baseline state. This is the case for evoked-potential work, as well as for the new era of brain imaging (PET, SPECT, and now fMRI). Alternatively, one evaluates performance-related issues, as in tests of reaction time or of cognitive function.

If we regard neurofeedback in its role as a probe of brain function, the trainee's brain is effectively serving in the role of detector, and the salient observables are the physiological shifts that can be noted in the trainee rather than the changes that may be seen in the signal. Since the brain is observing a correlate of its own activity, it comes to the task with an enormous advantage vis-à-vis a naive external observer—such as a neuroscientist, for example. The brain is performing a recognition task rather than a detection task. As soon as such recognition occurs, the brain is navigating on familiar ground.

And just as one can allow the brain to play the role of signal detection in the research on brain functional organization, one can allow the brain to inform us as to how it is best trained. Just as the French farmer uses a pig to find the truffles, we can let the brain lead us to its own most productive training procedures. This is largely a matter of skilled clinical observation.

Single-session effects are routinely seen in neurofeedback, and have in fact been documented by now using measurements of evoked potentials (Hill, 2013), of the contingent negative variation (CNV) (Magana et al., 2016), and of functional connectivity as revealed in fMRI data (unpublished). The effects of single sessions on the state of the system have been apparent since the early days of the field. (Indeed many published studies of biofeedback were based on results obtained after a mere one to three sessions.)

Once clinical objectives of neurofeedback became paramount, as occurred with SMR-beta training, the emphasis shifted toward "the behavior"', in the lexicon of operant conditioning, and talk of quick effects was dismissed in a general disparagement of "overclaiming". More particularly, the mere induction of state shifts was not seen as germane to the real objectives of the training. Once the issue was dismissed from the discussion, it became very difficult to reintroduce it. In fact, state shifts can be achieved within a matter of minutes, and such state shifts can be used to guide the training to its most propitious outcome. Cumulatively, these observations also illuminate the larger issue of how state regulation is organized.

Once the discussion is focused on the core issue of state regulation, one must have a schema to organize the findings. When the brain is regarded in its role as a control system, the first priority, as noted earlier, is to assure its own unconditional stability. As it happens, Sterman's SMR training was relevant to that core objective, but matters were not discussed in those terms at the time. The second tier is the regulation of state with respect to arousal, affect, autonomic function, and interoception—the self-monitoring of the state of the body. The objective here is homeostasis, or more appropriately a homeodynamic equilibrium. Autonomic regulation subsumes the brain's regulation of other bodily systems.

The final objective is regulating the brain's engagement with the outside world, the domain of executive function. Lubar's SMR-beta training explicitly trained attentional faculties while implicitly serving the purpose of arousal regulation (Lubar & Lubar, 1984). Independently of the hierarchy of regulation, in neurofeedback one is also concerned with containing behavioral disinhibition, as well as with rolling back learned behaviors, such as addictions, acquired fears and phobias, and other such specific issues.

Our own work in this field began in 1985 with the procedure pioneered by Sterman and first clinically deployed by Ayers. Sterman's sleep research involving cats had firmly established operant conditioning as the scientific model for EEG biofeedback (Sterman, Howe, & Macdonald, 1970). The cats had learned to produce SMR spindle bursts in greater abundance with the help of conditional food reward, and they benefited in terms of resistance to chemically induced seizures in consequence. The very first experiment was both blinded and fully controlled— by pure happenstance. And there was no placebo effect: they were cats, after all. It has even been said, tongue-in-cheek, that SMR-training for the management of epilepsy came into the world by immaculate conception. For a review, see Egner and Sterman (2006).

The same method used in cats was introduced to human subjects insofar as that was possible. The problem was that the large SMR spindle bursts that became apparent in cat sensorimotor cortex during resting-state conditions were not replicated in the human waking EEG. In the human EEG, we were faced with a smooth, well-behaved, continuous distribution of amplitudes over the low-beta frequency range. In line with the operant conditioning model, Sterman chose to set the threshold relatively high and continued to concern himself only with the threshold-crossing events. With such a high threshold, the rewards were sparse, as mandated in an operant conditioning design.

Ayers was the first to apply Sterman's method to a variety of clinical conditions (Ayers, 2004). Following Ayers, our first instrument design merely computerized Sterman's experimental design. The approach is illustrated in Fig. 8.5. The training signal is extracted from the raw signal with a filter of 3-Hz bandwidth. It is rectified and smoothed, and then a threshold is applied. Threshold crossings are signaled with a beep, and the beep is repeated (at half-second intervals) for as long as that status is maintained.

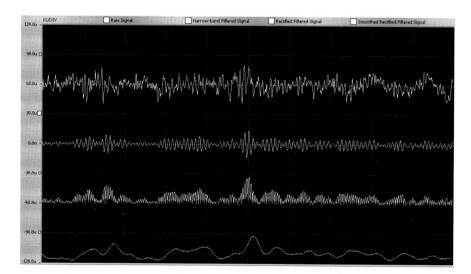

FIGURE 8.5 Conventional frequency-band neurofeedback is illustrated here.
The top trace is the raw EEG signal in 0.5–30 Hz bandwidth. The narrow-band filtered signal (3-Hz bandwidth, infinite impulse response filter, with 2nd-order roll-off) is shown in the second trace. The third trace shows the rectified signal, which is then smoothed with 0.5 Hz time constant to yield the fourth trace. A threshold is applied to the fourth trace to govern the discrete rewards.

Through her clinical work, Ayers found it helpful to set the threshold lower so that rewards would be much more plentiful, and we adopted that approach as well. With rewards no longer rare, the moorings were being loosened on Sterman's operant conditioning paradigm. The training was no longer event-focused, but rather had become state-focused. There would be runs of beeps followed by intervals of no beeps, effectively bringing lower-frequency modulatory influences into the picture. The thresholding strategy had shifted for the tactical purpose of enhancing engagement with the training task, but in fact the very nature of the process had been changed in consequence.

Our approach further differed from Sterman's in that we opted to display the entire real-time behavior of the SMR-band to the trainee, in the expectation that that would promote the brain's engagement with the process generally. This had become possible by virtue of computerization of the procedure, which we had accomplished by 1987. The brain derived far more information from the signal dynamics than we expected. This turned out to be the main event, with the threshold crossings a mere grace note. This consigned the operant conditioning aspect of the design, the threshold crossings, to an even further diminished role in the entire procedure.

DISCOVERY OF THE FREQUENCY SPECIFICITY OF NEUROFEEDBACK

With this much more dynamic approach to feedback, we observed an explicit dependence of arousal level on target frequency in the SMR-beta range. State sensitivity could be discriminated at the 0.5 Hz level (the frequency resolution we had available at the time). This seems quite surprising on its face, in that the EEG does not differ much over a 0.5 Hz range. This is shown in Fig. 8.6, where traces are given for three frequencies that differ by 0.5 Hz. It is indeed difficult to discern a difference in the time domain waveforms. Frequency domain data are also shown for the three bands, and these reveal at least a discernible

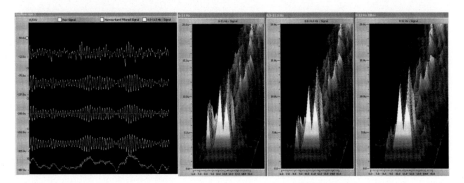

FIGURE 8.6 The EEG time course and corresponding spectral distribution are shown for three band-pass filters processing the identical signal with center frequencies that differ by 0.5 Hz around 11 Hz.

The time waveforms are difficult to distinguish; the spectrals reveal a more readily observable difference. The brain that is observing its own activity under such circumstances may very well respond differently to each of these signals.

difference. Parenthetically, it seems possible that the brain processes these signals as frequency-domain signatures rather than as time-domain waveforms, by analogy to our sense of hearing.

The previous discovery was actually somewhat fortuitous, as is often the case at the scientific frontier, and it was also long in coming. For years, we had been trying to establish the existence of a systematic difference between SMR and beta1 training, and were never able to do so. A Ph.D. dissertation was even devoted to the topic (Thorpe, 1997). Others were similarly occupied, even many years later, also without systematic success (Egner and Gruzelier, 2004). Once the coupling of arousal level and reward frequency was established firmly, however, there was no way back to standard bands. It became obligatory to tailor the training to each person in order to optimize the training.

The favored frequency was called the ORF, for Optimum Reward Frequency, or alternatively Optimum Response Frequency. The process involved moving incrementally to that frequency at which the person felt maximally calm, alert, and euthymic during the session. Symptom relief might also be experienced. It became a matter of maximizing the positive attributes and minimizing the severity of clinical complaints, and that process typically converged on a single frequency, the ORF. The process is illustrated in terms of a behavior surface in hyperspace in Fig. 8.7. The favored frequency then also consistently yielded the most propitious outcome for the training. The search for the ORF might take a number of sessions, but once it was identified, it was observed to remain rather stable over the course of training. Once the optimization strategy was adopted, it also became clear that the training of brain instabilities, such as seizures,

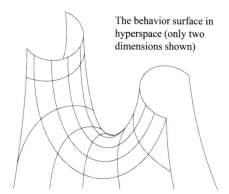

The behavior surface in hyperspace (only two dimensions shown)

FIGURE 8.7 The behavior surface is shown for the frequency dependence of two behavioral features, one that is being promoted and one that is being ameliorated.
Both the maximization and the minimization criteria are met at a single frequency, the Optimum Response Frequency. Training under these optimum conditions in state space maximizes the likelihood of a favorable outcome of the training.

migraines, panic, vertigo, and bipolar disorder was exquisitely sensitive to the choice of target frequency. Brain instabilities served as our canaries that drove the agenda for the further optimization of the training. They also presented the strongest argument for the optimization procedure itself.

The rest of the field did not follow our lead with regard to the individualization of reward frequencies, and this was for understandable reasons. Sterman had good physiological grounds for training the SMR-band of 12–15 Hz, to do so on the sensorimotor strip, and to adopt referential montage for the purpose. His ongoing work was essentially pinned to that protocol. Lubar's burden was to persuade an intransigent mainstream, and the best battering ram was to pursue a single claim with a single protocol. It was not helpful to have neurofeedback promoted as a panacea for every ailment in the mental health universe. Finally, there was nearly universal conviction that in order to get research results accepted, one had to be working with fixed protocols. The upshot was that by the time Egner and Gruzelier published in 2004 on the relative roles of SMR and beta1 bands, we had already been operating according to the ORF schema for some four years.

There was yet one more explanation for our discovery that led to our independent journey into adaptive training: We had gone back to bipolar montage, which had been universally employed in the early research of Sterman and Lubar. This likewise went against the grain of prevailing trends elsewhere within the field. From the early nineties on, there was a move to adopt QEEG-based targeting, and this came to play a primary role with respect to the inhibit aspect of neurofeedback protocols. With the attractions of QEEG-based training beckoning, there had been a corresponding shift toward the adoption of referential placement for neurofeedback, in the spirit of the reigning localization hypothesis of neuropsychology.

Initially, we responded to the appeal of this as well. In time, when it became of interest to move off the sensorimotor strip, we did so with a bipolar montage in order to keep one foot planted on familiar turf, and we observed that bipolar montage was giving us stronger effects. The brain found the relationship between two sites to be more salient than the activity at a single site. The greater level of discernment also made the training more frequency-specific, which then led to the identification of the extraordinary frequency-specificity of the response.

The concept of the ORF implied that there was an underlying frequency-based organization of brain function that was not necessarily apparent in the EEG. The implications of this are potentially huge, but we all understand that large claims demand good evidence. The only evidence that could be brought to bear in support of this concept was self-report by the trainee. Could such an edifice be constructed on the basis of mere subjective evidence? Skepticism was rampant. On the other hand, the clinical evidence was compelling, with brain instabilities in particular. Whereas a migraine might be expunged at one frequency, a nearby frequency might well evoke a migraine aura. The reproducibility of such phenomena turns anecdotal findings into evidence, and ultimately into publishable data. It is absurd for the purists to argue that once an anecdote, always an anecdote. On the contrary, the astute observations of patterns of consistency among disparate data are the very essence of good science.

By 2004, in our widening search for the ORF we had extended the range of target frequencies to cover the entire EEG band out to 40 Hz, our software limit. A substantial bias to the lower frequencies became apparent, however, and the range was gradually extended all the way down to 0–3 Hz with our 3-Hz signal bandwidth, so that the lowest target frequency was 1.5 Hz. The clinical strategy was to start the optimization procedure at 12–15 Hz, our traditional comfort zone, and to move up or down as necessary. The distribution we observed in target frequencies by 2005 is shown in Fig. 8.8. Our original default protocol of 15–18 Hz training had sunk into relative insignificance (Kaiser & Othmer, 2000). The distribution was essentially flat below the SMR range, but in fact, the highest peak was at the lowest frequency. This becomes apparent when one imagines the data plotted in terms of 1-Hz wide bins. Each of the bars in the graph represents three such bins, with the exception of the lowest, for which we have only one bin. Clearly the lowest frequency was somehow favored.

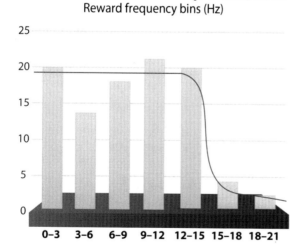

Distribution of reward frequencies, 2005

Reward frequency bins (Hz)

FIGURE 8.8 The distribution of reward frequencies that was observed in 2005 at the EEG Institute, just prior to entry into to the infra-low frequency (ILF) regime.

The most common single target frequency was actually the lowest, 1.5 Hz, which becomes apparent when one imagines this figure expressed in terms of 1-Hz bins (referring to the center frequency). This was the impetus for pursuing the further reduction in target frequency.

THE DISCOVERY OF INFRA-LOW FREQUENCY NEUROFEEDBACK

By 2006, the adoption of new software permitted the extension of the range to 0.05 Hz. This was the first venture into the training of the tonic slow cortical potential using a frequency-based approach. The clinical strategy of starting the optimization procedure at 12–15 Hz remained the same. The expectation was that the pile-up that had previously occurred at 1.5 Hz would distribute itself over the new range. Much to our surprise, the lowest frequency was once again favored. In fact, it was much more strongly favored than before, with about half the clients preferring the lowest frequency. This is shown in Fig. 8.9.

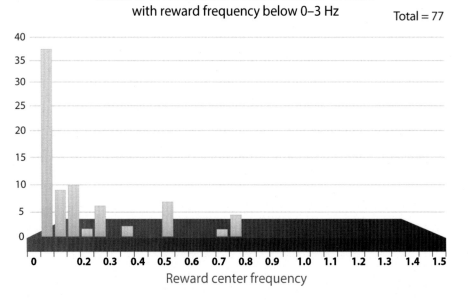

Number of EEG clients in last 6 months of 2006
with reward frequency below 0–3 Hz   Total = 77

Reward center frequency

FIGURE 8.9 The distribution of target frequencies (ORFs) is shown for the first 6-month period after the threshold to the infra-low frequencies was breached.
The lowest accessible frequency of 0.05 Hz was by far the most prevalent. This trend was further confirmed in 2007, indicating a need for the exploration of yet lower frequencies.

The venture into the infra-low frequency (ILF) region required an entirely new approach to training. The frequency was simply too low to permit threshold-based amplitude training, where the amplitude refers to the envelope of the spindle-burst activity. Instead, the trainee simply watches the unfolding low-frequency signal, which reflects the ebb and flow of (differential) cortical activation in its subtle fluctuations. One way or another, the brain appears to have no difficulty recognizing its connection to the displayed signal, and it responds accordingly. In fact, it does so with as much rapidity as at the higher frequencies. That would appear to violate expectations on the basis of signal processing theory. How can one explain quick responses to slowly changing signals? More specifically, how can such a response be so frequency-specific when an external observer would have to observe a good part of an entire period to be sure what the frequency is?

The answer is the obvious one: The brain cannot be keying on the basic low-frequency signal, which indeed is much too slow for feedback. Instead, it must be attending to the subtle fluctuations in that signal that relate to its own real-time operations. The basic rhythm being extracted by the filter software provides the context for what is being observed. It is not itself the observable.

As for the exquisite frequency sensitivity of this process, a mere change in perspective is required. Since the brain is both the agent and the observer of the unfolding signal, its sensitivity to the fluctuations is greater than that of an outside observer. The brain is not doing frequency detection. It is experiencing a process that intrinsically possesses high-frequency specificity. For both of these reasons, the trend toward lower frequencies did not lead to slower response in the training. On the contrary, by and large trainees were more responsive, and that responsiveness became apparent even earlier in the training than before. With more clinical experience, it was found that two-thirds of all clients preferred the lowest frequency. We obviously needed to move even lower to provide a wider range of options.

By 2008 new software allowed us to go down to 0.01 Hz in target frequency, with finer resolution. In short order, the peak in the distribution of ORFs moved down from 0.05 to 0.01 Hz. Quickly, it transpired that once again two-thirds of the population preferred the lowest frequency. Since the software allowed it, the target range was therefore extended to 1 mHz in late 2008. Once again, some two-thirds of clients preferred the new lowest frequency. With the bulk of the population now training at low frequency, the starting frequency was changed to 1.5 Hz. Yet the distribution of ORFs still covered the entire EEG spectrum. A number of people could not train at the low frequencies at all. The distribution is shown in Fig. 8.10.

In 2010 the range was further extended to 0.1 mHz, and after just a few months of clinical experience, the starting point of 0.1 mHz was adopted for everyone. With this fuller range available, very few clients remained who failed to optimize within the ILF range. It was apparent that clients had several preferred training frequencies over the spectrum, but of these, the lowest was always the most effectual.

In 2015 a final step was taken to extend the range even further to 0.01 mHz, and finally, the distribution of ORFs does broaden somewhat, as we had been expecting all along. On the other hand, the lowest frequency remains the dominant frequency in the distribution. With each downward step in range, the

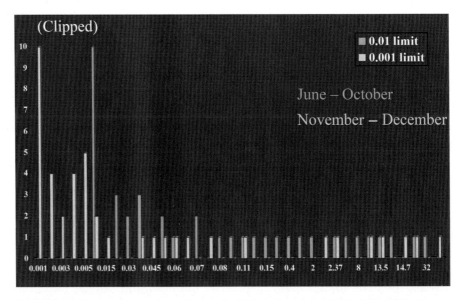

FIGURE 8.10 The distribution of target frequencies is shown for a four-month period in 2008 where the software limit had been extended to 0.01 Hz, along with the distribution obtained in the subsequent two-month period in which the software limit had been extended to 0.001 Hz, or 1 mHz. The distribution altered shape, and the pattern of preference for the lowest available frequency was sustained.

peak observed previously would be obliterated as trainees migrated toward the new low frequency. Those whose training optimized at ORFs other than the lowest were not affected.

It appears that there are two client populations. There are those whose state of dysregulation is dominated by brain instabilities, and there are those who primarily need calming of their agitated states. The former require very specific target frequencies, and the latter appear to gravitate toward the lowest target frequency that the software allows.

THE FREQUENCY RULES

Over the entire trajectory of development of this protocol, the consistent observation was made that when the left hemisphere was being trained with a lateralized placement, the optimum target frequency was twice that of the right hemisphere. This relationship held over the entire ILF range, and it held for all such placements. Interhemispheric placements required the same ORF as right-lateralized ones. We had previously established that in the EEG range of frequencies, the left hemisphere training optimized at 2 Hz higher than the right. The relationship is illustrated in Fig. 8.11. It has now been shown to hold over six orders of magnitude in target frequency.

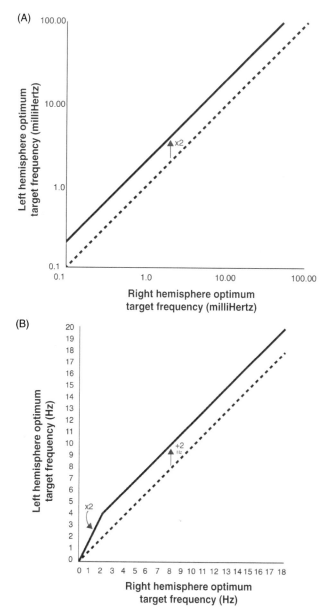

FIGURE 8.11 A consistent relationship was observed between the target frequencies (ORFs) for left- and right-lateralized placements in both the EEG spectral range and the ILF domain.
(A) The frequency relationship between left- and right-lateralized placements was geometric (harmonic) over the entire ILF region, as shown in a logarithmic plot; (B) The frequency relationship was arithmetic over the EEG frequency range, with the right-lateralized placements optimizing at 2 Hz lower than the left. The crossover between the two regimes is at 4 Hz with reference to the left hemisphere, 2 Hz with reference to the right. Historically, the relationship that prevails in the EEG range was discovered first.

Even though this relationship is entirely based on client report, it has the weight of cumulative evidence in support. First of all, it offers the virtue of consistency. The frequency rules are consistently observed across a large practitioner network that operates according to this formulation. For another, the model permits the prediction of ORFs as laterality is changed in training, and these predictions are consistently confirmed. Exceptions are relatively rare and explainable in terms of the occasional difficulty in identifying the ORF with finality. That being the case, one can use the existence of the frequency relation to argue for the validity of what has been found clinically and is being presented here. One can also make the case that an underlying order must exist according to which the brain organizes function in the frequency domain.

Once it is realized that the data presented above cannot be readily dismissed, a number of questions arise. The most obvious relates to the signal processing issues implied here. Anyone who freshly encounters this work can be counted upon to ask: "Just how is it possible to train at such low frequencies?" The second obvious question is whether any independent supportive evidence can be found.

The signal processing question must be approached in the spirit that we do have something to explain here. One cannot use concerns about signal processing to sweep the data off the table. The history of science is replete with examples of scientists ignoring data because they believed it to be impossible according to the models in their heads, and that has been the story of neurofeedback as well. The battle against rigidity of thought among the scientific elites has to be relentless. By now well over half a million people have been trained according to these frequency rules, and more than 300,000 of them in the ILF region. Thousands of practitioners are living by these rules daily. There has been ample validation.

To answer the first question, the signal processing analysis cannot rely on frequency-domain analysis exclusively. It must also take time-domain analysis into account (i.e., transient analysis). The signal that passes through the low-frequency filters still reflects the dynamics of life in real time, albeit heavily attenuated by the filter transfer function. The relevant question is then whether those dynamics are actually detectable, and that is a matter not of signal level but rather of signal-to-noise ratio. We have the empirical evidence that the signal-to-noise ratio is adequate for the purposes of feedback. The brain clearly recognizes its connection to the displayed signal, and it does so rather promptly and in nearly every case. By now, the vast majority of people respond within a first session, and if they were asked whether a placebo response could explain it, they would have little doubt that the answer is no. Their response to the training will have been unanticipated in its particulars. It cannot have been a matter of wish fulfillment.

We therefore move on to the more interesting questions raised by these findings. Uppermost is the question of whether any independent corroborative evidence exists that might support the existence of these special frequencies in the ILF regime.

IS THERE EVIDENCE FOR "SPECIAL" FREQUENCIES?

Can independent evidence be found that attests to the existence of "special" frequencies in the ILF region? It is challenging to get good data at these low frequencies because for these purposes conditions need to be stably maintained over the large sampling time. But such data do exist down to the frequency range of 0.01 Hz (Demanuele, James, & Sonuga-Barke, 2007). The authors took into account the dominant 1/f-dependence of signal amplitude in the ILF and EEG regime. Once the data are "normalized" with respect to this function, a characteristic frequency dependence emerges, one that is unique to the individual. The data are shown in Fig. 8.12. The data are shown in both linear and

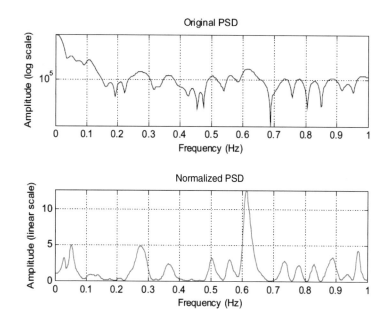

FIGURE 8.12 Persistent fine structure has been observed in the low-frequency region of the EEG and the upper end of the ILF spectrum, from 0.01 to 1.0 Hz.
This fine structure was rendered observable by first compensating for the secular 1/f–dependence of EEG and ILF spectral properties. Both logarithmic and linear plots are illustrated. The sharp nulls observed in the logarithmic plot testify to the stability of the observed spectrum at least over the data acquisition interval. It is appealing to suggest a correspondence between the ORF and one or another of these spectral peaks. *Taken from Demanuele, C., James, C.J., Sonuga-Barke, E.J.S. (2007). Distinguishing low frequency oscillations within the 1/f spectral behaviour of electromagnetic brain signals.* Behavioral and Brain Functions, 3*(62).*

logarithmic scales. The precise nulls that can be seen in the logarithmic plot demonstrate that the data are short-term stable in terms of frequency, that is to say, over the data acquisition interval at a minimum.

It is tempting to regard the peaks in this distribution as lining up with what would be identified as ORFs in the ILF training. The flat tops of the peaks then represent regions over which the response is most favorable. It is similarly tempting to suggest that the response characteristic has the shape of a resonance curve, as shown in Fig. 8.13. Both the real and imaginary components are illustrated. The real component reflects the dependence we see of the impact of the training on target frequency. The response is clearly stronger, as well as more distinctly favorable, at the resonance frequency and within its immediate vicinity (Othmer, 2009).

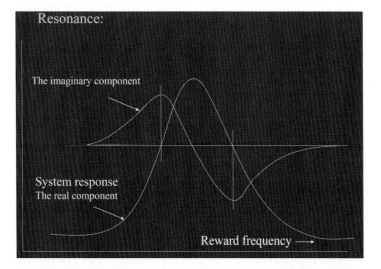

FIGURE 8.13 Collective clinical experience with the process of identifying the ORF in each case gives the impression that the response at these special frequencies is that of a resonant system. The response is both stronger and better at the resonance frequency than elsewhere. Additionally the response may be more equivocal within the immediate vicinity of resonance frequency, which could be attributed to the imaginary component of the resonant response.

What is the behavioral consequence of the imaginary component? Transient adverse responses to the training are often encountered as the resonance frequency is approached. It is as if the system is in some turmoil if it is challenged in the vicinity of the resonance frequency, and this could be identified with the imaginary component. The response is more ambiguous—it is complex. In practice, one just has to push through that zone in order to reach the desired resonance frequency. This behavior is reproducible, although for obvious reasons the deliberate testing of such reproducibility is not ethically permissible with clients. Instead, we have tested it on ourselves.

SKILL ACQUISITION

ILF neurofeedback clearly cannot be explained in terms of the standard operant conditioning model. There are no thresholds, and there are no rewards. There are not even any specific objectives. Instead, the process is directly analogous to ordinary skill learning, in which the brain witnesses the consequences of its own actions and on that basis refines its responses going forward. The brain is the creator of its own ongoing challenge in this kind of feedback. The act of recognizing its agency with respect to the signal intrinsically involves the brain's prediction of its subsequent trajectory. The brain then tries to bring about a convergence of the unfolding reality and its own prediction for the signal going forward. This engages its regulatory instrumentalities even while the brain simultaneously refines its interpretation of the signal. The process is entirely analogous to one whereby we learn to ride a bicycle or handle the steering wheel of a car. The significant differences are that in this case the skill is that of self-regulation, and the data provided relate directly to the process of self-regulation. Just as overt behavior is the "analog" of brain behavior employed in the learning of a motor skill, the EEG is the analog of brain behavior relevant to the learning of self-regulatory skills.

Ironically, the prediction model being invoked here to explain ILF neurofeedback is also required to make sense out of operant conditioning. The brain is organized as a correlation detector with a prospective orientation. Quite simply, it possesses intentionality, always projecting its immediate future. So it could even be said that the prediction-based skill-learning model described here also explains operant conditioning, rather than the other way around. As soon as correlation is detected by the brain in a standard operant conditioning design, it predicts its future course and anticipates its subsequent confirmation. Absent such confirmation, it loses interest in the hypothesis.

THE RESPECTIVE ROLES OF THE RIGHT AND LEFT HEMISPHERE IN NEUROREGULATION

Since this chapter is not a clinical dissertation, what can actually be accomplished with the ILF training has been given short shrift. For a clinical description of the process, the best resource is the Protocol Guide. It is presently in its fifth edition, which testifies to the rapid evolution of the clinical method (Othmer, 2015). A more detailed scientific description and clinical overview is to be found in Othmer, Othmer, Kaiser, and Putman (2013). With respect to clinical results, impulsivity is taken to be a good proxy for what can be accomplished with neurofeedback, in that it is largely functionally based. A report on improvements in impulsivity achieved with a mixed population of over 5000 has recently been published (Othmer & Othmer, 2016). This report is based on test results obtained in hundreds of clinics, and reflects more than 100,000 training sessions with ILF protocols. Large-scale validation of the protocols has therefore already been achieved.

On the basis of that extensive clinical experience, some top-level observations can be made that provide the context for more speculative thoughts to follow. If one surveys our collective clinical experience over the past 15 years, the dominant trend is one toward ever lower target frequencies in the drive toward greater clinical effectiveness in each case. This migration to lower frequencies bumped up against our software limits on several occasions, but was otherwise relentless.

The driver here must have been the migration toward the foundations of our regulatory hierarchy. Most certainly that holds in the frequency domain, but it also holds more generally. The training strongly impinges on core regulatory function of arousal, affect, autonomic balance, and interoception. All these are influenced by the higher-frequency training as well, but not as strongly or as directly. The matter of affect regulation and autonomic regulation with neurofeedback have rarely merited discussion in the literature. In the EEG range of frequencies, the signal is dominated by the specific function it supports, whereas in the ILF region, cortical activation is more directly reflected in the signal. Hence, it is more directly influenced in the ILF training process as well.

Within the ILF region, however, it remains counter-intuitive that the lower frequencies should turn out to have stronger effects in the training. Being stronger, they also make their appearance obvious sooner. It is possible that the lower frequencies offer a cleaner signal, one less compromised by extraneous factors. More than likely, however, the ILF training has revealed a hierarchy of regulation that extends all the way down to the circadian rhythm, and has demonstrated that training at the bottom of the hierarchy is more cogent.

Concurrently with the migration to ever lower frequencies there has been a migration toward right-side dominance in the allocation of training time. Every training starts either with right hemisphere training or with interhemispheric placement, or both. The standard placements are shown in Fig. 8.14. Since both placements train optimally at the same frequency, it is tempting to suggest that the right hemisphere plays a controlling role in both. In that early phase, left hemisphere training may not even be tolerated in the ILF regime. A kind of scaffolding model applies, in that the foundational protocols enable what follows. One has the strong sense that we are recapitulating the developmental hierarchy, refurbishing functional connectivity relationships sequentially as one progresses.

The necessity of allowing for two starting protocols traces to the two core issues. The first is that of calming of the nervous system, which is a priority for those who require the right parietal training. The second is brain stability,

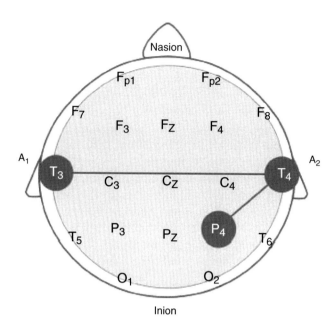

FIGURE 8.14 **The two principal starting placements are illustrated here. They target the two primary issues of the regulatory hierarchy, brain stability and arousal regulation.**
The interhemispheric placement targets brain stability, and the parietal placement targets arousal regulation. Both placements target regions where the Default Mode Network is accessible to us at the cortical surface. In a fraction of cases both starting protocols are required from the outset.

which likely determines the ORF whenever it is an issue. Those who need calming as a first priority tend to gravitate to the lowest frequency, whereas those who require stabilization distribute themselves over the spectrum. Some clients require both protocols from the outset. In these cases, we always have one ORF, not two.

The ILF training has been described as "a dance with the brain, but the brain gets to lead." The client's brain leads us to its priorities, provided one has the sensitivity to pay attention and the proper model to interpret the observations. The brain needs information on its own state to migrate out of its own constraints, and it guides us to the information it finds most useful to that end. It would indeed be the height of arrogance on our part to override the brain's wisdom that becomes apparent in this process.

The conclusion is inescapable: Sensitivity to the brain's own urgencies has led us to the lowest frequencies that are dynamically managed by the brain, to the foundations of the regulatory hierarchy (i.e., arousal and affect regulation; autonomic regulation and interoception), and to right-hemisphere priority. A consistent picture emerges when it is realized that the right hemisphere has priority with respect to the regulation of these core functions, our vegetative functions, and our internal homeostasis and sense of external safety.

Within the framework of the model of Intrinsic Connectivity Networks (ICNs), the quality of regulation is governed by the functional connectivity within the Default Mode Network (DMN) of brain function (Raichle et al., 2001; Raichle, 2010). Effectively we live our lives out of the Default Mode, the task-negative network. Task-positive networks are a mere perturbation on that core network organization. Since our concern is state regulation, our neurofeedback conversation is largely oriented to the DMN. Buckner has traced the functional connectivity of the hubs within the DMN and found the right hemisphere to be more tightly interconnected than the left, in particular, the mid-temporal regions that are our primary training sites (Buckner, Andrews-Hanna, & Schacter, 2008). Sridharan has shown that mediation between the DMN and the Central Executive Network is strongly dominated by the right fronto-insular cortex, the central hub of the salience network (Sridharan, Levitin, & Menon, 2008).

The Salience Network ranks second only to the Default Mode in its relevance to our task. It performs a global monitoring function to ascertain the integrity of internal regulatory function, as well as our safety with respect to external threat. More specifically, it is the right insula that principally bears this burden. The salience network is ranked with the task-positive networks, but in fact, its primary role is a relatively passive monitoring function. This is an "always-on" function that could equally well be ranked with the Default Mode in that regard. One could therefore alternatively understand the salience network as a perpetual partner to the DMN, seeing us safely through our periods of nonengagement and performing general risk management.

It is noteworthy that most of our clinical work to date has involved either lateralized training or interhemispheric placement. This is a good springboard to the next topic, the relationship between the two hemispheres as illuminated by our clinical history.

RELATIONSHIP OF THE LEFT AND RIGHT HEMISPHERES

It is of interest to inquire further into the relationship of the left and right hemispheres that is implied by the frequency rules. The following presents a speculative model that invites further inquiry. Whereas in the realm of computer technology the governing frequency of the system is the highest, namely the clock frequency, in biological systems it is the lowest frequency that sets the tone, so to speak. This suggests that the right hemisphere is playing a controlling role with respect to the core regulatory functions that are being engaged with the ILF training. The left hemisphere organizes itself on a harmonic of that activity. This implies that even when we do left-hemisphere training in the ILF domain, the right hemisphere still plays the controlling role, and our primary appeal is to those regulatory functions that are the principal responsibility of the right hemisphere. How then does the left hemisphere fold into this model?

The question of information flow between the two hemispheres was recently addressed in a study that elaborated on Lehman's microstate model in the context of what we now know about resting state networks (Lehmann et al., 2014). The study demonstrated that in the alpha and low beta range of frequencies, net information flow is strongly dominated by the left hemisphere. Much more information flows from the left hemisphere to the right than goes the other way. This leads readily to the supposition that the left hemisphere plays a controlling role with respect to the EEG bands that are most heavily involved in state regulation, namely the lower frequencies in the theta, alpha, and low beta bands.

Lehman's microstate model shows the brain toggling sequentially between four different microstates, two of which have a lateralized bias, and the others have an axial bias. What is observed in fMRI are the time-averaged effects of this activity. What the fMRI data also revealed is the hub structure, most particularly of the Default Mode. The current work revealed that all four microstates are organized around a common posterior cingulate generator, the parietal hub of the DMN. This hub is also the first to be discernible in imagery in early development. One of the microstates additionally has a left parietal hub; one has an additional right parietal hub; and a third has a frontal hub, the anterior cingulate. The fourth has solely the posterior hub. They are illustrated in Fig. 8.15.

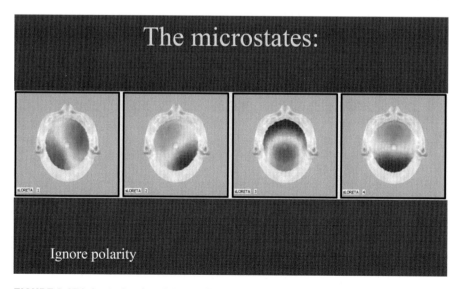

FIGURE 8.15 Lehmann's microstates are illustrated here.
One is left-lateralized, with a posterior bias; one is right-lateralized, also with a posterior bias. Two are axially oriented. All four microstates involve the posterior hub of the Default Mode, and only one of the four involves the frontal hub (Lehmann et al., 2014).

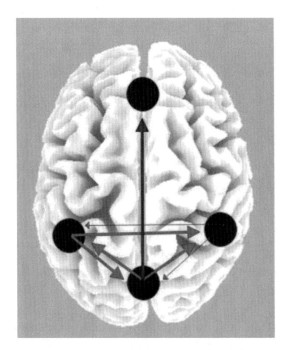

FIGURE 8.16 The hub structure underlying the microstates is illustrated here, along with the dominant information flow among the hubs that prevails in the alpha and low-beta regimes.
The predominant conversations are among the posterior hubs, but with a pronounced left-hemisphere bias. As for the front-back axis, the predominant information flow is from back to front. Firstly, this justifies giving the parietal region priority in the training of arousal regulation. Secondly, it supports the view that in the EEG frequency range the left hemisphere has priority with respect to regulatory functions subserved by the low to intermediate frequency EEG (Lehmann et al., 2014).

The corresponding hub configuration is shown in Fig. 8.16. The specific finding of the Lehmann study is that the information from the left parietal hub to the central posterior hub and to the right parietal hub vastly dominates the flow going the other way. Further, the flow from the posterior to the anterior hub dominates the reverse flow.

The following generalization, therefore suggests itself: In the ILF regime, the right hemisphere appears to be exercising a controlling role, with the left hemisphere slaved to it at a harmonic of the ORF. In the EEG range of frequencies, the left hemisphere is deemed to be playing the controlling role, but it organizes itself with respect to the right hemisphere at a frequency 2 Hz higher. The crossover between the two domains is at the only place where self-consistency is preserved, namely at 4 Hz on the left and 2 Hz on the right. The relationships are illustrated in Fig. 8.17. One may also conjecture that the ORFs across

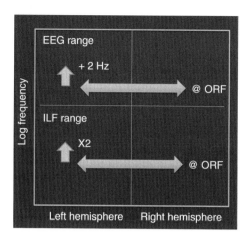

FIGURE 8.17 The proposed relationships between the left and right hemispheres in matters of state regulation are illustrated here.
The right hemisphere has priority with respect to core regulation, as reflected principally in the ILF region. Here the left hemisphere tracks the right at the second harmonic. In the EEG range of frequencies, by contrast, the left hemisphere assumes dominance, and the right follows at a lower frequency by two Hertz. Brain stability involves the relationship between the hemispheres most directly, and with respect to this core regulatory issue, the right hemisphere is dominant.

the spectrum are privileged frequencies at which this interaction is principally organized to effect mutual governance.

Within this general framework, when it comes to functions we associate with the limbic system we would look to train with right-hemisphere priority, and when it comes to functions we associate with the basal ganglia and the thalamocortical loop, we would look to train at some point with left-hemisphere priority. The context for this is provided by the regulatory hierarchy, the developmental hierarchy, the frequency hierarchy, and the scaffolding model (Chiron et al., 1997). The ILF training with right-hemisphere priority would therefore generally precede the training in the EEG range with left-hemisphere priority.

On the basis of the above, one may regard the neurofeedback training in the SMR and beta bands as involving some significant left-hemisphere priority, irrespective of the actual electrode placement. When it comes to left-hemisphere function, we recognize the terrain because the Sterman protocol and others have been occupying it for forty years already. We ourselves spent fifteen years there. Over that entire history, training protocols in the SMR-beta range have had a pronounced left-hemisphere bias.

THE RESPECTIVE ROLES OF NEUROFEEDBACK AND STIMULATION

With the above background, some commentary on the preferred roles for neurofeedback and stimulation is in order. Historically, stimulation methods have had the advantage of eliciting an almost immediate response, whereas neurofeedback protocols were typically slow to ramp up to where clinical benefit was unambiguous. Stimulation-based methods using audio-visual stimulation (AVS) were therefore the first to gain wide interest, and that interest was sustained through the (temporary) eclipse of alpha training in the late seventies.

The problem has been that the very accessibility of AVS systems caused them to be broadly adopted and used with insufficient intelligent guidance. With respect to clinical applications, the critical step is finding one's way toward the desired outcome, and that difficulty afflicts both kinds of technologies. Success in bringing the right remedy to bear on a particular problem sets a ceiling on what one may accomplish with both kinds of methods.

At higher frequencies, neurofeedback training becomes more problematic because gamma band activity only persists for short intervals. By the time feedback is given on prevailing activity, the brain has already moved on, by virtue of signal processing delays. Good work can still be done here, however. In this frequency region, the stimulation-based methods offer distinct advantages, at least in principle. For one thing, they can be used open-loop, on the principle of stochastic resonance. In this approach, the stimulation is applied randomly (with respect to ongoing EEG activity), on the assumption that occasionally it will be of such a phase as to induce a salutary response. At other times, the relative phase will be such that the brain is not particularly responsive. With respect to closed-loop operation, one may simply have lost some training efficiency.

It is in the conventional EEG range (say up to 40 Hz) where a good case can be made for both technologies. A choice, if one needs to be made, will likely be driven by nontechnical issues, such as practitioner preference. The bias of the Western mind is to place itself in charge of the process of neurorehabilitation and to prescribe the remedy. The Eastern perspective, out of which biofeedback and neurofeedback first arose, is more observational, reflective, and adaptive. The choice of tool will be more driven by the person in charge of the process than by inherent merits of either approach.

There is, however, one core issue that should be discussed in this connection. Some kinds of neurorehabilitation are facilitated by way of relaxation and others by way of activation. Clinicians tend to be oriented toward one of these or the other. ILF training unambiguously belongs with the relaxation-enabled technologies. The principal criterion for arrival at the ORF is that the person feels calm (while retaining alertness). The destressed and nonengaged state facilitates the functional renormalization that must take place within the intrinsic connectivity networks. Engagement, it is argued, commits the networks to a particular configuration and thus limits the brain's options.

On the other side, we have the potential of activation-induced plasticity, on the basis that the near neighborhood of any state of the brain is also available in state space. So if brain stimulation is used to push the brain into new terrain, the brain can migrate to a new state configuration more readily than otherwise. However, this distinction should not be overdone. Reinforcement methods also take the brain into terrain it would not otherwise occupy, and likewise enlarge the scope of neuronal activity that can be recruited into the rhythmic activity being reinforced. One should also not confuse the calmness that is sought in ILF training with stasis. If anything, the exposure to the ILF signal accelerates the migration of the state vector through state space, thus exposing the brain to more of the attractor landscape, and thus hazarding capture by adverse attractors. That is why the ILF training always needs to be attended by a knowledgeable and vigilant clinician. The training parameters may need to be adjusted on the fly to redirect the trajectory of the state vector.

In the general case, there is likely a need for both kinds of rehabilitation, but an obvious hierarchy exists. The relaxation-induced recovery should be accomplished before the brain is challenged to perform. If that is not done, then the process may well be limited in what it can accomplish because it is built on dysfunctional core network relations. The evidence for this proposition is cumulative from clinical experience, from the way clinical progression plays out presently in comparison to earlier higher frequency training. When the ILF training is done first, conditions often resolve that in an earlier day were addressed satisfactorily with conventional EEG band training. The converse, however, is not the case. Progress can be made with developmental delay, the autism spectrum, addictions, schizophrenia, epilepsy, migraines, bipolar disorder, and dissociative states that could not be achieved without ILF training—or at least not nearly as readily.

Finally, it must also be said that what has been discussed separately here can be readily combined in clinical practice. Each method would be deployed where it offers unambiguous advantages. In the ILF region, stimulation-based approaches are not applicable. In the high-frequency region, by contrast, stimulation-based approaches have natural advantages over training. In the intermediate region, one would ideally deploy both training and stimulation, as appropriate. Whereas prudence calls for introducing new approaches one at a time to new clients, there is no reason not to combine different methods into one training challenge incrementally.

It is in fact customary in clinical practice to combine ILF training with an inhibit strategy that covers the conventional EEG band. In a reinforcement paradigm, however, such inhibits are simply tokens of the brain's indiscretions. They carry no mandate for the brain. Stimulation-based methods can be used to actively disrupt such activity, evoking the brain's response more compellingly than can be achieved passively through mere alerting.

The inhibit-based protocols typically target dysfunctions that are observable in terms of excess EEG band amplitudes. These may be localized to particular sites or regions, and they tend to predominate in the lower part of the spectrum, particularly in the theta region and below. Similarly, there may be excess bispectral amplitudes, meaning excess correlations between spectral amplitudes of two frequencies at the same site. Also, there may exist excess coherences or even cross-frequency coupling between different sides, all exhibiting a bias toward the lower EEG frequencies. The watchword here is excess: when networks are in too intimate communication, are too closely or too persistently coupled, functional discrimination is lost. It must be said, however, that the discriminant here is organized for the recognition of excess.

At higher EEG frequencies, by contrast, functional deficits may arise from deficiencies in coupling between sites, manifesting in EEG amplitudes that are too low. One thinks in terms of the brain failing to rise to the challenge of cognitive function. This may point to an intrinsic problem in the gamma region of frequencies, or to a failure to coordinate the nesting activity at lower frequencies that govern the gamma range bursting activity. This traces back to a fundamental issue related to frequency. At high frequencies, it is a challenge to maintain common phase over the spatial domain at issue. The amplitudes are an index to the prevailing degree of local synchrony. At low frequency, one has the problem of discrimination among nearby frequencies.

A CAUTIONARY NOTE

The effects of both neurofeedback and stimulation are often so strong and so dramatic that the clinician may be drawn into a kind of benign tunnel vision that focuses on the obvious benefits of the training at the expense of attention to any adverse accompaniments. There is even the temptation to dismiss the adverse side effects as having anything to do with the neurofeedback or brain stimulation at all.

It must be recognized that both reinforcement-based neurofeedback and brain stimulation are essentially provocations that the brain has to contend with. In the systems perspective, the consequences likely extend beyond the matters of immediate clinical interest. The application of extrinsic constraints (through stimulation) cannot suffice to specify the optimal operational state for the brain. In short, self-regulation cannot be outsourced. The brain has to manage the dynamic frontier of self-organized criticality. Seen in that context, the most refined type of feedback involves the brain's tracking of its own states—the skill learning model.

The most comprehensive approach to brain rehabilitation and brain fitness training both begins and ends with skill learning. In between, both reinforcement- and stimulation-based methods can drive the agenda efficiently toward their appointed ends. Ultimately there should be no side effects of the training at all. Good self-regulatory competence should not have to be purchased at the cost of something else. This is a standard that all neurotherapy models should strive to achieve and maintain.

Implied in the foregoing is an essential kinship between neurofeedback and low-level stimulation. Although the evidence is not adduced here, the cumulative clinical experience demonstrates that the combination of the methods here discussed can effect substantial remediation across the entire spectrum of mental disorders, either in combination with pharmaceutical support or independently. Implicit also is the observation that there is no clinical objective for high-level stimulation, such as rTMS, that cannot also be met with low-level stimulation. The remedy lies in the altered behavior of neuronal assemblies, and the combination of feedback with low-level stimulation has shown itself to be sufficient for the purpose.

CONCLUSIONS

In this chapter, the thesis has been presented that the most critical failure modes of brain regulation involve the mass action of neuronal assemblies in the domain of timing and frequency. An understanding of the cerebral regulatory regime is therefore of keen interest. Neuronal networks are spatially organized into hierarchical configurations with small-world character. At every level, the distributions are scale-free. In the time domain, the neuronal assemblies are organized in terms of a frequency hierarchy that extends down to the circadian periodicity. The dynamics are observed to be scale-free; they drive the brain to a state of self-organized criticality.

The basis has been laid for the proposition that much of mental dysfunction is traceable to functional deficits in the domain of timing and frequency. As these are dynamically organized, they should be susceptible to systematic remediation with methods of reinforcement or stimulation, or simply through skill learning. The case is made for a regulatory hierarchy in the frequency domain that directly parallels the developmental hierarchy and the functional hierarchy of the cerebrum. Advantages accrue to organizing the training with cognizance of the regulatory hierarchy in the spatial domain as well as in the frequency domain. This means initiating the training in the ILF region with right-hemisphere priority, particularly for those afflicted with profound core dysfunction. That, in turn, lays the basis for the most productive deployment of reinforcement and stimulation-based methods to address residual functional deficits, accompanied by skill-learning approaches for the achievement of optimal function.

Finally, the clinical exploitation of neurofeedback has led to the discovery of frequency rules that govern brain organization in the frequency domain, and suggest that an underlying order prevails that governs interhemispheric coordination. These frequency rules support the hierarchical organization in the frequency domain, and they support right-hemisphere priority with respect to core regulatory function.

References

Alexander, A. R. (2014). *Infinitesimal: how a dangerous mathematical theory shaped the modern world.* New York: Scientific American/Farrar, Straus and Giroux.

Ayers, M. E. (2004). Neurofeedback for cerebral palsy. *Journal of Neurotherapy, 8*(2), 93–94.

Bak, P. (1997). *How nature works: the science of self-organized criticality.* New York: Springer-Verlag.

Barabási, A. (2002). *Linked: The new science of networks.* Cambridge, MA: Perseus Pub.

Berger, H. (1929). Über das Elektrenkephalogramm des Menschen. *Archiv für Psychiatrie und Nervenkrankheiten., 87,* 527–570.

Brown, B. B., & Klug, J. W. (1974). *The alpha syllabus: a handbook of human EEG alpha activity.* Springfield, IL: Charles C. Thomas.

Buckner, R. L., Andrews-Hanna, J. R., & Schacter, D. L. (2008). The brain's default network: anatomy, function, and relevance to disease. *Annals of the New York Academy of Sciences, 1124*(1), 1–38.

Chiron, C., Jambaque, I., Nabbout, R., Lounes, R., Syrota, A., & Dulac, O. (1997). The right hemisphere is dominant in human infants. *Brain.* 120, 1057–1065.

Demanuele, C., James, C. J., & Sonuga-Barke, E. J. S. (2007). Distinguishing low frequency oscillations within the 1/f spectral behaviour of electromagnetic brain signals. *Behavioral and Brain Functions, 3,* 62–72.

DeMott, D. W. (1970). *Toposcopic studies of learning.* Springfield: Charles C Thomas.

Egner, T., & Gruzelier, J. H. (2004). EEG Biofeedback of low beta band components: frequency-specific effects on variables of attention and event-related brain potentials. *Clinical Neurophysiology, 115,* 131–139.

Egner, T., & Sterman, M. B. (2006). Neurofeedback treatment of epilepsy: from basic rationale to practical application. *Expert Review of Neurotherapeutics, 6*(2), 247–257.

Engel, A., Konig, P., Kreiter, A., & Singer, W. (1991). Interhemispheric synchronization of oscillatory neuronal responses in cat visual cortex. *Science, 252*(5009), 1177–1179.

Fiser, J., Chiu, C., & Weliky, M. (2004). Small modulation of ongoing cortical dynamics by sensory input during natural vision. *Nature, 431*(7008), 573–578.

Hagmann, P., Cammoun, L., Gigandet, X., Meuli, R., Honey, C. J., Wedeen, V. J., & Sporns, O. (2008). Mapping the structural core of human cerebral cortex. *PLoS Biology, 6*(7.), .

Haken, H., & Stadler, M. (1990). *Synergetics of cognition: proceedings of the international symposium at Schloss Elmau, Bavaria, June 4–8, 1989*. Berlin: Springer-Verlag.

Hill, A.R. (2013). Measuring and Modulating Hemispheric Attention. UCLA: Psychology 0780. Available from: http://escholarship.org/uc/item/55c35139

Kaiser, D. A., & Othmer, S. (2000). Effect of Neurofeedback on Variables of Attention in a Large Multi-Center Trial. *Journal of Neurotherapy, 4*(1), 5–15.

Kelso, J. A. (1982). *Human motor behavior: an introduction*. Hillsdale, NJ: L. Erlbaum.

Lehmann, D., Ozaki, H., & Pal, I. (1987). EEG alpha map series: brain micro-states by space-oriented adaptive segmentation. *Electroencephalography and Clinical Neurophysiology, 67*(3), 271–288.

Lehmann, D., Pascual-Marqui, R.D., Milz, P., Kochi, K., Faber, P., Yoshimura, M., Kinoshita, T. (2014). The resting microstate networks (RMN): cortical distributions, dynamics, and frequency specific information flow. Avaiable from: http://arxiv.org/abs/1411.1949

Lubar, J. O., & Lubar, J. F. (1984). Electroencephalographic biofeedback of SMR and beta for treatment of attention deficit disorders in a clinical setting. *Biofeedback and Self-Regulation, 9*, 1–23.

Magana, V., Wirt, R., Pina, D., Escobar, O., Becerra, A., Garcia, S., & Abara, J.P. (2016). *Electrophysiological and Behavioral Evidence of Neurofeedback Enhancement of Attention and Preparatory Response*. Society for Behavioral Medicine, 37th Annual Conference. Washington DC.

Modha, D. S., & Singh, R. (2010). Network architecture of the long-distance pathways in the macaque brain. *Proceedings of the National Academy of Sciences, 107*(30), 13485–13490.

Othmer, S. (2009). Neuromodulation Technologies: An Attempt at Classification. *Introduction to Quantitative EEG and Neurofeedback: Advance Theory and Applications* (2nd ed.), 3–26.

Othmer, S., & Othmer, S. F. (2016). Infra-Low Frequency Neurofeedback for Optimum Performance. *Neurofeedback and Biofeedback Treatments for Advances in Human Performance Case Studies in Applied Psychophysiology*. Los Angeles: The EEG Institue.

Othmer, S. F. (2015). *Protocol Guide for Neurofeedback Clinicians*. Los Angeles: EEG Info.

Othmer, S., Othmer, S. F., Kaiser, D. A., & Putman, J. (2013). Endogenous neuromodulation at infra-low frequencies. *Seminars in Pediatric Neurology, 20*(4), 246–260.

Plenz, D., & Niebur, E. (Eds.). (2014). *Criticality in neural systems*. Germany: Weinheim.

Raichle, M. E. (2010). Two views of brain function. *Trends in Cognitive Sciences, 14*(4), 180–190.

Raichle, M. E., Macleod, A. M., Snyder, A. Z., Powers, W. J., Gusnard, D. A., & Shulman, G. L. (2001). A default mode of brain function. *Proceedings of the National Academy of Sciences, 98*(2), 676–682.

Sherrington, C. (1906). *The integrative action of the nervous system*. New Haven, Conn: Yale University Press.

Sridharan, D., Levitin, D. J., & Menon, V. (2008). A critical role for the right fronto-insular cortex in switching between central-executive and default-mode networks. *Proceedings of the National Academy of Sciences, 105*(34), 12569–12574.

Sterman, M. B., Howe, R. D., & Macdonald, L. R. (1970). Facilitation of spindle-burst sleep by conditioning of electroencephalographic activity while awake. *Science, 167*, 1146–1148.

Thorpe, T. (1997). EEG biofeedback training in a clinical sample of school age children treated for attention-deficit/hyperactivity disorder. *California School of Professional Psychology* (unpublished).

Zhang, K., & Sejnowski, T. J. (2000). A universal scaling law between gray matter and white matter of cerebral cortex. *Proceedings of the National Academy of Sciences, 97*(10), 5621–5626. doi: 10.1073/pnas.090504197.

Further Reading

Freeman, W. J. (2000). *How Brains Make up Their Minds*. New York: Columbia UP.

Freeman, W., & Kozma, R. (2010). Freeman's mass action. *Scholarpedia, 5*(1), 8040.

Kandel, E. R. (2006). *In search of memory: the emergence of a new science of mind*. New York: W.W. Norton & Company.

Kelso, J. A. (1995). *Dynamic Patterns; the Self-Organization of Brain and Behavior*. The MIT Press, Cambridge, Mass.

Raichle, M. E. (2011). The restless brain. *Brain Connectivity, 1*(1), 3–12.

CHAPTER 9

Cultural Factors in Responses to Rhythmic Stimuli

Udo Will

Ohio State University, Columbus, OH, United States

In recent decades the cognitive sciences have seen some remarkable developments that reflect shifting and broadening perspectives of the discipline. One is the recognition that cognition is shaped and formed by an intricate interplay of biological, environmental and experiential, that is, cultural factors. This realization has profited from a notable tradition of cultural and cross-cultural research in psychology (Berry, Poortinga, Bruegelmans, Chasiotis, & Sam, 2011; Berry, 2011) with an initial research emphasis on social cognition, personality traits, cognitive styles, and so on. It has produced multiple lines of evidence for cultural differences engendering differences in psychological behavior (Han & Northoff, 2008; Kitayama & Cohen, 2007), and, following the introduction of modern imaging methods (fMRI, EEG,MEG), has uncovered influences of cultural factors on various brain functions and activations of specific neural substrates, including low-level perceptual processes (Han et al., 2013; Chiao & Immordino-Yang, 2013).

Another 'discovery', that of music as a relevant domain for cognition research, expanded the purview of the cognitive sciences, which since their inception, had mainly used language processing as their main reference model for complex human cognition. The 'new' field promised to deliver valuable and informative insights into the functioning of the human brain via the analysis of commonalities and differences in the domains of music and language. It also offered, at least in principle, plenty of opportunities for cross-cultural studies as music is shaped and influenced in multiple ways by cultural conditions (Trehub, Becker, & Morley, 2015). However, following the tradition of music psychology, research was initially devoted to Western tonal art music, using stimuli derived from this art form and participants familiar with it. One of the main reasons for this bias was the musical background of the researchers, who

Rhythmic Stimulation Procedures in Neuromodulation. http://dx.doi.org/10.1016/B978-0-12-803726-3.00009-2

rarely had experience with and knowledge of nonwestern music. The exclusive focus on Western music was and is problematic because it puts into question any generalizing conclusion drawn from such research: music develops within a specific sociocultural context, and concepts and practices of Western music follow different principles than those from nonwestern societies (Will, 1999). With the mounting realization of the influential role of cultural factors in the ontogenetic formation of human cognitive capacities, there was a growing need for a switch from a so-called culture-neutral framework, which in effect had a clear bias toward Western culture, to a culture-specific one. Consequently, since a couple of years we see an increasing number of studies investigating cultural variability of cognitive processing of music (Stevens, 2011). Culture-specific research projects have for instance been able to show that enculturation affects one's understanding of musical structures even before adulthood (Morrison, Demorest, & Stambaugh, 2008), and that it shapes our musical memory (Demorest et al., 2010).

However, this kind of research has so far mainly focused on general musical abilities and pitch and melody processing. Only a few cross-cultural studies have been conducted that investigate cultural effects on rhythm processing, even though this is likely an especially pertinent field for cross-cultural cognition studies due to specific aspects of time and rhythm: As I have pointed out elsewhere, a remarkable fact concerning time and rhythm is that, though temporal experience and exploration of the world are essential components of our being in this world and of the interactions with our physical and social environment, humans have no special sense organs for time or rhythm—nor does any other organisms for that matter (Will, 2016). What we experience as time and rhythm is not something given, preexisting in the outside world. It is a cognitive construct, the components for which are extracted from and mediated by our actions, in addition to perception of other types of information, and as a construct it is shaped and formed not only by physiological and psychological processes, but also by sociocultural factors, ideas, and concepts.

It may be useful to recall that the significance of our temporal processing capabilities is not primarily, as suggested by many a laboratory context, to allow us to make 'time measurements', that is, determine durations or interval lengths per se. Their significance arises much more from the fact that they enable us to detect temporal similarities and regularities, rhythmic pattern, and periodicities in the world around us, especially our social environment, and to adjust our behavior accordingly. Such a view on time and rhythm processing is guided by an action oriented approach to cognition, one in which cognition is understood as enactive, as a form of practice. A key premise of it holds that cognition cannot be adequately understood as merely serving to construct hypotheses about the world, but rather as subserving action through the identification of patterns and regularities in the dynamic sensorimotor coupling between agents and their environment. Accordingly, cognitive states and their associated neural activity patterns need to be studied primarily with respect to their functional role in action generation. As will be shown in the following, such an inactive paradigm is not merely a hypothetical and conceptually sustainable one. It is grounded in plenty of experimental evidence and supported by theoretical approaches that demonstrate the action-relatedness of temporal processing.

Within this enactive framework it is the dynamic sensorimotor coupling of agent and environment that is at the basic of the cultural influences on cognition. However, a notable challenge for research investigating these influences is the conceptualization and operationalization of culture. Culture is frequently conceived of as an observable, empirical reality, an irreducible entity formed by an enumerable set of components, with clearly identifiable boundaries and stability across time. Such a reified notion is problematic as it leads to considerable conceptual and logical contradictions (Kuper, 1999; Will, 2007). Apart from the practical impossibility to identify the complete set of components, such a notion is also difficult to reconcile with empirical data as it flouts the historic character of human societies and the contingency of their constantly changing boundaries. Human interaction and communication networks are not limited by politically defined administration units (Amselle & M'Bokolo, 1999; Amselle, 2001). Hence, to operationalize culture as nation, state or similar geopolitical units, makes little sense from an explanatory perspective (Frijda & Jahoda, 1966).

Because the notion of culture as either a reified, empirical system or as a fixed collection of enumerable units does not fit well the empirical evidence (d'Andrade, 2001), there is a pronounced trend to move away from the use of general cultural dimensions and to focus on more specific explanations (Breugelmans, 2011). This can be done on the basis of a reconceptualization of the notion of culture as an attribute characterizing socially learned behavior and socially attained and transmitted knowledge and values, as it has, for example, been proposed by Tomasello (1999). We can then employ these cultural factors, that is, the defined specific practices, and investigate the effect they have on cognitive processes. Furthermore, as there is no a priori defined set of such practices for any social group, we can also examine the effects of different sets of cultural factors across groups, even within larger human societies, thereby avoiding the pitfalls of the traditional reified notion of culture. For example, two of the more frequently studied cultural factors are 'being a musician', that is, having received special training in and exposure to a specific music, and 'being speaker of a particular language', that is, having exposure to and practical experience with a specific language. If cultural factors are used in intragroup studies they have the advantage, given all other cultural factors are the same for all group members that they permit to explore the influence of individual factors. However, despite their conceptual advantages, when using them in cross-cultural research there are still considerable methodological challenges to overcome. For example, a factor needs to be operationalized in a way it designates equivalent learned behaviors in each of the groups to be compared. Additionally, in practice, cultural factors are frequently not sufficiently specified to guarantee they denote equivalent behaviors across groups. In cases like these the interpretation of results can be rather limited, often restricted to the mere assertion of group differences, and with no further analysis of how the differences relate to specific aspects of the factors investigated. Such difficulties and limitations will be considered in the studies reviewed later.

Before examining the research that has investigated cultural variability of rhythm processing, it seems apt to present a short overview of the current views on models and mechanisms of time and rhythm processing that will give us the necessary background to gauge the place of those studies within the current research context.

CURRENT MODELS OF TIME AND RHYTHM PROCESSING

Some models of time perception propose that timing is accomplished by a single, a model process. These models, also called 'dedicated models' (Ivry & Schlerf, 2008), suggest that temporal processing is done by specialized and unique neural structures that are employed across different tasks and modalities. The problem with such models is that there is little support for them from clinical studies. No specific brain structures dedicated to temporal processing have so far been identified: patients with cerebellar degeneration, Parkinson's disease, or prefrontal lesions, all show similar dysfunctions in temporal perception. In addition, an increasing number of studies report results that are difficult to reconcile with models of dedicated, amodal temporal processing. This situation has led to the development of alternative, intrinsic models (for review: Ivry & Schlerf, 2008; Wiener, Turkeltaub, & Coslett, 2010), which assume that various parts of the brain can fulfill such functions because of the inherent temporal character of neural processing. Models of intrinsic timing mechanisms propose that not only different task and different timespans but also different sensory modalities employ different neural structures and mechanisms. On the basis of such models one would expect modality-specific temporal processing, and even within one modality different mechanisms should be identifiable because different contexts and tasks are expected to recruit different neural structures and/or processes.

In a metaanalysis of 41 imaging studies on time perception, Wiener et al. (2010) found that, indeed, different neural structures seem to be involved in time processing depending on stimulus and task characteristics of the studies. Main differences in activation patterns were related to different stimulus durations and to tasks that involved either perceptual or motor timing. For example, the study by Lewis and Miall (2003) tested brain activation patterns for processing of durations of 0.6 and 3 s. Their results showed that, although many brain regions were active during both sub- and suprasecond duration perceptions, there were specific differences for the two time spans. The frontal operculum, left cerebellar hemisphere and middle and superior temporal gyri, all showed significantly greater activity during processing of the shorter intervals, supporting the hypotheses that the motor system is preferentially involved in the measurement of sub-second intervals. Additionally, a neurochemical study by Wiener, Lohoff, and Coslett (2011b) indicates that different dopaminergic systems are involved in subsecond and suprasecond timing processes.

Various studies have advanced evidence for sensory specific mechanisms in temporal processing. For example, duration discrimination thresholds have been shown to be lower for auditory than for visual stimuli (Grondin, Roussel, Gamache, Roy, & Ouellet, 2005; Merchant, Zarco, & Prado, 2008). Furthermore, the involvement of primary sensory cortices (Shuler and Bear, 2006), and the fact that certain brain areas are only involved in temporal processing of one modality (Bueti, Bahrami, & Walsh, 2008) are further indications of sensory specificity. In line with predictions of intrinsic models, differential intramodal processing has also been identified. In same/different decision tasks on auditory rhythms Hung (2011) found significant behavioral (reaction time, accuracy) and brain activation (fMRI) differences for vocal and instrumental rhythms, with activation differences in auditory sensory areas (primary auditory cortex, and anterior and posterior superior temporal gyrus). Differences between vocal and instrumental rhythms are, however, not limited to sensory input processing. They have also been found in short-term memory tasks (Klyn, Will, Cheong, & Allen, 2016), with faster reaction times and more rapid accuracy depreciation for vocal rhythm. Klyn et al. (2016) also report that in rhythm reproduction tasks memorization of vocal rhythms recruits the participation of the articulatory loop (the articulatory loop is a working memory mechanism that prevents verbalizable memory content from decaying through repeated articulation), but memorization of instrumental rhythms does not, indicating that different memory mechanisms are available for the two rhythm types. These reports of sensory specific components of temporal processing indicate an additional source of variability in rhythm perception and pose a clear challenge for approaches that, implicitly or explicitly, consider rhythms as abstract, disembodied cognitive representations of temporal relationships of external events and rhythm processing as a quintessentially amodal process (Deutsch, 1986; Povel & Essens, 1985).

Support for the hypothesis that multiple distinct procedures underlie timing is provided by the study of Portugal, Wilson, and Matell (2011), showing that different environmental and behavioral contexts elicit distinct patterns of activity variation in the striatum. A study by Grahn and McAuley (2009) reports that participants' subjective strategies can influence the selection of neural circuits. They suggest that different neural networks are employed in a tempo discrimination task, depending on whether subjects implicitly used a beat-based or a duration-based strategy. Subjects' prior experience with music has also been found to affect temporal processing. Musical training seems to contribute to faster extraction of timing information (Kraus & Chandrasekaran, 2010) and also improves auditory processing in some brain areas (i.e., planum temporale and dorso-lateral prefrontal cortex) that are also used in language processing (Franklin et al., 2008; Ohnishi et al., 2001). On the other hand, their training may enable professional musicians to employ different forms of mental representation of rhythms than nonmusicians (Aleman, Nieuwenstein, Böcker, & de Haan, 2000; Schaal, Banissy, & Lange, 2014; Palmer & Krumhansl, 1990). Due to musical training musicians can use symbolic representation systems that may be effective in extracting rhythms quickly and storing them in more stable and easier to recall representations (Brodsky, Kessler, Rubinstein, Ginsborg, & Henik, 2008). Musicians have been found to be more accurate than nonmusicians in rhythm memory tasks. Performance differences between these two groups have been found to vary with experimental tasks and were much larger in reproduction than in discrimination tasks (Klyn et al., 2016). As rhythm reproduction relies on both on the extraction and memorization of a rhythm and on the ability to accurately reproduce it, musical training likely confers a general advantage on the latter, which could explain at least part of the difference between the reproduction and discrimination experiments.

In their review on timing models, Ivry and Schlerf (2008) point out that some of the effects taken as evidence for intrinsic mechanisms could be due to the influence of nontemporal factors on temporal perception tasks and they advise that future research should be cautious about marrying temporal and nontemporal mechanisms. While I agree that a distinction between theses mechanisms is important, it would be ill conceived for time studies to exclude the latter because in real world situations organisms do not evaluate temporal aspects in isolation. It has for example been demonstrated that in the striatum, temporal information is embedded within contextual and motor related neural activity (Portugal et al., 2011). In response to events or event sequences, we simultaneously consider temporal, intensity, spectral, and spatial dynamics to identify sources and causes, and the interaction between the dynamics of these features shapes our experience and guides us in organizing appropriate actions. In addition, nontemporal mechanisms like memory and attention are essential parts of our temporal experience, so essential, indeed, that there would be no time or rhythm experience at all if we had no memory.

MECHANISMS OF TIME AND RHYTHM PROCESSING

The development of intrinsic timing models required a reformulation of the underlying neural mechanisms. The most frequently postulated mechanism for dedicated models, the already mentioned clock-accumulator module, is based on the idea that a neural clock (pacemaker) emits a stream of pulses that are accumulated in a separate unit (the accumulator), and the number of pulses accumulated during a certain time interval corresponds to the duration of that interval. In addition to being a largely hypothetical model, it is unlikely that a mechanism like this could be implemented in such a varied and flexible way as required by intrinsic timing models. Alternative and more flexible mechanisms that have been proposed are, for example, the clock-less mechanism based on neural networks with a range of synaptic time constants and short-term plasticity of Karmarkar and Buonomano (2007). Another model, the energy readout model, based on the idea that the magnitude of neural activity could encode duration, was proposed by Pariyadath and Eagleman (2007), and this model has been shown to be able to predict the interaction of duration perception and attention.

Available evidence suggests that temporal processing may make use of not just one but multiple mechanisms, and we need to consider the specific temporal process in question when looking for possible underlying mechanisms. For example, in our interactions with the environment we can distinguish discrete actions, like pointing or reaching, from continuous actions, like walking, dancing or music making, and one can ask whether periodic continuous activity can simply be understood as a concatenation of discrete movements or whether the two are based on different types of movement organization and therefore require distinct control mechanisms. With computational simulations of movement models based on dynamical systems theory, Huys, Stukenda, Rheaume, Zelaznik, and Jirsa (2008) were able to demonstrate that distinct control mechanisms may underwrite continuous and discrete movements. The latter are non-autonomous and governed by fixed-point dynamics; their timing cannot originate from their own dynamics and therefore requires external timing control arising from neural structures that are not involved in the realization of their dynamics. In contrast, continuous movements—if they are not too slow, are governed by limit cycle dynamics, they are autonomous and their timing emerges from the movement dynamics.

This idea of distinct mechanisms for discrete and periodic movements has gained support through an fMRI study (Schaal, Sternad, Osu, & Kawato, 2004) that examined brain activation pattern for discrete and periodic (continuous) wrist movements. For continuous movements significant activations were found in contralateral areas including primary sensorimotor areas, premotor area and supplementary motor areas, cingulate cortex and in the ipsilateral cerebellum. Activations for discrete movements were more widely distributed, they included the same areas and, in addition, selectively activated dorsal premotor, pre-SMA, prefrontal and parietal (Brodman 6, 47, 7, 40) and contralateral cerebellum. Lewis and Miall (2003) pointed out that discrete movements may have more widely distributed activations because they seem to place additional demands (decide start, control the discreteness, select movement direction for each gesture; timing to control discrete movements) and they noted that, for the discrete condition of the Schaal et al. (2014) study, activated areas were similar to those found in studies of discrete interval timing. Hence it is likely that at least part of specific activations in the discrete condition reflect the necessary 'external' timing control, and that this 'external' timing control may be realized not in a single neural structure but in a rather distributed network.

INTERNAL PERIODICITY

The temporal characteristics of periodic movements have been explored in a study by MacDougall and Moore (2005). In an analysis of long-term spectra of human movements they demonstrated that the power spectrum of periodic locomotor activity in everyday contexts (walking, running, cycling) exhibits a dominant spectral peak at 2Hz with smaller harmonic side peaks above and below this frequency. This peak frequency did not correlate with gender, age, height, weight, or body mass index, and is could not be explained as an active or passive mechanical resonance. Therefore it is likely that the sharply tuned 2 Hz acceleration of head and limbs during active locomotion reflects a central, movement related, neural "resonant frequency" or internal periodicity. This resonance frequency of 2Hz corresponds to the spontaneous tempo of finger tapping (Kay, Kelso, Saltzman, & Schoner, 1987; Collyer, Broadbent, & Church, 1994) and the preferred metronome tempo (Fraisse, 1982) identified in separate studies. Interestingly, a similar resonance frequency has also been reported for other movement related neural processes like the vestibuloccular and vestibulocollic reflexes, which act to coordinate vertical head and eye movements during walking to maintain dynamic visual acuity (MacDougall & Moore, 2005). Further, periodic auditory stimulation has been shown to produce brainwave synchronization that is maximal at a stimulation rate of 2Hz (Will & Berg, 2007). These studies suggest that periodic movements of body, head, and limbs are under the control of oscillatory neural circuits with a resonance frequency of 2 Hz, and that these circuits are influenced by and interacting with related networks with a similar resonance frequency. It is worth noting that this resonance frequency is not shared by all periodic processes of our body, and several have been shown to work at distinctly different frequency ranges. For example, our circadian cycle is slightly longer than 24 h, and, more relevant in the present context because it is related to voice production, the brainstem controlled breathing cycle has a period of ca. 0.25 Hz at rest.

There are suggestions that the internal periodicity that guides periodic movements may be involved in temporal perception. A study by Press, Berlot, Bird, Ivry, and Cook (2014), reporting an interaction between duration perception and movements, indicates such involvement, and a study by Denner, Wapner, McFarland, and Werner (1963) suggests that the internal periodicity may act as a time reference in duration perception. Though Denner et al. (1963) only used single interval stimuli for their experiment, such a function of the internal periodicity seems all the more pertinent in the case of temporal judgment of event sequences or rhythms. In particular, it is a characteristic of periodic limb and body movements that, even though their timing is an autonomously emerging property, the controlling oscillatory circuits can nevertheless be influenced by sensory inputs or volition, in order to assure that the movements performed are appropriate for whatever the organism is doing at the time (e.g., Marder, 2001). The interaction between internal periodicity and external stimuli has been characterized as follows (Will, Clayton, Wertheim, Leante, & Berg, 2015): periodic limb movements that are performed spontaneously and in the absence of any external stimuli are executed with frequencies corresponding to the internal periodicity. However, the presence of external stimuli, no matter whether they are irregular or quasiperiodic, leads to modifications of internal periodicity and associated motor sequences activations. The resonance of the internal periodicity is perceived as pulse and finds its expression in corresponding periodic motor behavior. In the case of regular periodic external stimuli, internal periodicity and movements become entrained to the external stimuli, that is, pulse and music align (Will et al., 2015). The modifiability of the internal periodicity, and hence of the felt pulse, through external stimuli and volition, appears to be a central mechanism that allows for a flexible adaptation to various contextual conditions, contributing to the cultural variability of rhythm experience.

The idea that motor adjustments to, and temporal judgments of, event sequences are made on the basis of a comparison of external and internal periodicities has first been developed within the framework of entrainment theory. It is based on the notion of attention cycles that are synchronized with the internal periodicity and permit synchronization between external event sequence and the internal periodicity (Jones, 1976; Large & Jones, 1999; Jones & McAuley, 2005). McAuley, Frater, Janke, and Miller (2006) contrast such a timing mechanism, which they call beat or pulse based timing in reference to the internal periodicity that is perceived as pulse, with an interval based timing mechanism, and suggest that these may be alternative temporal processing strategies that can be recruited depending on the content and context of a given situation.

In terms of a neural substrate for the internal periodicity, several studies point towards the implication of the basal ganglia. In a study involving patients with metabolic dysfunction of the basal ganglia Wittmann, von Steinbüchel, and Szelag (2001) found that these patients have problems tapping steadily at their preferred tempo (~2 Hz), most tending to speed up towards the maximum movement rate (around 5 Hz). This is consistent with the findings of fMRI study on beat perception by Grahn and Rowe (2012), who found significant activation of the striatum for tapping continuation, though not for beat finding tasks. The authors suggest a central role of striatum in beat prediction. It is also reported that the involvement of the striatum was not found to be dependent on musical training (Grahn & Rowe, 2012; Grahn & Brett, 2007), which would account for the observation that pulse perception is done with ease by most people and does not require special training.

This brief review of mechanisms involved in temporal and rhythm processing already contained several hints about how cultural factors might interact and shape them. As mentioned above, there are relatively few crosscultural comparative studies on temporal processing, but there is a considerable number of studies that investigate the influence of factors like context, environment (including cultural artifacts), and music background and training, on these processes from an intracultural perspective. The following sections examine both types of studies and their findings concerning the effects of cultural factors on rhythm processing in more detail.

PERCEPTUAL VARIABILITY OF THE BASIC COMPONENTS OF TEMPORAL EXPERIENCE

Perception of events, simultaneity, temporal order and duration form the basic components of our temporal experience. The first three of them are known to be shaped basically by stimulus features and the physiological characteristics: The perceived onset of events can be regarded as a function of the duration and the temporal envelope of the sounds. The fusion threshold (2–3 ms for auditory stimuli, 27–30 ms for touch, and 40–50 ms in the visual modality) is conditioned by physiological characteristics of the sensory modalities, and a central component that takes account of stimulus features (e.g., stimulus length) leading to the phenomenon that stimuli with considerable onset asynchrony are still considered simultaneous (e.g., Brenner & Smeets, 2010). The determination of temporal order, which entails event distances larger than the fusion threshold, is shaped by contextual factors, and requires separation of events by at least 30 ms for event pairs and up to 100–300 ms for event sequences.

Stimulus features and stimulus context have also been found to influence duration perception, with filled and empty intervals being processed differently. The term 'filled interval' is used to describe the length of a state change as demarcated by an onset and an offset, whereas 'empty interval' designates the separation between two successive short events, that is, the temporal distance between events. Single filled intervals are perceived as longer (Wearden, Norton, Martin, & Montford-Bebb, 2007) and more accurate (Goldstone & Goldfarb, 1963; Rammsayer & Lima, 1991) than empty intervals. These accuracy data for single intervals, however, do not seem to generalize to rhythmic sequences as the accuracy for decisions on empty interval sequences has been found to be better than for filled interval sequences (Klyn et al., 2016), an effect that is probably due to the improved temporal order identification if sound events are separated by silent intervals (Warren & Obusek, 1972).

The sensitivity for duration contrasts, for example, the difference limen, has been shown to be influenced by cultural factors like musical training and language background. Jeon and Fricke (1997) have reported that the discrimination threshold for duration is a factor of two smaller in musicians than in nonmusicians, while other studies have shown that both musicians and speaker of Finnish, a language in which phonemic duration is linguistically relevant, have a higher sensitivity for duration differences that speakers of French (Marie, Kujala, & Besson, 2012) or German (Tervaniemi et al., 2006), and nonmusicians.

Our perception of duration, however, is not only informed by temporal features of events. A study by Press et al. (2014) showed that events were perceived as longer and judgments were more consistent when tactile or visual stimuli were congruent with concurrent action. The authors suggest that this may be a consequence of preactivated action expectancy (events perceived to begin before action onset), and that action preparation activates representations of anticipated effects. These findings suggest that when we interact with our physical or social environment our actions are not only based on expectations and predictions built on past experiences, but we immediately evaluate the consequences of our actions so that adjustments can be made if necessary.

Numerous studies have also demonstrated an interaction of pitch and duration perception in speech, as well as nonspeech sounds—a phenomenon relevant for rhythm perception in both music and speech. Again, different effects have been reported for filled and empty intervals: filled higher pitched sounds are perceived as longer than lower pitched ones, while empty intervals marked by lower pitched (flanker) tones are perceived as longer than those marked by higher pitched flanker tones (Lake, LaBar, & Meck, 2014; Fan & Will, 2016). Furthermore, it has been demonstrated that durations of speech and nonspeech sounds with dynamic pitch (varying pitch contours) are perceived as longer than those with fixed (level) pitch (Lehiste, 1976; Cumming, 2011), with larger pitch changes resulting in longer duration perception, and rises judged as longer than falls (Rosen, 1977).

In addition, various studies have demonstrated a substantial language specific dependency of the size of the duration-pitch interaction effect [e.g., Lehnert-LeHouillier, 2007 (Spanish, German, Thai, Japanese); Gussenhoven &

Zhou, 2013 (Dutch, Mandarin); Simko et al. 2015 (Finnish, Estonian, Swedish, Mandarin)]. These studies suggest that pitch of a sound influences judgment of its duration for speakers of different languages in different ways. For example, speakers of languages that use pitch to cosignal phonological quantity (e.g., Estonian and Finnish) show a greater influence of pitch level on duration estimates, whereas speakers of tonal Mandarin, which does not use length contrasts, were found to be less precise in durational judgments, and to a smaller degree influenced by the interaction between duration and pitch. The language specificity of the reported effect sizes can be seen as support for a, probably production based, hypercorrection hypothesis, according to which duration judgments are influenced by regularities in speakers' language environment.

Finally, the interaction of duration and pitch for empty interval has also been shown to be influenced by factors like musical training and language background. Duration judgments of tone language speakers (Mandarin) and musicians were found to be significantly less influenced by pitch of the marker tones than nontone language speakers (English) and nonmusicians (Fan & Will, 2016).

The dependencies of pitch-duration interactions on our socio-cultural environment and active experiences (training) have advanced our understanding of rhythm processing and production in both speech and music. By growing up within specific linguistic and/or musical contexts we not only become familiar with their associated temporal regularities and absorb them in the production of our linguistic and musical behavior, the perceptual and production regularities are also reinforced neurally. Although the neural mechanisms underlying cultural influences on duration processing and the pitch-duration interaction are currently not fully understood, studies indicate that plasticity of brainstem neurophysiology during language acquisition (Krishnan, Xu, Gandour, & Cariani, 2005) and music learning (Wong, Skoe, Russo, Dees, & Kraus, 2007) play a role in this process, and changes at the brainstem level probably lead to further adjustments at the perception and production level. There is also evidence for top–down influences along the descending auditory pathway that have been interpreted as improving the encoding of complex sounds that are important for behavior (i.e., Perrot et al., 2006; Norena, Micheyl, Durrant, Chery-Croze, & Collet, 2002).

MICROTIMING AND INTERVAL RATIOS

Western music psychology has for a long time assumed that human rhythms tend to be based on simple integer ratios, as they are easier to produce and discriminate than complex ratio rhythms. Clarke (1987) performed a series of experiments in which listeners had to judge three tone series with temporal interval ratios varying between 1:1 and 1:2. Participants tended to hear simple integer ratios (1:1 or 1:2) whatever the actual ratios of the tones. These results can be explained as categorical rhythm perception that is strongly influenced by the Western music theory and its notational system that conceptualize and represent rhythms in terms of small integer ratios, complex ratios being interpreted simply as the result of expressive timing or accidental inaccuracy.

Support for such an interpretation comes from studies of non-Western music showing that rhythm production in not universally based on small integer ratio intervals. An analysis of interval ratios in percussive accompaniments of about 400 Australian Aboriginal songs performances did not reveal any preference for small integer intervals (Will, 2011). Even the distribution peak for the most frequently occurring interval, that close to a 1:1 ratio, turned out to be bimodal, that is, this interval is performed with an unequal subdivision (a ratio of c. 0.992). Interval ratios were found to vary with geographic location of the music and different areas show preferences for different ratio combinations, but noninteger interval ratios dominated in rhythm accompaniments of all regions analyzed (Will, 2004, 2011). Though most songs have only one type of rhythm accompaniment, more than a third have two or more nonisochronic rhythmic patterns. The rhythms are linked with distinct sections of the songs (for example mean interval ratios of 0.97 may be found in the opening and final section, while the middle section has mean ratios of 0.68), sometimes alternating multiple times (Will, 2004, 2016). Both the regional differences and the systematic and deliberate use of non-integer ratio intervals questions the general validity of Repp (1998) assertion that microtiming, that is, the deviation from simple small interger ratios, is not the result of musicians' intentional behavior but rather a necessary consequence of expressive performances of certain musical structures in Western music. Iyer (2002) has already pointed out that expressive timing in Western music, comprising phenomena like rubato, ritardando, and accelerando, is very different from microtiming in non-Western music, which for him is characterized by asynchronous unisons, subtle separation of rapid consecutive notes, asymmetric subdivisions of a pulse, and microscopic delays (Iyer, 2002). The noninteger ratio intervals of Australian Aboriginal music would clearly fall into this category of microtiming, and the deliberate, structured utilization of such intervals in Aboriginal music is an unmistakable indication that they result from intentional behavior. This argument can be extended to other non-Western musics that have been shown to make use of microtiming and noninteger ratio intervals, like Jazz (Friberg & Sundström 2002; Iyer, 2002) or Malian drumming (Polak, 2010; Polak & London, 2014), although the question whether specific noninteger ratio intervals in these musics operate as distinctive features for songs, genres or styles—as they appear to do in Australian Aboriginal music—still remains to be explored.

Though instrumental rhythms with unequal, asymmetric, or noninteger beat subdivisions are found in many non-Western music traditions, few attempts to explain the underlying mechanisms have been made. A promising recent approach by McGuiness (2015) applied a pulse-coupled oscillator model to beat microtiming variations and demonstrated that its behavior is capable of emulating timing characteristics of a nonisochronic sequence taken from a studio recording of Afro–American music (C. Stubblefield's 'The Funky Drummer' break). The model is based on the idea of intrapersonal interactions of multiple oscillators (two, in this case) that mutually entrain at different frequencies—an approach that could, at least in principle, also work for unequal beat subdivisions. However, it is currently not clear whether more elaborated versions of such a model are capable of emulating the whole range of systematically and consistently used 'microtiming' intervals that have been described in the literature. In particular, the dynamics of transition from one stable state to another, as for example, found in Australian Aboriginal song accompaniments with different rhythmic ratios for different song section, have yet to be modeled and explained.

ENTRAINMENT

The way oscillatory processes influence each other is determined by their coupling, and hence dependent on nature and strength of their coupling force or forces. What constitutes a coupling force is not something pre-given but depends on the nature of the interacting oscillatory processes. There are indications for a variety of coupling factors in musical context that range from biological and physiological to psychological and cultural factors. It has, for example, been shown that the biological factor age can act as regulators of coupling strength (McAuley et al., 2006). Other factors that have been proposed to influence our bodily responses to music are familiarity, musical training, cultural practices, belief systems, and aesthetic values associated with these factors (Clayton, Sager, & Will, 2005; Will & Turow, 2011). Even intention has been suggested as a factor influencing the intrinsic coordination dynamics (Kelso, 1995). Although only relatively few studies have so far attempted a more detailed examination of coupling factors in musical entrainment, they have given us important insight into the mechanisms involved.

Cultural variability has been demonstrated at the level of intraindividual entrainment, when, for example, a singer accompanies her/himself on an instrument. Clayton et al. (2005) showed tight interlocking of vocal and instrumental rhythms for a Ghanaian, and weak synchronization with fluctuating alignment for an Australian Djyrbal performer, and their analysis indicates that the degree of coherence in synchronization is affected by both musical and social factors.

As mentioned above, in pulse-based listening, when we listen to music or speech, the way our internal pulse aligns with what we hear shapes our experience of the rhythms. The way external stimuli and internal periodicity couple is based on the interaction of acoustical, motoric, prosodic, and conceptual processes, and the ways in which the processes interact are shaped by our enculturation. They are therefore likely to differ not only according to, for example, language or music background but also to genre and performance context. An example at the genre level would be the listener unfamiliar with jazz and tapping on the 'wrong' beat.

The contribution of specific interactions of different factors to culture specific experiences of rhythms is illuminated by a study on rhythms in northern Potosí (Bolivia) by Stobart and Cross (2000). The authors describe melodies of several Quechua and Aymara song genres that, depending on how they align with the listener's internal pulse, can be heard as anacrustic (starting with a lead-in or off-beat) or as starting with a down beat (on-beat). In a listening experiment they found that Western European participants perceived the Bolivian test melodies as anacrustic, whereas Bolivian participants perceived then as starting on-beat. The authors also note that, although the sound durations of the Bolivian melodies had complex ratios, the European listeners tended to perceive them recategorized in terms of simple integer ratios that may have helped them in their anacrustic interpretation of the rhythms. Further analyses of musical performances supported the results of the listening experiment. They showed that the footfall of Bolivian musicians, marking rhythmic accents, corresponded to the first melody sound starting on the beat, and that the downstroke of the charango (a small guitar-like instrument) was synchronized with first syllables and footfalls (Stobart & Cross, 2000).

In their discussion of the perceptual differences, Stobart and Cross point out a possible significant contribution of the local languages. For example, the Quechua language has a primary stress that occurs on the penultimate syllable and a secondary stress that is fixed to the first syllable. They suggest, the fact that the first syllable of a word has a privileged status as a referential feature—carrying the secondary stress in speech and song—may help to explain why Bolivian subjects tend to perceive the first sound as marking the pulse.

In support for a pulse-based theory of rhythm perception the Stobart and Cross (2000) study can be seen to highlight that experience of rhythms is not simply based on an analysis of relevant features of the stimulus sequence, but on an evaluation of the alignment of the stimulus sequence with our internal pulse, our motor resonance, which in turn is shaped by our interactions with the environment. In this study both language experience and musical training can be considered coupling factors that shape the way stimulus sequences and motor resonance align.

Effects of musical training and listening experience on synchronization to external musical stimuli have been examined in a study on motor responses to North Indian alaap performances (Will et al., 2015). Though no differences between professional Hindustani musicians and professional North American musicians without training and experience in Hindustani music were found in terms of the number of subjects that showed synchronization, there were significant group differences in the way they aligned with the stimuli. Notably, overall group differences were significant in the more regular jor and jala sections, but not for the alap sections, suggesting that listeners' experience has a larger effect on synchronization to regularly than to irregularly timed music. Interestingly, the nonexperienced group showed a higher degree of synchronization (i.e., smaller variance in the phase distribution) than the experienced Hindustani musicians. However, a subject-by-subject analysis revealed that this is not due to weaker phase synchronization of the professional Hindustani musicians but to their larger intersubject variability. The more uniform response of the nonexperienced listeners suggests a stronger reliance on stimulus features, while the Hindustani musicians show a larger set of response options and strategies due to their individual musical training and experience (Will et al., 2015).

A remarkable example of socio-cultural influences on alignment phase in interindividual synchronization has been reported by Lucas, Clayton, and Leante (2011). Their analytical field study shows that in the Afro–Brazilian Congado ritual, which involves multiple music groups, an encounter of two groups can lead to a maximization of their phase differences. The authors argue that it is group identity that leads to tight temporal within-group coordination and minimization of coordination between groups. The ethnographic information they supply, however, suggests that it is not so much group identity as such, but the specific function that group identity has in specific performance situations that determines whether this factor acts as a coupling factor or not: It is only in an encounter of two groups from different communities that there is an intent to mark their respective group identity and this intent, in turn, shapes the temporal alignment between the groups.

Studies of synchronization in jazz and Cuban dance ensembles (Doffman, 2009, 2013; Poole, 2012) show that specific phase alignments—and the resulting rhythmic feel or groove - are also considered as indicators of group identities in other musical traditions. Specific phase alignments may also be indicators of other socio-cultural factors that shape interaction and communication within ensembles, like song ownership or shared leadership. For example, in one of their Australian case studies Clayton et al., 2005 demonstrate that the status of 'song owner/leader' (based on age, musical knowledge, socio-cultural legitimacy, etc.) of one of the performers is reflected in the specific phase alignment between the rhythms of the performers involved. In contrast, Western string quartets are marked by a balance between stable leadership and flexible and effective group interaction, and show more evenly distributed leadership roles (Glowinski, Badino, Ausilio, Camurri, & Fadiga, 2012; Badino, D'Ausilio, Glowinski, Camurri, & Fadiga, 2013).

The above studies advance our understanding of the complex nature of coupling in musical entrainment. Interactions between oscillatory processes are not only influenced by biological or physiological factors relating to the implementation of these processes, they are also shaped by cognitive, socio-cultural and emotional factors. The reviewed studies effectively highlight the special status of the concept of coupling that links physiological, psychological and socio-cultural domains within the framework of entrainment theory. At the same time they demonstrate convincingly that identification of these factors and the exploration of their functioning are only possible when detailed ethnographic and entrainment analyses are combined and can complement each other.

A different kind of approach is found in the study of Cameron, Bentley, and Grahn (2015), who examined rhythm processing in Canadian and Rwandan participants. As stimuli they used rhythms transcribed from East African recordings (from Uganda and Kenya) and especially composed Western rhythms, both resynthesized with sine-tone sequences. Participants were asked to perform three experimental tasks, a same/different comparison, a rhythm reproduction task and a beat tapping task. Though accuracy differences between groups were found for the comparison task these could not, for methodological reasons, be attributed to cultural differences. East African rhythms were better reproduced by both groups than Western rhythms. The difference between the two rhythm types was larger for African participants, though their accuracy was significantly lower than that of the Canadians. In the tapping task both groups were more accurate (showed lower absolute asynchrony) in tapping to familiar rhythms, and East Africans tapped at a larger number of metrical levels than Canadian participants. Unfortunately, due to methodological difficulties and problems (differences in testing conditions and familiarity with experimental procedures between the groups, unspecified cultural variables, and use of artificial stimuli), the study only permits to conclude that there were differences in processing of rhythm and beat between the two participant groups, but contributes little to an understanding of the reasons for these differences.

ENTRAINMENT AND EMOTIONS/ALTERED STATES

Effects of music and musical rhythms on emotions and altered states of attention, arousal, and mood, or even of trance states, reported for a plethora of ritual contexts, have been subject of a long standing debate among anthropologists and ethnomusicologists. A most comprehensive review of the anthropological debate was given by Gilbert Rouget in his book 'Music and Trance:

A Theory of Relations between Music and Possession' (Rouget, 1990). Rouget gives an overview on ritual trances involving various types of music, and reviews the anthropological literature on their neurophysiological foundations. He reports on the work of Charles Pidoux, a physician and ethnopsychiatrist who studied possession cults in Mali, and who, in the 1950s was one of the first to hypothesize that drumbeats might act upon "different levels of the neural axis". Rouget then turns to Andrew Neher's laboratory study of auditory driving (Neher, 1961) and his hypothesis that ritual trance states are induced through physiological changes caused by auditory driving in response to periodic drum sounds (Neher, 1962)—a hypothesis that was the basis for much speculation concerning the biological mechanisms of trance in the years between 1962 and 1980, and exerted considerable influence on the discourse on possession and trance ritual of prominent anthropological scholars. In his discussion Rouget raises a series of objections against Neher's theory and its interpretation. Some of his objections seem rather unqualified and inappropriate, others are absolutely crucial when laboratory findings concerning physiological reactions to music are used to (re-)interpret ethnographical research. Rouget's main points are: (1) the lab experiment shows certain physiological reactions, but as subjects did not fall into trance during the lab experiment, they do little to clarify the link between physiological reactions and trance states described by ethnologists. (2) The analogy Neher constructs between his auditory driving results and reports from photic driving experiments (where altered/epileptic states have been reported), is unconvincing on the basis of the data Neher reports. (3) Ethnographic evidence clearly indicates that drum rhythms are *not a sufficient* criterion to evoke trance: in ceremonies that use drum rhythms many if not most participants do not enter into trance states, though all are exposed to the same music. Those who do fall into trance are thought to have a physiological predisposition that distinguishes their responses from that of other participants. (4) Ethnographic reports also indicate that drum rhythms are *not a necessary* element to produce trance states: instead of drums various societies use for examples gong instruments or hand clapping in trance ceremonies (Rouget, 1990).

Subsequent research has consolidated and supported these points. For example, in her research on religious ecstatics and deep listeners J. Becker (2011) reports that physiological responses to music obtained in the lab were markedly different from those obtained in a familiar social setting of the participants. She tested several of her Pentecostal ecstatics in both context and found that her physiological measures (heart rate and galvanic skin response) indicated increased emotional arousal in the social compared to the lab setting (Becker, 2011). In her research Becker also found that religious ecstatics show a different physiological response profile than other participants and she describes this profile as a predisposition to ecstasy that combines a strong physiological response to all music with special response sensitivity to one's favorite music. This means, though participants showed physiological responses to certain features of the music, the strong emotional responses cannot be explained by particular features of the music but require the consideration of participants' physiological response profile as well as their individual relationship with the music (Becker, 2011; Penman & Becker, 2009). The music can be said to function more like a catalyst, not as the cause for the strong emotional response, especially as it is the musical content that engenders the listener's relationship to it in the first place (s. Sloboda & Juslin, 2001; Levitin, 2006). Similar to Rouget, Becker points out that (religious) trance can and does happen without music, but, in addition to the performance context, she considers the presence of deeply arousing music as helping devotees attain a state of religious ecstasy.

A recent trance related study, and one of the very few neurophysiological field studies, is that of Oohashi et al. (2002), who performed an EEG study with Balinese dancers of the ritualistic Calonarang drama. They report that only the performer that entered a trance state (as indicated by eye movement, facial expression, muscle stiffness and tremor during the trance state and the way of recovery from that episode) showed a significant increase in theta and alpha band activity, but no paroxysmal (epileptic) discharges. Dancers that did not enter a trance state showed no comparable activity changes. Unfortunately, a concurrent musical analysis was not within the purview of this study. Nevertheless, as both trancing and nontrancing dancers were exposed to the same music and the power spectra in the pretrance phase were not significantly different, this study is in line with the argument that the music acts as a facilitator rather than the cause of the trance state.

However, the study leaves open whether and to what extent the observed EEG band changes are characteristic for the specific ritual studied or whether they are general indicators of entering a trance state. Questions like these can only be answered through a comparison of studies performed on different rituals, but such studies are not yet available and Oohashi and colleagues give us an excellent idea about the difficulties of performing such studies in the field: Because of the religious character of rituals and the sanctity of the performance space foreigners are rarely given access to them. If they are, it may take a long time to establish mutual trust with the local community, obtain consent from local authorities, and gain the collaboration of the performers involved. Oohashi et al. (2002) report that it took them 11 years—no wonder then that we have not seen more of such studies.

METER

Another level of temporal organization that had been shown to be shaped by cultural factors is that of meter (in contrast to the temporal fine structure— the rhythm—of a sound sequence, meter refers to its periodic accent pattern). Developing as an oral tradition that enhanced memorization (Rubin, 1995) in both performers and listeners, it was initially a special form of heightened speech—an oral style between conversation and recitation, with more rhythmic speech patterns and more clearly enunciated words—temporally restructured through reorganization of syllable sequences on the basis of contrastive prosodic features. This 'poetic meter' arranges syllables in patterned sequences that are defined (that is, 'measured' in the sense of counting) by distinct number of syllables and structured by length, tone and/or other accent patterns. For example, in poetry of some Australian Aboriginal language groups where syllable length is a contrastive language feature, syllables are organized in patterns of short/long contrasts, but without assignment of fixed durational proportions (Will, 2004, 2016). Early Chinese poetry used the contrasts of level and deflected tones, which also implied length contrast (level tones having a longer duration than deflected tones), and later developed into forms with fixed arrangements of tonal patterns (Watson, 1971). In these cases the length contrasts are relative and flexible; they are not based on absolute, isochronic duration units. Poetry in China and Australia, but also in Africa (Agawu, 1995), additionally makes use of dynamic accents, though these are not linked to fixed positions within a meter, they can be freely used and serves as creative means of expression. In various traditions, the temporal structure of heightened speech and song poetry is distinctly different from that of accompanying instrumental music (for Aboriginal Australian music: Will, 2004, 2016), which is based on the internal periodicity of periodic movements (pulse based) and whose rhythms are created from quasi-isochronous unit intervals and subdivisions or multiples thereof.

In Europe, with the development of (rhythmic) modal notation by the 12/13th century composers of the Notre Dame School, we see a gradual replacement of the even and unmeasured rhythm of early polyphony and plainchant with patterns based on the metric feet (units or rhythmic groups) of classical poetry. This was an important step towards the development of the notation for vocal polyphonic music in the 13th–16th century (mensural notation; Hoppin, 1978) and the application of metrical principles in the temporal organization of instrumental music. However, the current concept of meter in Western music only developed since the 17th century and is very different from the concept of meter in poetry, in earlier European and in non-Western music. This modern Western meter, which fuses the formerly distinct concepts of tactus (Latin for touching, striking, referring to the musical 'beat') and of measure (an assemblage or sequence of beats), is based on the assumption of isochronous time units, beats or pulses, and refers to relatively short cyclical pattern of strong and weak pulses or beats. It acts as a framework for the actual (melodic) rhythms that may or may not be congruent with the underlying meter, and as such it is rather an abstract concept than a musical structure. Non-Western musical (i.e., instrumental) meter, e.g. those found in South–Eastern Europe, the

Middle East, or South Asia, are also based on countable pulse units, but show a range of more or less complex patterns and cycle lengths, using a variety of sound features (duration, intensity, spectra and sound quality, sound dynamics) to structure these additive patterns in a memorable (and countable) way, and, as such, exhibit perceptual qualities lacking in Western meter.

A couple of electrophysiological studies have compared processing of rhythm and meter, with meter referring to the periodic structure of the musical beat and rhythm to the temporal fine structure of the acoustic event sequence (not beat or periodicity related). They did not find any large topological differences, in contrast to predictions of some theories, for example, Lerdahl and Jackendoff (1983), who suggested rhythm might be processed by the left and meter by the right brain hemisphere, and Ivry and Robertson (1998), who proposed a local (rhythmic)/global (metric) dichotomy. The EEG study of Kuck, Grossbach, Bangert, and Altenmueller (2003) found that processing of both rhythm and meter leads to cortical activations over frontal and temporal regions with right hemispheric dominance, and responses in the rhythm condition being located slightly more centroparietal than in the metric condition. Processing of rhythm and meter does not seem to differ in terms of global contrasts, but appears to be performed in largely overlapping networks. An EEG study by Geiser, Zaehle, Janckce, and Meyer (2009) reported differences in the time course of processing during rhythm and meter processing and differences between musically trained (professional musicians) and nontrained subjects. These differences were larger in the 'meter' condition; a specific EEG component was identified for both attended and unattended stimuli in the rhythm condition, but only for attended stimuli in the meter condition. This suggests that, in contrast to rhythm processing, meter processing is more strongly influenced by musical training and requires special attention.

The role of native cultural experience on meter perception was explored in a study by Hannon and Trehub (2005) that examined discriminability of metrical variants of simple (4/4) and complex (7/8) meter in Native American and Native Bulgarian/Macedonian listeners. Their results showed that both groups could distinguish variants of simple meter, but only the Bulgarian/Macedonian group could reliably discriminate complex meter variants. They also found that a group of 6–7 month old American infants were able to distinguish variants of both simple and complex meter. The researchers took the results to suggest that, though infants are able to discriminate simple as well as complex meter variants, adults' discriminative abilities are shaped by enculturation processes; growing up in a culture dominated by simple meter diminishes the ability to distinguish variants of complex meter. In addition, a study by Soley and Hannon (2010) found that already infants younger than 9 month exhibit clear preferences for rhythmic pattern of their own culture—despite their undiminished discriminative abilities.

Although the latter two studies do not permit to reconstruct the ontogenetic development of meter perception, the results are compatible with the suggestion that human infants are initially capable of distinguishing with equal ease between variants of simple and complex meter. By growing up in a specific cultural environment, that is, through exposure to specific types of music and musical activities, their discrimination abilities develop and adjust to metrical features of their musical environment. However, even before their metrical discrimination abilities are tweaked, that is, before the age of 9 month, infants already show a pronounced preference for metrical features of their culture. Both phenomena appear to reflect changing attentional sensibilities as meter perception has been shown to require special attention (see Geiser et al., 2009; aforementioned)—a feature that clearly sets it apart from beat perception.

CONCLUSIONS

The studies reviewed here demonstrate that the idea that all humans perceive and experience temporal periodic structures like the rhythms of music or speech in the same way is untenable. Linguistic and musical practices and traditions differ across socio-cultural groups in a significant number of aspects, including structural and functional features and relationships with other bodily activities like movement and dance. The exposure to a specific speech or music environment leads to culture-specific perceptual reorganization and fine-tuning of cognition that affects cognitive processes ranging from duration perception to meter perception and affective responses. Even individual preferences and habits are shaped by affordances of our socio-cultural environment.

Due to the array of socio-cultural factors that contribute to its formation, cognition, and in particular rhythm cognition, is to a remarkable extent socially modulated. Then the fact that cognition is also a social phenomenon helps to explain why cognitive processes studied in natural settings are frequently found to work significantly different from processes studied in an abstract, desituated setting like a laboratory.

It has been highlighted how intimately processing and experience of rhythms are connected to our body situated in and interacting with an environment, with bodily action, motion, and motor resonance. Rhythms engage us because they evoke a motor-based bodily resonance. The act of listening to rhythmic music activates many of the same neural processes that underlie our body's exploration of the environment. The sensory and motor processes involved, perception and action, evolve conjointly in ontogeny and the emergent neural connections between the sensory and the motor system form the basis for perceptually guided action, that is, for cognition.

Effects of cultural factors, socially learned behaviors as well as cultural artifacts, on cognitive processing of temporal and rhythmic information at various levels of the processing hierarchy has been unambiguously demonstrated in numerous studies. However, the culture-specificity of these influences is more difficult to assess, and this is mainly due to methodological difficulties. In crosscultural studies it is not meaningful to operationalize 'culture' as nationality or residency in a country or city because the implied homogeneity of cultural traits within such geopolitical units does not correspond to the reality. There is unquestionably need for more comparative, crosscultural studies of rhythm processing that employ clearly specified cultural factors as experimental variables in order to advance our understanding of cultural effects on processing and experience of time and rhythm.

References

Agawu, K. (1995). *African Rhythm: A Northern Ewe Perspective*. Cambridge, UK: Cambridge University Press.

Aleman, A., Nieuwenstein, M. R., Böcker, K. B., & de Haan, E. H. (2000). Music training and mental imagery ability. *Neuropsychologia, 38*(12), 1664–1668.

Amselle, J. -L. (2001). *Branchments. Anthropologie de l'universalité des cultures*. Paris: Flammarion.

Amselle, J. -L., & M'Bokolo, E. (1999). *Au Coeur de l'ethnie: ethnie, tribalisme et état en Afrique*. Paris: La Découverte.

Badino, L., D'Ausilio, A., Glowinski, D., Camurri, A., & Fadiga, L. (2013). Sensorimotor communication in professional quartets. *Neuropsychologia*. Available from: http://dx.doi.org/10.1016/j.neuropsychologia.2013.11.012i

Becker, J. (2011). Rhythmic Entrainment and Evolution. In J. Berger, & G. Turow (Eds.), *Music, Science and the Rhythmic Brain. Cultural and Clinical Implications* (pp. 49–72). New York, London: Routledge.

Berry, J. W. (2011). The ecocultural framework: a stocktaking. In F. J. R. van de Vijver, A. Chasiotis, & S. M. Breugelmans (Eds.), *Fundamental questions in cross-cultural psychology* (pp. 95–114). Cambridge: Cambridge University Press.

Berry, J. W., Poortinga, Y. H., Bruegelmans, S., Chasiotis, A., & Sam, D. L. (2011). *Cross-cultural psychology: Research and applications* (3rd Ed.). New York: Cambridge University Press.

Breugelmans, S. M. (2011). The relationship between individual and culture. In F. J. R. Vijver, A. Chasiotis, & S. M. Breugelmans (Eds.), *Fundamental questions in Cross-Cultural Psychology* (pp. 135–162). Cambridge, NewYork: Cambridge University Press.

Brenner, E., & Smeets, J. B. J. (2010). How well can people judge when something happened? *Vision Research, 50*, 1101–1108.

Brodsky, W., Kessler, Y., Rubinstein, B. S., Ginsborg, J., & Henik, A. (2008). The mental representation of music notation: notational audiation. *Journal of Experimental Psychology: Human Perception and Performance, 34*(2), 427.

Bueti, D., Bahrami, B., & Walsh, V. (2008). Sensory and association cortex in time perception. *Journal of Cognitive Neuroscience, 20*, 1054–1062.

Cameron, D. J., Bentley, J., & Grahn, J. A. (2015). Cross-cultural influences on rhythm processing: reproduction, discrimination, and beat tapping. *Frontiers in Psychology, 6*, 366.

Chiao, J. Y., & Immordino-Yang, M. H. (2013). Modularity and the cultural mind: contributions of cultural neuroscience to cognitive theory. *Perspectives on Psychological Science, 8*(1), 56–61.

Clarke, E. (1987). Categorical rhythm perception: an ecological perspective. In A. Gabrielsson (Ed) *Action and Perception in Rhythm and Music*. (Stockholm: Publications of the Swedish Academy of Music, no. 55, 1987), pp. 19–33.

Clayton, M. R., Sager, R., & Will, U. (2005). In Time with the Music: The Concept of Entrainment and its Significance for Ethnomusicology. *European Meetings in Ethnomusicology, 11,* 3–75.

Clayton, M., Dueck, B., & Leante, L. (2013). *Experience and meaning in music performance.* Oxford: Oxford University Press.

Collyer, C. E., Broadbent, H. A., & Church, R. M. (1994). Preferred rates of repetitive tapping and categorical time production. *Perception in Psychophysiology, 55,* 443–453.

Cumming, R. (2011). The effect of dynamic fundamental frequency on the perception of duration. *Journal of Phonetics, 39*(3), 375–387.

d'Andrade, R. (2001). A Cognitivist's View of the units debate in cultural anthropology. *Cross-Cultural Research., 35*(2), 242–257.

Demorest, S. M., Morrison, S. J., Stambaugh, L. A., Beken, M., Richards, T. L., & Johnson, C. (2010). An fMRI investigation of the cultural specificity of music memory. *Social Cognitive and Affective Neuroscience, 5*(2–3), 282–291.

Denner, B., Wapner, S., McFarland, J. H., & Werner, H. (1963). Rhythmic activity and the perception of time. *The American Journal of Psychology, 76*(2), 287–292.

Deutsch, D. (1986). Recognition of durations embedded in temporal patterns. *Perception & Psychophysics, 39*(3), 179–186.

Doffman, Mark (2009). Making it groove! Entrainment, participation and discrepancy in the 'conversation' of a jazz trio. *Language and History, 52*(1), 130–147.

Fan, Y., & Will, U. (2016). Musicianship and tone-language experience differentially influence pitch-duration interaction. In *Proceedings of the 14th ICMPC 2016*San Francisco, USA. .

Fraisse, P. (1982). Rhythm and tempo. In D. Deutsch (Ed.), *The Psychology of Music* (pp. 149–180). New York: Academic.

Friberg, A., & Sundström, A. (2002). Swing ratios and ensemble timing in jazz performances: evidence for a common rhythmic pattern. *Music Perception, 19*(3), 333–349.

Frijda, N. H., & Jahoda, G. (1966). On the scope and methods of cross-cultural research. *International Journal of Psychology, 1,* 109–127.

Franklin, M. S., Moore, K. S., Yip, C. -Y., Jonides, J., Rattray, K., & Moher, J. (2008). The effects of musical training on verbal memory. *Psychology of Music, 36*(3), 353–365.

Geiser, E., Zaehle, T., Jankcke, L., & Meyer, M. (2009). The neural correlate of speech rhythm as evidenced by metrical speech processing. *Journal of Cognitive Neuroscience, 20*(3), 541–552.

Glowinski, D., Badino L., Ausilio A., Camurri A., & Fadiga L. (2012). Analysis of Leadership in a String Quartet. *Third International Workshop on Social Behaviour in Music* at ACM ICMI.

Goldstone, S., & Goldfarb, J. L. (1963). Judgment of filled and unfilled durations: intersensory factors. *Perceptual and Motor Skills, 17*(3), 763–774.

Grahn, J. A., & McAuley, J. D. (2009). Neural bases of individual differences in beat perception. *Neuroimage, 47,* 1894–1903.

Grahn, J. A., & Rowe, J. B. (2012). Finding and feeling the musical beat: striatal dissociations between detection and prediction of regularity. *Cerebral Cortex, 23,* 913–921.

Grahn, J. A., & Brett, M. (2007). Rhythm perception in motor areas of the brain. *Journal of Cognitive Neuroscience, 19,* 893–906.

Grondin, S., Roussel, M. E., Gamache, P. L., Roy, M., & Ouellet, B. (2005). The structure of sensory events and the accuracy of time judgments. *Perception, 34,* 45–58.

Gussenhoven, C., & Zhou, W. (2013). Revisiting pitch slope and height effects on perceived duration. *INTERSPEECH,* 1365–1369.

Han, S., Nordhoff, G., Vogeley, K., Wexler, B. E., Kitayama, S., & Varnum, M. E. W. (2013). A cultural neuroscience approach to the biosocial nature of the human brain. *Annual Review of Psychology, 64,* 335–359.

Han, S., & Northoff, G. (2008). Culture-sensitive neural substrates of human cognition: a transcultural neuroimaging approach. *Nature Reviews Neuroscience, 9,* 646–654.

Hannon, E. E., & Trehub, S. E. (2005). Metrical categories in infancy and adulthood. *Psychological Science, 16,* 48–55.

Hoppin, R. H. (1978). *Medieval Music.* New York: W.W. Norton.

Hung, T. H. (2011). *One music? Two musics? How many musics? Cognitive Ethnomusicological, Behavioral, and fMRI Study on Vocal and Instrumental Rhythm Processing (Doctoral dissertation).* Columbus OH: The Ohio State University.

Huys, R., Stukenda, B. E., Rheaume, N. L., Zelaznik, H. N., & Jirsa, V. K. (2008). Distinct timing mechanisms produce discrete and continuous movements. *PLoS Computational Biology, 4*(4), e1000061.

Ivry, R. B., & Robertson, L. C. (1998). *The two sides of Perception.* Cambridge, MA: MIT Press.

Ivry, R. B., & Schlerf, J. E. (2008). Dedicated and intrinsic models of time perception. *Trends in Cognitive Sciences, 12,* 273–280.

Iyer, V. (2002). Embodied mind, situated cognition, and expressive microtiming in African–American music. *Music Perception, 19*(3), 387–414.

Jeon, J. Y., & Fricke, F. R. (1997). Duration of perceived and performed sounds. *Psychology of Music, 254,* 70–83.

Jones, M. R. (1976). Time, our lost dimension: toward a new theory of perception, attention, and memory. *Psychological Review, 83*(5), 323–355.

Jones, M. R., & McAuley, J. D. (2005). Time judgments in global temporal contexts. *Perception & Psychophysics, 67,* 298–317.

Karmarkar, U. R., & Buonomano, D. V. (2007). Timing in the absence of clocks: encoding time in neural network states. *Neuron, 53,* 427–438.

Kay, B. A., Kelso, J. A., Saltzman, E. L., & Schoner, G. (1987). Space-time behavior of single and bimanual rhythmical movements: data and limit cycle model. *Journal of Experimental Psychology: Human Perception and Performance, 13,* 178–192.

Kelso, J. A. S. (1995). *Dynamic patterns: the self-organization of brain and behavior.* Cambridge, MA: MIT Press.

Kitayama, S., & Cohen, D. (Eds.). (2007). *Handbook of Cultural Psychology.* New York: Guilford.

Klyn, N. A. M., Will, U., Cheong, Y. J., & Allen, E. T. (2016). Differential short-term memorization for vocal and instrumental rhythms. *Memory, 24*(6), 766–791.

Kraus, N., & Chandrasekaran, B. (2010). Music training for the development of auditory skills. *Nature Reviews Neuroscience, 11*(8), 599–605.

Krishnan, A., Xu, Y., Gandour, J., & Cariani, P. (2005). Encoding of pitch in the human brainstem is sensitive to language experience. *Cognitive Brain Research, 25*(1), 161–168.

Kuck, H., Grossbach, M., Bangert, M., & Altenmueller, E. (2003). Brain processing of meter and rhythm in music. *Annals of the New York Academy of Sciences, 999,* 244–253.

Kuper, A. (1999). *Culture. The antrhopologists' account.* Cambridge: Harvard University Press.

Lake, J. I., LaBar, K. S., & Meck, W. H. (2014). Hear it playing low and slow: How pitch level differentially influences time perception. *Acta Psychologica, 149,* 169–177.

Large, E. W., & Jones, M. R. (1999). The dynamics of attending: how people track time-varying events. *Psychological Review, 106*(1), 119–159.

Lehiste, I. (1976). Influence of fundamental frequency pattern on the perception of duration. *Journal of Phonetics, 4,* 113–117.

Lehnert-LeHouillier, H. (2007). The influence of dynamic f0 on the perception of vowel duration: Cross-linguistic evidence. In *Proceedings of the 16th International Congress of Phonetic Sciences* (pp. 757–760). .

Lerdahl, F., & Jackendoff, R. (1983). *A generative theory of tonal music.* Cambridge, MA: MIT Press.

Levitin, D. J. (2006). *This is your brain on music: The science of a human obsession.* New York: Dutton.

Lewis, & Miall (2003). Brain activation patterns during measurement of sub- and supra-second intervals. *Neuropsychologia, 41,* 1583–1592.

Lucas, L., Clayton, M., & Leante, L. (2011). Inter-group entrainment in Afro–Brazilian Congado ritual. *Empirical Musicology Review, 6*(2), 75–102.

Marie, C., Kujala, T., & Besson, M. (2012). Musical and linguistic expertise influence pre-attentive and attentive processing of non-speech sounds. *Cortex, 48,* 447–457.

MacDougall, H. G., & Moore, S. T. (2005). Marching to the beat of the same drummer: spontaneous tempo of human locomotion. *Journal of Applied Physiology, 99/3,* 1164–1173.

Marder, E. (2001). Moving Rhythms. *Nature, 410,* 755.

McAuley, J. D., Frater, D., Janke, K., & Miller, N. S. (2006). Detecting changes in timing: Evidence for two modes of listening. In *Proceedings of the 9th ICMPC* (pp. 566–573). .

McGuiness, A. (2015). Modelling microtiming beat variations with pulse-coupled oscillators. *Timing & Time Perception, 3*(1–2), 155–171.

Merchant, H., Zarco, W., & Prado, L. (2008). Do we have a common mechanism for measuring time in the hundreds of millisecond range? Evidence from multiple-interval timing tasks. *Journal of Neurophysiology, 99,* 939–949.

Morrison, S. J., Demorest, S. M., & Stambaugh, L. A. (2008). Enculturation effects in music cognition. The role of age and music complexity. *Journal of Research in Music Education, 56*(2), 118–129.

Neher, A. (1961). Auditory driving observed with scalp electrodes in normal subjects. *Electroencephalography and Clinical Neurophysiology, 13,* 449–451.

Neher, A. (1962). A physiological explanation of unusual behavior in ceremonies involving drums. *Human Biology, IV,* 151–160.

Norena, A., Micheyl, C., Durrant, J., Chery-Croze, S., & Collet, L. (2002). Perceptual correlates of neural plasticity related to spontaneous otoacoustic emissions? *Hearing Research, 171*(1/2), 66–71.

Ohnishi, T., Matsuda, H., Asada, T., Aruga, M., Hirakata, M., Nishikawa, M., Katoh, A., & Imabayashi, E. (2001). Functional anatomy of musical perception in musicians. *Cerebral Cortex, 11*(8), 754–760.

Oohashi, T., Kawai, N., Honda, M., Nakamura, S., Morimoto, M., Nishina, E., et al. (2002). Electroencephalographic measurement of possession trance in the field. *Clinical Neurophysiology, 113*(3), 435–445.

Palmer, C., & Krumhansl, C. L. (1990). Mental representations for musical meter. *Journal of Experimental Psychology: Human Perception and Performance, 16*(4), 728.

Pariyadath, V., & Eagleman, D. (2007). The effect of predictability on subjective duration. *PLoS ONE, 2,* e1264.

Penman, J., & Becker, J. (2009). Religious ecstatics, "deep listeners," and musical emotion. *Empirical Musicology Review, 4*(2), 49–70.

Perrot, X., Ryvlin, P., Isnard, J., Guenot, M., Catenoix, H., Fischer, C., et al. (2006). Evidence for corticofugal modulation of peripheral auditory activity in humans. *Cerebral Cortex, 16*(7), 941–948.

Polak, R. (2010). Rhythmic feel as meter. Non-isochronous beat subdivision in jembe music from Mali. *Music Theory Online, 16*(4), .

Polak, R., & London, J. (2014). Timing and meter in Mande drumming from Mali. *Music Theory Online, 20*(1), .

Poole, A. (2012). *Learning, Performance and Entrainment in Cuban Music.* PhD thesis, Open University, UK.

Portugal, G., Wilson, A. G., & Matell, M. S. (2011). Behavioral sensitivity of temporally modulated striatal neurons. *Frontiers in Integrative Neuroscience, 5,* 30.

Povel, D. J., & Essens, P. (1985). Perception of temporal patterns. *Music Perception, 2*(4), 411–440.

Press, C., Berlot, E., Bird, G., Ivry, R., & Cook, R. (2014). Moving time: the influence of action on duration perception. *Journal of Experimental Psychology: General, 143*(5), 1787–1793.

Rammsayer, T. H., & Lima, S. D. (1991). Duration discrimination of filled and empty auditory intervals: cognitive and perceptual factors. *Perception & Psychophysics, 50*(6), 565–574.

Repp, B. H. (1998). Obligatory "expectations" of expressive timing induced by perception of musical structure. *Psychological Research, 61*, 33–43.

Rosen, S.M. (1977). The effect of fundamental frequency patterns on perceived duration. In *Speech Transmission Laboratory—Quarterly Progress and Status Report* volume 18. Stockholm, Sweden: KTH, pp. 17–30.

Rouget, G. (1990). *La musique et la transe*. Paris: Gallimard.

Rubin, D. C. (1995). *Memory in oral traditions. The cognitive psychology of epic, ballads, and counting-out rhymes.* New York Oxford: Oxford University Press.

Schaal, S., Sternad, D., Osu, R., & Kawato, M. (2004). Rhythmic arm movement is not discrete. *Nature Neuroscience, 7*, 1136–1143.

Schaal, N. K., Banissy, M. J., & Lange, K. (2014). The rhythm span task: comparing memory capacity for musical rhythms in musicians and non-musicians. *Journal of New Music Research, 44*(1), 3–10.

Shuler, M. G., & Bear, M. F. (2006). Reward timing in the primary visual cortex. *Science, 311*(5767), 1606–1609.

Simko, J., Aalto, K.D., Lippus, P., Włodarczak, M. & Vainio, M.T. (2015). Pitch, perceived duration and auditory biases: Comparison among languages. *Proceedings of the 18th International Congress of Phonetic Sciences.* (ICPhS 2015).

Sloboda, J. A., & Juslin, P. N. (2001). Psychological perspectives on music and emotion. In J. A. Sloboda, & P. N. Juslin (Eds.), *Music and emotion: Theory and research* (pp. 71–104). Oxford: Oxford University Press.

Soley, G., & Hannon, E. E. (2010). Infants prefer the musical meter of their own culture: a cross-cultural comparison. *Developmental Psychology, 46*(1), 286–292.

Stevens, C. J. (2011). Music perception and cognition: a review of recent cross-cultural research. *Topics in Cognitive Sciences, 4*, 653–667.

Stobart, H., & Cross, I. (2000). The andean anacrusis? Rhythmic structure and perception in Easter songs of Northern Potosi, Bolivia. *British Journal of Ethnomusicology, 9*(2), 63–94.

Tervaniemi, M., Jacobsen, T., Rottger, S., Kujala, T., Widmann, A., Vainio, M., et al. (2006). Selective tuning of cortical sound-feature processing by language experience. *European Journal of Neuroscience, 23*(9), 2538–2541.

Tomasello, M. (1999). *The cultural origins of human cognition*. Cambridge: Harvard University Press.

Trehub, S. E., Becker, J., & Morley, I. (2015). Cross-cultural perspectives on music and musicality. *Philosophical Transactions of the Royal Society B, 370*, 20140096.

Warren, R. M., & Obusek, C. J. (1972). Identification of temporal order within auditory sequences. *Perception & Psychophysics, 12*, 86–90.

Watson, B. (1971). *Chinese lyricism: Shih poetry from the second to the twelfth century*. New York: Columbia University Press.

Wearden, J. H., Norton, R., Martin, S., & Montford-Bebb, O. (2007). Internal clock processes and the filled duration illusion. *Journal of Experimental Psychology: Human Perception and Performance, 33*, 716–729.

Wiener, M., Turkeltaub, P. E., & Coslett, H. B. (2010). The image of time: a voxel-wise meta-analysis. *Neuroimage, 49*, 1728–1740.

Wiener, M., Lohoff, F. W., & Coslett, H. B. (2011b). Double dissociation of dopamine genes and timing in humans. *Journal of Cognitive Neuroscience, 5*, 31.

Will, U. (1999). La baguette magique d'ethnomusicologie. Repenser la notation et l'analyse de la musique. (The magic wand of Ethnomusicology. Re-thinking notation and its application in music analyses) In *Cahiers de Musiques Traditionelles*, Vol. XII. Georg, Genève, pp. 9–34.

Will, U. (2004). Oral memory in Australian song performance and the Parry-Kirk debate: a cognitive ethnomusicological perspective. In E. Hickmann, & R. Eichmann (Eds.), *Studies in music-archaeology* (pp. 161–180). (vol. IV). Leidorf: Rahden.

Will, U. (2007). In the garden of cultural identities silk flowers quickly grow roots: On the logic of culture, race and identity in postmodernist discourse. *EME, 12*, 18–36.

Will, U. (2011). Perspektiven einer neuorientierung in der kognitiven musikethnologie (prospects for a reorientation in cognitive ethnomusicology). In W. Steinbeck (Ed.), *Selbstreflexion in der Musikwissenschaft, Kölner Beiträge zur Musikwissenschaft* (pp. 193–211). (Vol. 16). Kassel: Bosse Verlag.

Will, U. (2016). Temporal processing and the experience of rhythm: a neuro-psychological approach. In A. Hamilton, & M. Paddison (Eds.), *The Nature of Rhythm: Aesthetics, Music, Dance and Poetics*. Oxford University Press.

Will, U., & Berg, E. (2007). Brainwave synchronization and entrainment to periodic stimuli. *NeuroScience Letter, 424*, 55–60.

Will, U., & Turow, G. (2011). Introduction to entrainment and cognitive ethnomusicology. In J. Berger, & G. Turow (Eds.), *Music, science and the rhythmic brain. Cultural and clinical implications* (pp. 3–30). New York, London: Routledge.

Will, U., Clayton, M., Wertheim, I., Leante, L., & Berg, E. (2015). Pulse and entrainment to non-isochronous auditory stimuli: the case of North Indian alap. *PLoS ONE, 10*(4), e0123247.

Wittmann, M., von Steinbüchel, N., & Szelag, E. (2001). Hemispheric specialisation for self-paced motor sequences. *Cognitive Brain Research, 10*, 341–344.

Wong, P. C., Skoe, E., Russo, N. M., Dees, T., & Kraus, N. (2007). Musical experience shapes human brainstem encoding of linguistic pitch patterns. *Nature Neuroscience, 10*, 420–422.

Further Reading

Doffman, Mark (2013). Groove: temporality, awareness and the feeling of entrainment in jazz performance. In Clayton Martin, Dueck Byron, & Leante Laura (Eds.), *Experience and Meaning in Music Performance*. Oxford: Oxford University Press.

London, J., & Polak, R. (2014). The influence of task and context on complex rhythm production: Evidence from Malian Drumming. *Proceedings of the 13th International Conference for Music Perception and Cognition*, ICMPC, p. 386.

Merriam, A. P. (1964). *The anthropology of music*. Evanston, IL: Northwestern University Press.

Nozaradan, S., Peretz, I., & Mouraux, A. (2012). Selective neuronal entrainment to beat and meter embedded in musical rhythm. *The Journal of Neuroscience, 32*(49), 17572–17581.

Snyder, J. S., & Large, E. W. (2005). Gamma-band activity reflects the metric structure of rhythmic tone sequences. *Cognitive Brain Research, 24*, 117–126.

Wiener, M., Matell, M. S., & Coslett HB (2011a). Multiple mechanisms for temporal processing. *Frontiers in Integrative Neuroscience, 5*, 31.

Will, U. (2012). Coupling factors, visual rhythms, and synchronization ratios. *Empirical Musicology Review, 6*(3), 180–185.

CHAPTER 10

Speculation on the Nature and Future of Rhythmic Stimulation

James R. Evans
Sterlingworth Center, Greenville, SC, United States

MECHANISMS OF HEALING

"This treatment changed my life". "I tried everything, but this was all that worked". "Why isn't this wonderful treatment more widely known? These are statements my business associates and I hear quite regularly regarding neurofeedback (NFB). However, almost certainly they also are statements heard by professionals in other areas of health care, not only those of conventional medicine using new treatments for cancer, etc, but by practitioners of all alternative/supplementary procedures, including those covered in this book. Despite lack of definitive scientific research support for greater efficacy of one procedure as compared to another, very often there is heated argument between practitioners (and among those within any one profession) regarding which, or whose, specific procedure is most efficient, safe, or cost-effective. Yet, they all seem to "work" if one uses as evidence the thick files of client/patient testimonials kept by many professionals—why? One of the purposes of this chapter is to speculate on possible answers to that question.

Skeptics generally focus on the explanation that positive effects of all alternative procedures are due to belief, expectation or placebo. It is likely that most professionals today would agree that those are very important factors in all treatments, both conventional and alternative. However, it has been the writer's experience that most persons who have discovered a treatment that "works" for them, whether music therapy, pEMF stimulation, an antidepressant medication, or some other, strongly reject placebo as the main factor. Some reason that if it is "only placebo", why weren't any of the various other procedures they had tried equally effective, even those recommended by their highly trusted and respected physician, minister, or friends? And, some may cite evidence that the procedure has been found effective with animals, which

Rhythmic Stimulation Procedures in Neuromodulation. http://dx.doi.org/10.1016/B978-0-12-803726-3.00010-9

presumably have no expectations regarding the treatment. Within the field of NFB with which the writer is most familiar it is not uncommon to find rapid changes in attitude of persons initially skeptical of the effects which modifying one's brain electrical activity can have on cognition and emotion. That is, after even a single short session a client or curious volunteer often will make a remark, such as, "I am surprised. Something was happening there, and it wasn't my imagination".

So, if not all placebo, then what? Could there be other general mechanisms underlying efficacy? Some might invoke divine intervention as the main factor in an individual's " just happening" to find an effective procedure leading to cure or symptom alleviation. Having been influenced as a child by a Christian Scientist maternal grandmother, and having witnessed examples of apparent response to individual or group prayer, and the mysterious occurrence of "synchronistic" events and spontaneous healings which seemingly were "meant to be", the writer does not negate the possibility of such intervention as a factor, but believes there may be more "earthly" forces at work.

Other variables in the mix of potential mechanisms underlying success (or failure) of alternative procedures relate to concepts such as human consciousness and unconsciousness, will, ego, "I", me, myself, motivation and intention; perhaps best summarized as the "I Factor". Definitions of these terms vary widely, and the neural mechanisms mediating consciousness remain one of the greatest mysteries in neuroscience. Yet, we sense that the phenomena to which such terms refer do exist. We often use related phrases, such as "I am conscious, therefore I exist", "just do it", "yes I can" or "no, I won't", "I want (don't want) to change", "I am what I think", "that's my attitude (belief)", "I intend (don't intend) to", and "that was a "Freudian (unconscious) slip I made". Although such phenomena may be considered major aspects of the placebo response, they are mentioned here because, along with various social and cultural variables, the writer and his associates regularly suspect they have significance for facilitating (or inhibiting) NFB progress (see Evans, Dellinger, Guyer, & Price, 2016). Individual differences in such "I" variables, along with individual differences in susceptibility to the specific rhythmic features of the stimulation involved, likely are reasons for common research findings of no significant group effect of a specific treatment, yet powerful effects for individual participants.

There are persons who profess to believe that mental illness is a myth, and/ or that persons facing adversity, can and should just "man-up", "carry-on", etc. Some such persons appear to have more ability than others to do so, and may rarely or never seek treatment, conventional or alternative. And, if they do receive treatment, their "will"- and "I"-related factors may facilitate speed and quality of responding. Cognitive behavior therapy (CBT) in its various forms recognizes the power of conscious thought, has a well verified track record regarding treatment success, and often is used successfully in conjunction with other procedures. The "power of positive thinking", even when not considered within a spiritual context or as part of formal CBT, may facilitate positive outcomes. And, some therapies, such as hypnotherapy, attempt to bypass "blockages" of positive conscious thought, and/or recruit unconscious thought processes in the service of healing. When used along with procedures such as those covered in this book, remissions of symptoms, or cures, may be facilitated.

THE PRIMACY OF RHYTHM

In the writer's opinion, still another factor needs to be considered: the rhythmicity inherent in nearly all alternative approaches to healing (whether or not explicitly recognized or addressed by proponents). But, if so, how and why? Certainly it is true that oscillations, vibrations and their organized patterns and interactions, i.e. rhythms as defined in this book, are very basic phenomena. In the book, *The Elegant Universe* (2003) physicist Brian Greene discusses string theory which some believe "provides a single explanatory framework capable of encompassing all forces and all matter" (Green, 2003). A principle of that theory is that particles consist not just of atoms, electrons, protons, neutrons and quarks, but, even more basically, of a "vibrating, oscillating, dancing filament" that physicists have named a string. In other words, "everything at its most microscopic level consists of combinations of vibrating strands" (or "strings") (pp. 14–15). Considered from such a point of view, some might assign religious/spiritual connotations to oscillatory activity and rhythm, that is, an alpha–omega status, with life's beginnings involving rhythmic sexual activities, and its earthly ending signalled by "brain death", that is, cessation of rhythmic brain electrical activity. A related anecdote: My son, Kent, at the innocent age of six, and apparently in a pensive mood while contemplating a rainbow, exclaimed, "Maybe God is a rainbow". Since the arc of the rainbow contains colors reflecting the full frequency spectrum of light in a sequentially organized (visually rhythmic) manner, some might say this was a valid insight.

The term "primordial" could be applied appropriately to rhythm. Consider, for example, the vibrations of atomic particles and molecules, planetary movements, and even the rhythmic activity involved in natural human conception, that is, the rhythmic movements of intercourse leading to rhythmic release of sperm cells at orgasm, followed by their rhythmic "swim" up the vaginal canal aided by the rhythmic contractions of the latter. And then there are the life–death and day–night cycles, the cyclic seasonal changes, and the well known natural rhythms of the human body, such as female menstruation, beating of the heart and breathing, as well as the more recently discovered and less obvious rhythms of brain electrical activity, and saccadic eye movements.

Of probable relevance here is the topic of cymatics. Dating back at least to the 17th century it has been demonstrated that when certain materials, such as layers of sand are placed on a metal surface which has been set to vibrating by a source of oscillation (usually sound), geometric images will form in the sand with their shapes determined at least in part by the frequency spectrum of the oscillations—a sort of "sight of sound" production. Shapes produced often are of similar form to mandalas used in some types of meditation or as seen in nature. Some have perceived these as reflecting an invisible, primordial force field with vibrational properties, and healing potential. Perhaps not surprisingly "cymatic therapy" practitioners now can be found, often making extreme claims, such as using sound frequencies which engender specific shapes to restore youthfulness.

Dysregulations (arhythmias or dysrhythmias) of some natural rhythms, such as that of the heart, are known to be associated with discomfort, decreased efficiency, disease or risk of death, and their reregulation often is the goal of both conventional treatment and some of the stimulation procedures described in this book. As seen from such perspective, it may not be surprising that within the last thirty years or so concepts, such as "vibrational medicine" and "energy medicine" have been coined and popularized in best-selling books, and "sound" therapies and "light" therapies touted as " medicines of the future". Perhaps, with only moderate variations on this theme, the present book can add terms, such as "rhythm medicine", "rhythm treatment", "rhythmization", or "rhythmizers" to this movement.

For most persons it likely goes unnoticed, but frequency, vibration and rhythm-related terms are common in the English language, attesting to the importance and relevance we implicitly place on such phenomena. Examples include: being "in tune with" or "resonating with" an idea or another person; being "on the same wavelength" as another; sensing "good (or bad) vibrations" concerning something or someone; having an "harmonious" relationship with or being "in sync" with a situation or person.

THE MUSIC-EEG CONNECTION

Interestingly, most of the aforementioned terms also relate to concepts in music which have been considered by others to be analogous to or metaphors for brain function. For example, a twentieth century English neurophysiologist and pioneer in the study of brain electrical activity, W. Grey Walter, often is quoted, (e.g., Robbins (2008), as saying, "We are dealing essentially with a symphonic or orchestral composition, but one in which the performers may move about a little, and may follow the conductor or indulge in improvisation—more like a jazz combo than a solemn philharmonic assembly". The analogy with music is elaborated upon further by Robbins, and implied in the title of his book, *A Symphony in the Brain*, one of the earliest books published solely on the topic of NFB.

The writer also has found the concept of brain (or even whole body) function analogous in many ways to that of a symphony orchestra. Environmental *location* is important, e.g., an orchestra may function better in one hall than another due to variations in temperature or acoustical quality; and efficiency of processing of specific perceptual and cognitive functions varies with location in the brain and spatial patterns of neural networks. The relative role to be played by the string section as compared to the wood wind section, etc during an orchestral production is critical, as are the relative roles to be played by specific brain regions and neural networks for efficient processing of specific brain functions. And, of special relevance to this chapter, *timing* between and among the conductor and individual musicians is critical for an orchestral production, as is *neural timing* between, among, and within brain regions and networks. Neural timing research recently has been emerging as one of the most important topics in neuroscience. The music—brain function analogy could be carried still further. For example, if an individual musician is ill, poorly motivated or had an out-of-tune or damaged instrument it would be expected to impair the quality of the orchestral performance to varying degrees unless a way could be found to accommodate for it, e.g. decrease or eliminate that musician's or instrument's role, repair or tune the defective instrument. Somewhat similarly, a dysregulated, damaged or malnourished brain would be expected to preclude optimal performance, and likely lead to symptoms of mental, emotional and /or physical disorder. Happily, due largely to the recently demonstrated and accepted fact of neural plasticity of the human brain, effective treatments for brain dysregulation and damage are becoming increasingly available. Many are described briefly in Chapter 1, and in greater detail in other chapters of this book.

RHYTHM AND THE BRAIN

The study of relationships between rhythm and the brain has grown very rapidly in the past decade, with it now often being taken for granted by many writers that a very close relationship exists. The rapidly growing literature on the topic varies widely in terms of detail and sophistication, ranging, for example, from detailed, sophisticated texts, such as Rhythms of the Brain (Buzsaki, 2006) praised by many neuroscientists, to those written for popular consumption, using simplified language, and often making extreme and unsubstantiated claims.

The brain-rhythm connection also is a fast-growing topic of scientific research. In 2005 a book by Michael Thaut titled Rhythm, Music and the Brain was published (Thaut, 2008). And, in 2006 and 2007 Stanford University's Center for Computer Research in Music and Acoustics presented symposia titled, respectively, Brainwave Entrainment to External Rhythmic Stimuli, and Music, Rhythm and the Brain.

A path of reasoning followed by some serious researchers of rhythm and the brain, as well as by the majority of persons "discovering" and/or advocating rhythm/brain related methods with claimed potential for curing all manner of disorders, proceeds more or less as follows: (1) findings from physics noting that all objects from atoms to everything else in the universe (including brains, of course) vibrate (oscillate; have wave-like character) at natural and unique frequencies; (2) normally, these vibrations (oscillations; waves) resonate with each other in a harmonious manner facilitative of health and efficient cognitive function; (3) factors, such as brain damage, stress, negative thinking, poor diet and impact of incompatible oscillations from external sources can disrupt this harmony, causing "disease", malfunction of the immune system, etc.); (4) exposure to rhythmic external stimulation (or facilitation of internally generated stimulation) of appropriate frequencies can resonate with the normal frequencies and entrain them back to normal, thus restoring health and function.

Of course this is a greatly oversimplified "bare-bones" description. Even if basically accurate, the actual process certainly would be complicated by the almost infinite number of interacting vibrations and wave forms of widely varying natural (or resonant) frequencies and their harmonics, the "fragility" of resonances and the need for resonance relationships to be flexible as processing needs change. Given such complication, it is difficult to comprehend how exposure to relatively simple, apparently one dimensional types of stimulation, such as earlier versions of auditory-visual stimulation, rhythmic massage, or many of the "sound therapies" could be effective; yet apparently they are in many cases. Perhaps the answer lies in statements, such as those by Cole (1985) that "the synchrony of many small pushes add up to a much larger one" (p. 264), and that although whatever resonates responds to one frequency above all others, it also responds to a range of sympathetic vibrations; that is, there is no need for perfect resonance. This appears to be in line with both the notion of the "butterfly effect" mentioned in Chapter 1 where the flapping (rhythmic pulsing) of a butterfly's wings at one location may lead to hurricane force winds at another, and with the effect the tiny electromagnetic pulses of pEMF can have on brain function.

MUSINGS

In this section the writer speculates on some "far out" (some readers may say "even farther out") rhythm-related notions which may resonate with certain readers who find them intellectually stimulating, but eliciting nausea in others who may label some of them nonsense and overly simplistic with no place in a book with any claims to scientific credibility. Such were the reactions of two different reviewers of a book on rhythm this author once coedited with Manfred Clynes (Evans & Clynes, 1986). One reviewer, an American experimental psychologist, rated the writer's speculative chapter "dysrhythmia and disorders of learning and behavior" as the poorest of the 10 chapters, while a reviewer in France rated the work as an interesting blend of science and "new age" thinking. The term "musings" seems appropriate, and is a term my medically trained co-editor used to refer to my rambling, emailed thoughts about NFB during the not-long-ago days when he was debating whether to incorporate such treatment into his neurology and epileptology practice (which he did, with much success). So here are some rambling, random, rhythm-related "musings" (alliteration intended). It is unlikely that any are original, but all relate to thoughts which have entered the writer's consciousness over several decades of pondering the concept of rhythm and its relationships to human behavior and brain function. While rather disorganized, and somewhat biased toward NFB, all relate either directly or tangentially to the topic of this chapter, that is, rhythmic stimulation. Hopefully at least some will contribute toward the purposes of this book.

- The concept of a biofield appears to have wide potential. It is an integral part of the theory behind development of the Neurofield equipment and procedure which is gaining strong acceptance as a supplement to NFB (see Chapter 1). As defined by Gerber (2001), the term biofield refers to the "energy field which surrounds and interpenetrates the physical body" (p. 555), and is made up of magnetic and electromagnetic energies generated by living cells, as well as subtle energetic fields. To some, one's biofield represents, contains or even "is" the true person. The concept of "aura" often is considered very similar or identical. It is involved in many explanations of the means by which Reiki and healing touch (see Chapter 1) achieve results, that is, healing "energy" from a practitioner's body supposedly penetrates, interacts with, or otherwise has a positive effect on the client. Although some have claimed to measure such hypothetical energy, to the writer's knowledge no reliable, valid and generally accepted measure exists.

In a way this situation is reminiscent of the status of "germ" theory in the early to mid-1800's when there was circumstantial evidence for an unknown "something" that appeared related to observed connections between washing hands and other cleaning measures and observations of decreased mortality and complications from surgery and childbirth. However, it remained for the development of improved microscopes revealing the organisms we now call germs to discover the true connection. Today, many would agree there is circumstantial evidence for a construct, such as a biofield in the many reports of efficacy of Reiki and therapeutic touch, and perhaps in other situations.

In the writer's opinion the latter situations include the fascinating phenomena of the Theramin instrument, wherein music is produced by a person moving hands over an instrument without ever actually touching it. It certainly seems that some form of energy emanating from the "musician's" hands is interacting across space with the instrument, causing it to emit tones. If it were possible to analyze the music so produced, perhaps in a similar manner to that now done with brain electrical activity, it might prove to be a quantifiable reflection and representation of the player's biofield, and have huge implications for many things ranging from improved diagnosis and treatment of disease, to improved matching of persons on dating web sites. It might be referred to as a "quantified electromagnetic biogram" (QEBG) and compete with QEEG for relevance. Who knows, it might even prove to be a "key" to ESP phenomena as the discoverer of the human EEG, Hans Berger, had hoped to find in brain electrical activity!

Or, perhaps just slightly more conservatively, it might be discovered that "true" or "genuine" healers exist, and have distinct QEBGs which distinguish them from "wanna-be" healers. Considered as roughly analogous to "skeleton key" capabilities or to the universal donor capability of type O blood, the biofields of such healers might be capable of penetrating, interacting with, harmonizing with and regulating those of a wide variety of persons. Conceivably, this could explain rather common reports of apparent "experimenter effect" in medical research where predicted results occurred for one treatment group, but not for another even though there were no known treatment differences—only different researchers or assistants applying, or being present during, treatment. In other contexts this might help explain the special "healing presence" attributed to some physicians, nurses and other health care professionals, and the exceptional success of certain psychotherapists and counselors. Two anecdotes may help support such reasoning.

The writer once had a very professionally successful chiropractor speak as a guest lecturer to a university class on alternative medicine. After his lecture to the class on the theory and methods of spinal manipulation and correction of subluxations, he confided to me that he suspected the main reason for his many therapeutic successes was that he transferred some type healing "energy" from his hands to the patient. And then, there is the story of "Bonesetter Reese" (more can be found via Google). As a child growing up in Western Pennsylvania, the writer often heard remarks from family and family friends, such as, "Sure wish Bonesetter Reese was still around, he could fix this back pain (headache, etc.)". It seems he was what today might be called a sort of "practical chiropractor" with no medical education who traveled to many steel mills and other businesses in the region, helping alleviate symptoms stemming primarily from industrial injuries. It has been reported that a wealthy person from the area, having experienced or observed his healing powers, offered to send him to medical school. However, after enrolling he soon dropped out, apparently perceiving little or no need for training. Did he have "naturally" healing hands, or a unique biofield capable of interacting with and regulating the fields of others? Whatever, the dynamics, the story has relevance to practice in the various fields of alternative medicine.

There is little or no regulation within the vast majority of those fields regarding who may or may not practice. For example, within the field best known to the writer, NFB, few states regulate its practice. In many states a person with a high school education or less, with money to purchase equipment and with a weekend of training, and who may not know a cortex from a cerebellum, can legally open a practice. And this is true for most of the procedures covered in this book. For many, this situation seems ludicrous. Statements often are made, such as, "How can this be? How can they be allowed to tamper with brain function?", or "Such people are dangerous; they interfere with people going for medical treatment for conditions which may be life-threatening", or "Why should I incur huge debt and spend seven (eight–twenty) years of university study and internship only to have to compete for clients with such minimally trained persons?" Yet very heated arguments often arise over this topic, with some emphasizing the lack of negative side effects, the high cost of conventional medicine which often precludes persons from receiving any help at all, and the existence of gifted healers who, it is argued, should be allowed to practice even if not be able to afford, or need, a university degree and be licensed as a health care provider. This situation might at least be somewhat alleviated if an objective means of identifying "natural healers" existed.

- A phenomenon often observed is the closeness in time of death of couples who have been married for a great many years. Of course, obvious explanations are that they usually are very old, so death of the widow or widower was to be expected soon anyway, or that many times the survivor essentially committed suicide as by failing to eat or sleep properly while grieving, or that the survivor died from the same contagious disease. However, it can be speculated that when two people live closely together for many years, a rhythm-related bonding occurs, perhaps involving entrainment of one person's biological rhythms by those of the other. This would be similar to the often-reported synchronizing of menstrual rhythms among college women living in the same dorm room. In the case of a very long term marriage, the death of one would mean the cessation of his/her rhythm, perhaps causing disruption of the "syncopated rhythm" of the survivor and generating life threatening dysrhythmias in his or her body.

- If, as skeptics claim, powerful positive results of most, if not all, alternative medicine procedures are due to placebo, should courses in how to enhance it be mandatory for all health care training programs, perhaps taught by business school faculty with advertising skills! And, related to this, if it has such powerful effects, could those effects also turn negative for treatment efficacy ("nocebo" effect)? This might occur if, for example, a trusted physician or friend tells a patient that a procedure is useless, or the patient reads a news release stating that some treatment procedure, conventional or alternative, has been scientifically demonstrated to be ineffective. Perhaps that is an equally valid reason for professional training in managing placebo effects.

- Within the various alternative/complementary medicine fields the question often is asked, "Why do we have so much difficulty being accepted by mainstream medicine and the general public? Common, obvious responses include: (1) conventional medical practitioners do not want competition for patients; (2) we are a potential threat to the multi-billion dollar pharmaceutical industry; (3) we have not demonstrated efficacy through enough (or any) well replicated, randomized, double blind, placebo controlled scientific studies; (4) insurance plans generally do not cover our practices. While there very likely is some validity to these responses, the first two appear to the writer to be less powerful at this point in time than often is claimed. Although the situation may be changing, most alternative approaches remain "under the radar" and, to date, probably have had minimal or no significant impact on demand for, or income from, conventional medical services or pharmaceutical needs. Because of major individual differences in the variables under study, the "gold standard", group comparison research of the type cited in #3 is extremely difficult to complete. Standardization, rather than individualization, of treatment specifics is demanded in most research, and that tends to prevent intended treatment effects. And, in any event, where "gold standard" type research has been done with results favorable to some alternative approach, mainstream scientists often find ways to reject the findings, seemingly demanding higher standards than those required for research involving more conventional treatments. In fact, many conventional medication-based treatments have little or no scientific research support. Therefore, it seems likely that reason #3 has had minimal effect on acceptance of alternative approaches. In the writer's opinion, however, there are other more important, but less often considered, causal factors as cited later.

The variety and rapidly growing numbers of alternative treatment approaches, with their often exaggerated and highly advertised claims of efficacy for treatment of all manner of disorders by practitioners with very limited or no formal training in relevant fields has to be very confusing to the general public, and a "turn-off" for even broad-minded conventional health-care practitioners. While some such self-styled, "wanna-be" healers may have genuine skills to offer, due to their common use of advertising ploys, such as colorful web sites, initials after their names suggesting strong medically-related qualifications, and inclusion of "buzz words" (e.g., clinically proven, evidence-based, quantum-physics derived), they often are perceived as charlatans out to fleece the public. Furthermore, and perhaps especially in regard to the rhythmic stimulation procedures covered in this book, emphasis on terms, such as "waves", "frequencies, and "vibrations" in practitioners' descriptions of their methods creates for many persons an automatic association with often-ridiculed "far-out, new-age" ideas from the 1960's and 1970's used by "hippies" who were believed to be more apt to engage in dreaming than rigorous scientific reasoning. This is likely to be true especially when such terms are used in conjunction with terms, such as "soul" or "spiritual".

A rarely mentioned possibility for slow progress in acceptance could be the huge implications for society which conceivably could occur if certain ideas proved true and applicable. For example, consider the upheavals which would follow if it were proven definitively to be true, as some of the more extreme advocates claim, that accurate medical diagnoses can be made with aura reading, or cancer is readily cured and ravages of time on the body reversed by exposure to the correct electromagnetic or sound frequencies, or certain interactions among biofields enable what now are referred to as paranormal experiences. Repercussions likely wound rival or surpass the discovery of atomic energy. Little wonder that many persons reject such type thinking which, for them, may include most or all of alternative medicine.

In the writer's opinion, these variables have more potential for slowing broad acceptance, and preventing insurance coverage than any of the more commonly cited ones listed above. In the writer's opinion some sort of regulation concerning truth in advertising, and type and quality of training necessary for license to practice is badly needed if broader acceptance and insurance coverage are to be attained.

■ Within the rhythm-related approach to healing and health most familiar to the writer, NFB (aka neurotherapy; EEG biofeedback), there is much disagreement among practitioners regarding not only who should be doing it, but also mechanisms of action and most effective methods. A great deal of intradiscipline conflict occurs, often involving outbursts of anger and occasional lawsuits. Over about thirty years of involvement in the field, the writer often has observed positive effects of use of several different approaches by self or others, including relatively simple one- and two-channel training and 19 channel amplitude training (QEEG-guided or not, with or without reference to a database, and with or without ancillary treatment, such as vitamin supplements or counseling.), infralow frequency training, and use of low resolution electromagnetic tomography (LORETA) software for QEEG and subsequent NFB. My associates and I at the Sterlingworth Center, for various reasons have chosen to use the LORETA approach for clients with more complicated symptom pictures, and use two channel, QEEG-guided and database-referenced treatment with clients with apparently less complex presentations. In some cases concurrent personal counseling is provided. This decision, however, was based on clinical experience and logical reasoning by the writer, more than on controlled research demonstrating superiority of this approach over others. For other NFB practitioners choice of approach may be determined by financial limitations, the experience and choice of persons under whom the practitioner trained or the advertising skills of vendors seeking to sell their particular equipment. This situation will persist unless and until controlled research can clearly demonstrate superiority of one method over others. Incidentally, this is true as well for making decisions about efficacy of all the procedures covered in this book. Until then it certainly is a "buyers beware" state of affairs.

■ Another musing regarding NFB in particular has special relevance to neural timing. Knowledge concerning specific brain *locations* of neural dysregulation (e.g., Brodmann areas) undoubtedly is extremely important when attempting neuromodulation, and may be determined through QEEG findings referenced to a quality database. However, considering the critical importance of neural timing within and among brain regions and networks, it would seem that identifying and targeting abnormalities of frequency-related measures of timing, such as coherence, phase relationships, phase shift and phase lock duration, are equally important. Relevant to that are what the writer considers "milestones" in the history of NFB involving training of EEG phase and/or coherence in addition to amplitude. These included early use of bipolar rather than unipolar electrode placement (montages) by some practitioners who reported better results. Some explained that as due simply to involvement of more cortical regions, but others noted that bipolar placement involves training of EEG power differences between two sites which are related to phase of the two EEG signals; hence it can indirectly modulate phase relationships. At another turning point one of the NFB pioneers, Joe Horvat, began reporting greatly improved treatment results using coherence training, earning him the nickname "Dr. Coherence", and stimulating broad interest in training that measure of neural timing. A third development which provided strong evidence that addition of timing-related stimulation to conventional NFB of the time was invention and use of the Roshi NFB instrument. This involved presentation of light stimulation via goggles of a type commonly used for auditory-visual entrainment (AVE) of EEG frequencies, but with the additional capability of modifying phase relationships among brain waves. The Roshi soon became recognized by many in the field as having greater training effectiveness than conventional NFB and was adopted by many practitioners. A fourth development which many in the field would consider a milestone is the low energy neurofeedback system (LENS). Its exact mechanisms of action are unknown, and it is described officially as involving NFB, but not stimulation. Its healing efficacy is believed by its inventor to be due, in part, to its ability to disrupt chronic, maladaptive cortico-thalamic phase settings, and thus allow new and better phase relationships to develop. Currently LENS is considered by many in the field to be one of the best forms of NFB treatment. Both the Roshi and LENS are described more fully in Chapter 1. Taken as a whole, these milestones appear to the writer to provide very strong evidence for the critical role neural timing (with its inherent rhythmicity) plays in normal and abnormal brain function, and thus its central role in stimulation procedures for neuromodulation. Neural timing now is broadly assessed and targeted for training in the LORETA approach to NFB which involves multiple phase-related variables operating among Brodmann areas and neural networks.

THE FUTURE

Although, as noted earlier, practitioners of alternative medicine frequently lament a lack of greater acceptance, the fact is that a huge amount of out-of-pocket money is spent each year on alternative and complementary medicine. This is true especially if professions are included which are not covered in this book because of lack of obvious emphasis on rhythmic stimulation (such as Tai chi, yoga, chiropractic, acupuncture and nutrition). Due largely to increasing costs of medical care and prescribed medications, the almost inevitable negative side effects of medication and the managed care-driven decreasing time permitted for personal attention from medical personnel, this situation can be expected to continue and expand.

Many alternative providers overly emphasize the (often unsubstantiated) value or their methods, while negating the value of conventional medicine in apparent attempts to replace or "cut into" the popularity (and profits) of the latter. Such attempts at negation overlook the many and continuing major life-saving and life-extending developments of conventional medicine regarding heart disease, cancer, stroke, infectious disease, and in the long run may more often encumber rather than facilitate acceptance. Hopefully, the near future will bring effective regulations containing reasonable and fair criteria for acceptability of the advertising and practicing of the various alternative approaches. Given its increasing popularity, greater funding of research on alternative and complementary (integrative) medicine may occur, results of which should help in developing regulations, and expedite acceptance of the more valid procedures.

Considering the rationale and evidence for primacy of rhythm and its role in regulating and organizing neural timing, it can be predicted that rhythmic stimulation procedures will play an increasingly greater part in future developments in techniques for neuromodulation. It is reasonable to predict that even as improved pharmaceutical products become available, side effects and cost will inhibit their popularity, while new, noninvasive, but increasingly effective neuromodulation techniques with very few, if any, negative side effects and lower cost will increase in popularity and use. This will bring safer and more enduring relief for the ills of individuals. And, at some point insurance companies will become aware of this, and find it to their financial advantage to provide coverage. As such a scenario clearly unfolds, however, expect very strong opposition from the pharmaceutical industry, perhaps even in the form of "dirty tricks," such as faked data suggesting no positive value of, or great harm resulting from what, for now, are considered non-conventional treatments.

If speculating on which of the many rhythm-related neuromodulation techniques *should* become the future leaders, the writer at this point in time would nominate LORETA-type NFB guided by QEEG and neuropsychological test results, and practiced in conjunction with a stimulation technique, such as Neurofield (or another which is pEMF or music-based) and access to personal counseling and medical consultation should the need arise. This approach allows for quite accurately determining and targeting specific regions of brain dysregulation. However, it is recognized that this is a relatively expensive approach, and perhaps necessary only in cases of brain damage, complicated learning handicap and/or multiple diagnoses. In less complex cases, single or two-to-four channel approaches may suffice. Therefore, it is likely that the latter *will* become most popular in the future. The future may bring clarification of a present debate within the field of NFB concerning whether or not operant conditioning is an essential mechanism of its efficacy, or whether, given appropriate feedback information regarding its functioning, the brain can and will modify itself without external reward. If the latter should prove to be true, it would seem, offhand, that more automated, and perhaps less costly NFB could be provided.

NFB was nominated over other approaches covered in this book not only because of the writer's many first-hand positive experiences with it, but because it deals *directly* with the rhythms of the central nervous system, can target for modification not just global brain function, but function in and among specific regions, may be more likely to elicit enduring results since it includes learning by the client, and provides opportunity to view specific changes in the EEG, thus enabling objective assessment of ongoing neurological modification, and not having to rely solely on client or parental reports of desired behavioral change.

If considering more speculative predictions for the future, one could foresee technological developments leading to discovery of exact neural timing mechanisms underlying, for example, not only disorders, such as depression, ADHD, autism or substance abuse, but also phenomena, such as the placebo effect, or motivation (and lack thereof), thus perhaps allowing for precise modulation of these mechanisms to the benefit of individuals, as well as society in general. Of course major moral and ethical problems could arise, if, for example, technology enabled recognition of one's specific thoughts via analysis of EEG patterns unique to each. Even now that is possible in a rough sort of manner, for discriminating an exciting from a relaxing thought, or a spoken lie from a true statement. Imagine the repercussions (tongue-in-cheek): no more need for WikiLeaks; many private detectives out of work; more work for divorce lawyers, EEG technicians and those in a new profession of "thought police"; creation and use of "thought scrambling" drugs to prevent accurate thought detection; and a new psychiatric diagnosis of "chronic, drug-induced scrambled thought". Seriously, however, such future development is not necessarily science fiction.

If rhythmic stimulation as described in this book is capable of facilitating normal (or enhanced) function and harmony among neurons and networks of neurons within individual brains, then, as increasing numbers of individuals are helped by this technology, positive effects on social groups from marriages, to families, businesses, and perhaps even political parties could be predicted. That is, persons with enhanced sleep, cognitive function, empathy, and control of attention and emotion as commonly reported following exposure to rhythmic stimulation might promote interpersonal harmony as group participants. Such could be an *indirect* social and cultural effect of individuals' positive responses to rhythmic stimulation. However, there can be powerful effects when such stimulation is applied *directly* to groups. One has only to think of the group cohesion, harmony, and "like-mindedness" which, for better or worse, often occurs during religious ceremonies, at sports events, or when a group is "entranced" by a charismatic speaker at a political rally. In such cases it seems obvious that the rhythmic music, as of a national anthem, or the rhythm of a speaker's voice can facilitate interpersonal harmony and group cohesion. In these fractious times we may need an "international anthem" to promote world unity. One is reminded here of a televised Coca Cola advertisement several years ago which included the words, "I'd like to teach the world to sing in perfect harmony." Perhaps music, with its universally powerful rhythmic effects, actually could do it (see Chapters 6 and 9). Or is that such a tall order that the music soon would become only an echo in the breezes of time?

Based on many of the tenets put forth in this book regarding harmonizing and organizing qualities of various forms of rhythmic stimulation, when there is no need for professional treatment, one may wish to consider the following relatively common and inexpensive means of reducing internal chaos and promoting health: rhythmic walking on a mountain trail or rocking in a porch chair while listening to favorite relaxing music and viewing the visual rhythms of mountain ranges in the distance; sitting or lying on a beach absorbing the full spectrum of sunlight and Schumann resonances emanating from the earth beneath, while simultaneously listening to favorite relaxing music with rhythms of waves in the background. No wonder mountains and beaches are popular for vacation getaways! And, of course, for in-between times one could get a rhythmic massage while listening to rhythmic, but relaxing music and rhythmically breathing the aroma of a relaxing essential oil. And, if no time for all that, or it proves to be sensory overload, many find that simply engaging in rhythmic deep breathing can suffice.

References

Buzsaki, G. (2006). *Rhythms of the Brain*. New York: Oxford University Press.

Cole, K. C. (1985). *Sympathetic Vibrations: Reflections on physics as a way of life*. New York: William Morrow and Company.

Evans, J., & Clynes, M. (Eds.). (1986). *Rhythm in psychological, linguistic and musical processes*. Springfield, Illinois: Charles C Thomas.

Evans, J. R., Dellinger, M. B., Guyer, M. A., & Price, J. (2016). Intervening variables and neurotherapy success/failure. In T. Collura, & J. Frederick (Eds.), *Handbook of clinical qeeg and neurotherapy*. New York: Routledge.

Gerber, R. (2001). *Vibrational medicine: The #1 handbook of subtle-energy therapy* (3rd ed.). Rochester, Vermont: Bear & Company.

Green, B. R. (2003). *The elegant universe*. New York, New York: W.W. Norton & Company, Inc.

Robbins, J. (2008). *A symphony in the brain: The evolution of the new brain wave biofeedback*. New York: Grove Press.

Thaut, M. H. (2008). *Rhythm, music and the brain: Scientific foundations and clinical applications*. New York: Routledge.

Index

Printed in the United States
By Bookmasters